PRINCIPLES AND PRACTICE
OF BIOANALYSIS

PRINCIPLES AND PRACTICE OF BIOANALYSIS

Richard F. Venn
Pfizer Central Research, Sandwich, UK

Taylor & Francis
Taylor & Francis Group
New York   London

Published in 2000 by
CRC Press
Taylor & Francis Group
6000 Broken Sound Parkway NW, Suite 300
Boca Raton, FL 33487-2742

International Standard Book Number-10: 0-7484-0843-6 (Softcover)
International Standard Book Number-13: 098-0-7484-0843-6 (Softcover)

Library of Congress Cataloging-in-Publication Data

Catalog record is available from the Library of Congress

Taylor & Francis Group
is the Academic Division of T&F Informa plc.

**Visit the Taylor & Francis Web site at
http://www.taylorandfrancis.com**

**and the CRC Press Web site at
http://www.crcpress.com**

CONTENTS

List of contributors   xvi
Preface   xvii

1   **Physico-chemical properties of drugs and metabolites and their
    extraction from biological material**   1
    HUGH WILTSHIRE

    1.1   Introduction 1
          1.1.1   Metabolite isolation 1
          1.1.2   Bioanalysis 1
          1.1.3   Enrichment of drugs and metabolites 1
          1.1.4   Differences between metabolite isolation and drug analysis 2
    1.2   Physico-chemical properties of drugs and solvents 2
          1.2.1   Energy changes on solution 2
          1.2.2   Molecular phenomena behind solubility/miscibility 3
          1.2.3   Water miscibility and water immiscibility 8
    1.3   Partition 9
          1.3.1   Extraction efficiency 9
    1.4   Ionisation and its effect on the extraction of drugs 11
          1.4.1   Ionisation, pH and pK 11
          1.4.2   Titration curves 12
          1.4.3   Henderson–Hasselbach equation 13
          1.4.4   Buffers 16
          1.4.5   Distribution coefficient 17
    1.5   Solvent extraction 18
          1.5.1   Choice of solvent 18
          1.5.2   Mixed solvents 20
          1.5.3   Dealing with plasma proteins and emulsions 21
          1.5.4   Choice of pH for solvent extraction 21
          1.5.5   Artefacts arising during the extraction of drugs and
                  metabolites 21
          1.5.6   Modification and derivatisation of drugs and metabolites 24

1.5.7   Ion-pair extraction 26
1.5.8   Recoveries 26
1.6   The 'first law of drug metabolism' 26
1.7   Bibliography 27

2   **Solid-phase extraction**   28
CHRIS JAMES

2.1   Introduction 28
2.2   General properties of bonded silica sorbents 30
2.3   Sorbent/analyte interactions 30
       2.3.1   Solvation 30
       2.3.2   Non-polar 31
       2.3.3   Polar 32
       2.3.4   Ion exchange 32
       2.3.5   Covalent 33
       2.3.6   Mixed-mode interactions 34
       2.3.7   Polymeric sorbents 35
       2.3.8   Miscellaneous 36
2.4   Sample pretreatment of different biological matrices 36
       2.4.1   Liquid samples 36
       2.4.2   Protein binding 36
       2.4.3   Solid samples 37
2.5   Developing SPE methods 37
2.6   Example of an SPE method 38
2.7   Disc cartridges 38
       2.7.1   Potential advantages 39
       2.7.2   Disadvantages 40
2.8   96-Well format (e.g. Porvair Microsep™ system) 40
2.9   Direct injection of plasma 41
2.10 Other new developments 41
       2.10.1 Fines 41
       2.10.2 A cartridge in a pipette tip? 41
2.11 Conclusions and future perspectives 42
2.12 Bibliography 42

3   **Basic HPLC theory and practice**   44
ANDY GRAY

3.1   Origins 44
3.2   Applications 44
3.3   Apparatus 45
       3.3.1   Column 45
       3.3.2   Plumbing 47

3.3.3   Pumps 48
3.3.4   Injectors 49
3.3.5   Column ovens 50
3.3.6   Detectors 50
3.4   The chromatographic process 50
3.4.1   Basic principles 50
3.4.2   Molecular forces 51
3.4.3   Distribution 51
3.4.4   Theoretical plates 52
3.5   The chromatogram 54
3.5.1   Retention 54
3.5.2   Resolution 54
3.5.3   Peak shape 58
3.5.4   Effect of temperature 61
3.5.5   Effect of flow rate and linear velocity 62
3.5.6   Effect of sample volume 64
3.6   Separation mode 64
3.6.1   Normal phase 64
3.6.2   Reverse phase 65
3.6.3   Gradient reverse phase 68
3.6.4   Ion suppression and ion pairing 69
3.6.5   Ion exchange 72
3.6.6   Others 72
3.7   Column care 73
3.8   Bibliography 74

4   **HPLC optimisation**   75
DAVID BAKES

4.1   Objective 75
4.2   System parameters 75
4.3   Reverse-phase HPLC 76
4.4   Ion-pair HPLC 82
4.5   Ion-exchange HPLC 85
4.6   Normal-phase HPLC 87
4.7   Chiral HPLC 90
4.7.1   Chiral columns 91
4.7.2   Diastereoisomers 94
4.7.3   Chiral complexing agents 96
4.7.4   Chiral summary 96
4.8   Column switching in HPLC 97
4.9   Gradient reverse-phase HPLC 100
4.10   Column conditions 101
4.11   Computerised optimisation of HPLC 103

4.12 Conclusions 104
4.13 Glossary 104
4.14 References 105

5 **HPLC detectors**   **106**

RICHARD F. VENN

5.1 Introduction 106
5.2 Principles of detection 107
    5.2.1 Solute-property detectors 107
    5.2.2 Bulk-property detectors 108
5.3 Selectivity in detectors 108
5.4 Detector response 109
    5.4.1 Linearity 109
    5.4.2 Time constant 110
5.5 Detector types 111
    5.5.1 UV–visible detectors 111
    5.5.2 Fluorescence detectors 115
    5.5.3 Electrochemical detectors 120
    5.5.4 Multifunctional detectors 122
    5.5.5 Radiochemical detectors 123
    5.5.6 Other detectors 124
5.6 Sensitivity considerations 126
    5.6.1 Irradiation 126
    5.6.2 Pre-column derivatisation 126
    5.6.3 Post-column derivatisation 127
5.7 Selectivity 127
5.8 Detector problems 128
    5.8.1 Noise due to bubbles 128
    5.8.2 Spurious peaks 128
    5.8.3 Baseline instability 128
5.9 Appendix 129
    5.9.1 Buying a detector 129
    5.9.2 Which detector to use? 129
5.10 Bibliography 129

6 **Gas chromatography: what it is and how we use it**   **131**

PETER ANDREW

6.1 Why gas chromatography works 131
6.2 Factors that affect the chromatography 132
6.3 Choices in GC 133
    6.3.1 Stationary phase 133
    6.3.2 Mobile phase 135

6.3.3 Column length 136
6.3.4 Column diameter 136
6.3.5 Film thickness 136
6.3.6 Flow rate 137
6.3.7 Temperature 137
6.3.8 Some rules of thumb 138
6.4 GC hardware 138
6.4.1 Pneumatics 139
6.4.2 Sample introduction 140
6.4.3 Detectors 145
6.5 Derivatisation for GC 147
6.6 A GC strategy for bioanalysis 148
6.7 Bibliography 148

7 Thin-layer chromatography   149
HUGH WILTSHIRE

7.1 Introduction 149
7.2 Uses of TLC 150
7.2.1 Preparative TLC 150
7.2.2 Metabolic profiling 151
7.2.3 'Rules of thumb' 155
7.3 Some recommended solvent systems 158
7.4 Detection of compounds on TLC plates 158
7.5 Bibliography 159

8 Capillary electrophoresis: an introduction   160
PETER ANDREW

8.1 Introduction 160
8.2 How capillary electrophoresis works 160
8.3 Why capillary electrophoresis works 162
8.3.1 Electro-osomotic flow 162
8.3.2 Free-solution capillary electrophoresis 162
8.3.3 Micellar electrokinetic capillary chromatography 164
8.3.4 Electrochromatography (electrically driven HPLC) 165
8.4 CE hardware 166
8.4.1 The capillary 166
8.4.2 Sample introduction 166
8.4.3 Detectors in CE 168
8.4.4 Sensitivity in CE 169
8.5 Use in bioanalysis 169
8.6 Bibliography 170

9 Immunoassay techniques   **171**

RICHARD F. VENN

9.1   Introduction 171
9.2   Definitions 171
9.3   Theory 172
    9.3.1   Mass action 172
    9.3.2   Competitive assays 173
    9.3.3   Non-competitive assays 174
9.4   Requirements for immunoassay 175
    9.4.1   Antibody 175
    9.4.2   Label 175
    9.4.3   Separation 177
9.5   Practical aspects 178
    9.5.1   Preparation of hapten–carrier protein conjugates 178
    9.5.2   Immunisation 179
    9.5.3   Antibody detection 180
    9.5.4   Antibody titres 180
    9.5.5   Calibration curves 180
    9.5.6   Matrix effects 182
9.6   Data handling 182
    9.6.1   Standard curves 182
    9.6.2   Fitting 182
    9.6.3   Precision profile 182
9.7   Advantages of immunoassay 184
    9.7.1   Sensitivity 184
    9.7.2   Throughput 184
    9.7.3   Selectivity 184
    9.7.4   Ease 184
    9.7.5   Automation 184
9.8   Disadvantages of immunoassays 185
    9.8.1   Time: how long does it take? 185
    9.8.2   Selectivity 185
    9.8.3   Matrix effects 185
9.9   What can go wrong? 185
    9.9.1   Matrix effects 185
    9.9.2   Concentration effects 187
9.10 Immunoassay strategy 187
9.11 Example 187
    9.11.1 Sampatrilat 187
9.12 Affinity chromatography 187
    9.12.1 Immobilisation techniques and media 190
    9.12.2 Elution techniques 191
    9.12.3 Re-use/reconditioning 192

9.12.4 The interface between affinity chromatography and analysis 193

9.13 The future 193

    9.13.1 Phage libraries for antibodies 193

    9.13.2 Monoclonal antibodies 193

    9.13.3 Molecular imprinting 193

    9.13.4 Non-competitive assays for small molecules 194

    9.13.5 Use of low-specificity immunoassay for discovery compounds 194

    9.13.6 Indwelling optical fibre probes 194

9.14 Summary 194

9.15 Bibliography 194

10 Automation of sample preparation   196

CHRIS JAMES

10.1 Introduction 196

10.2 Approaches to automation 197

    10.2.1 SPE 197

    10.2.2 Protein precipitation methods 198

    10.2.3 Multi-well plate technology 198

    10.2.4 Liquid-handling procedures 198

    10.2.5 Avoiding evaporation 199

10.3 Simple automation 199

10.4 Column switching 200

10.5 Prospekt and Merck OSP-2 202

10.6 Benchtop instruments – sequential sample processing 202

    10.6.1 Zymark BenchMate 203

    10.6.2 Gilson ASPEC XL 203

    10.6.3 Hamilton MicroLab 204

10.7 Benchtop instruments – parallel sample processing 205

    10.7.1 Zymark RapidTrace 205

    10.7.2 Gilson ASPEC 4 205

    10.7.3 Multiple probe liquid-handling robots 205

10.8 Gilson ASTED 206

10.9 Full robotic systems 207

10.10 When to automate? 207

10.11 Example methods 208

10.12 Conclusions and future perspectives 208

10.13 Bibliography 209

11 Fundamental aspects of mass spectrometry: overview of terminology   **211**

MIRA V. DOIG

11.1   Introduction 211

11.2   Inlets 211

      11.2.1   Septum inlet 211

      11.2.2   Direct probe inlet 212

      11.2.3   GC inlets 212

      11.2.4   LC inlets 213

11.3   Ion sources 216

      11.3.1   Introduction 216

      11.3.2   Electron impact ionisation 216

      11.3.3   Chemical ionisation 218

      11.3.4   Atmospheric-pressure chemical ionisation 219

      11.3.5   Fast atom bombardment 220

      11.3.6   Thermospray 221

      11.3.7   Electrospray 223

      11.3.8   Other desorption techniques 225

11.4   Analysers 226

      11.4.1   Single-focusing magnetic instruments 226

      11.4.2   Double-focusing instruments 228

      11.4.3   Quadrupole analysers 229

      11.4.4   Time of flight (ToF) analysers 231

      11.4.5   Ion-trap mass analysers 231

11.5   Detectors 233

      11.5.1   Electron multipliers 233

      11.5.2   Negative-ion detection 234

11.6   Data acquisition and processing 234

      11.6.1   Instrument control 234

      11.6.2   Data acquisition/preliminary data processing 234

      11.6.3   Secondary data processing/data presentation 235

11.7   Bibliography 239

12 Applications of mass spectrometry: quantitative mass spectrometry   **240**

MIRA V. DOIG

12.1   Quantification 240

      12.1.1   Gas chromatography–mass spectrometry (GC–MS) 240

      12.1.2   Liquid chromatography–mass spectrometry (LC–MS) 241

      12.1.3   Quantitative API LC–MS and its contribution to the drug development process 241

12.2   Internal standardisation 242

12.3   Data acquisition 243

      12.3.1   Selected ion versus mass chromatogram 243

      12.3.2   Mass analysis 243

12.3.3   Calculation of the mass of the selected ion 244

12.3.4   Data storage and processing 245

12.4   Developing a quantitative method 245

12.5   Analysis of prostanoids by GC–MS 246

12.6   An example of thermospray LC–MS 249

12.7   Examples of API LC–MS 250

12.8   The future 253

12.9   Bibliography 254

13   Mass spectrometric identification of metabolites   **255**

JANET OXFORD AND SORAYA MONTÉ

13.1   Objectives 255

13.2   Introduction 255

13.3   Tandem mass spectrometry (MS–MS) 256

    13.3.1   Theory 256

    13.3.2   Instrumentation 256

    13.3.3   MS–MS scans and their application to metabolite identification 258

13.4   Isotopically labelled compounds in metabolite identification 265

13.5   Practical aspects for the identification of metabolites by mass spectrometry 266

    13.5.1   Introduction 266

    13.5.2   Electron impact ionisation and chemical ionisation 268

    13.5.3   Fast atom bombardment 270

    13.5.4   Thermospray LC–MS 271

    13.5.5   Electrospray LC–MS 273

    13.5.6   Ion-trap mass spectrometry coupled to external atmospheric-pressure ionisation sources 274

    13.5.7   Summary 275

    13.5.8   Overall comments 276

13.6   Bibliography 277

14   Nuclear magnetic resonance in drug metabolism   **278**

PHIL GILBERT

14.1   Introduction 278

14.2   Basic theory of the NMR phenomenon 278

14.3   Parameters of the NMR spectrum 280

    14.3.1   Chemical shift 280

    14.3.2   Spin–spin coupling 281

    14.3.3   Intensity 286

14.4   Practical considerations 286

    14.4.1   Types of spectrometer 286

    14.4.2   Sample preparation 287

14.5   NMR applications in drug development 288
    14.5.1   No sample preparation 288
    14.5.2   Solid-phase extraction sample preparation 288
    14.5.3   HPLC fractions 291
    14.5.4   Fluorinated compounds 291
    14.5.5   Stable isotope labelling 293
14.6   Plasma metabolites 294
14.7   Biochemical changes 294
14.8   Summary 294
14.9   Appendix: fourier transform and some multi-pulse techniques 294
    14.9.1   Why use pulse NMR? 294
    14.9.2   The pulse 296
    14.9.3   Time and frequency 296
    14.9.4   Multi-pulse experiments 296
    14.9.5   Conclusion 301
14.10 Bibliography 301

15   Strategy in metabolite isolation and identification   302
HUGH WILTSHIRE

15.1   Stage 1: radiochemical synthesis 302
    15.1.1   Choice of label 302
    15.1.2   Position of ^{14}C label 303
15.2   Stage 2: animal experiments 303
    15.2.1   Routes of excretion 304
    15.2.2   Formulation and route of administration 304
    15.2.3   Collection of urine and bile 304
15.3   Stage 3: metabolite isolation and characterisation 304
    15.3.1   Enrichment 304
    15.3.2   Analysis 306
    15.3.3   Separation 307
    15.3.4   Purification 315
    15.3.5   Characterisation 315
15.4   Stage 4: identification of metabolites 321
    15.4.1   Mass spectrometry 323
    15.4.2   NMR 325
    15.4.3   Degradation, derivatisation and comparison with
           authentic material 327
    15.4.4   Ambiguities 330
15.5   Stage 5: quantitative aspects of metabolism 330
    15.5.1   Quantification of excretion balance studies 330
    15.5.2   Quantitative aspects of metabolite isolation 331
    15.5.3   Quantitative measurement of metabolic profiles 331

15.6   *In vitro* studies 333
    15.6.1   Isolation of metabolites from *in vitro* incubations 334
    15.6.2   Cross-species comparisons of metabolic profiles 336
    15.6.3   Mechanistic studies 337
15.7   Identification of plasma metabolites 337
15.8   Good laboratory practice 339
15.9   Conclusions 341

16   **Strategy for the development of quantitative analytical procedures**   342
DAVID BAKES

16.1   Introduction 342
16.2   Preliminary requirements 343
16.3   Detection 345
16.4   Separation 348
16.5   Sample preparation 349
16.6   Solid-phase extraction 349
16.7   Extraction sequence 350
16.8   Liquid/liquid extraction 352
16.9   Quantification 354
    16.9.1   Rule of one and two 354
    16.9.2   Standardisation 354
    16.9.3   Peak height and area 355
    16.9.4   Calibration check 355
16.10   Validation 356
16.11   Support work 356
    16.11.1   Matrix substitution 356
    16.11.2   Stability 357
    16.11.3   Metabolites 358
16.12   Conclusions 358

Index   359

CONTRIBUTORS

Peter Andrew
Glaxo-Wellcome, Park Road, Ware, Herts SG12 0DP, UK

David Bakes
Servier Research and Development Ltd, Fulmer Hall, Windmill Road, Fulmer, Slough SL3 6HH, UK

Mira V. Doig
Advanced Bioanalytical Service Laboratories, Wardalls Grove, Avonley Road, London SE14 5ER, UK

Phil Gilbert
Celltech Chiroscience, Granta Park, Abington, Cambridge CB1 6GS, UK

Andy Gray
AstraZeneca R&D Charnwood, Bakewell Road, Loughborough, Leics LE11 5RH, UK

Chris James
Pharmacia & Upjohn, S.p.A., 20014 Nerviano (MI), Italy

Soraya Monté
SmithKline Beecham Pharmaceuticals, New Frontiers Science Park North, Third Avenue, Harlow, Essex CM19 5AW, UK

Janet Oxford
Celltech Chiroscience, Granta Park, Abington, Cambridge CB1 6GS, UK

Richard F. Venn
Pfizer Central Research, Department of Drug Metabolism, Ramsgate Road, Sandwich, Kent CT13 9NJ, UK

Hugh Wiltshire
Roche Discovery Welwyn, Broadwater Road, Welwyn Garden City, Herts AL7 3AY, UK

PREFACE

We hope you enjoy this book and find it helpful and informative.

Analysis – and in particular, bioanalysis – is a vital part of life today. It is fundamental to our increasing understanding of the complexities we meet in our universe (from microscopic to cosmic). It is also fundamental to our daily lives as members of a regulated society. Almost every industry or activity is subject to regulation: analysis of compounds (pollutants, toxins, endogenous substances, drugs (medicinal or recreational), growth agents, herbicides, pesticides . . . the list is endless) in the relevant matrix (soil, plasma, tissues, cerebrospinal fluid, urine, meat, cereal, water, etc.) is an essential part of complying with these regulations.

While the examples used in the book are drawn from the pharmaceutical world, the analytes and matrices discussed will not be very different from those being studied elsewhere. The book may be used as a reference to specific techniques, as a guide to assist the analyst in achieving the desired end result (a sufficiently sensitive and selective analytical method suited to the analyte and matrix under investigation) or as a textbook for the student of bioanalysis or metabolism. As such it will be of much wider use than merely for those involved in the pharmaceutical industry. Almost every chemistry and biochemistry student (undergraduate and postgraduate) will need to use the techniques described in this book at some time in their student life; some will use them throughout their careers. Thus we hope that this book will fill a gap and provide some basic instruction in what is a central part of our world today.

The genesis of this book lies in the history of the Drug Metabolism Discussion Group (DMDG), an umbrella organisation set up by those involved in drug metabolism within the British pharmaceutical industry. Major aims of the DMDG are to encourage co-operation and training in drug metabolism within the industry. One incentive for this is that there is no university course in drug metabolism; graduates (and, to a lesser extent, postgraduates) coming into drug metabolism departments require specialist training. Out of this grew the DMDG and a number of courses, for example the basic training course, the clinical and pre-clinical pharmacokinetics courses, the immunoassay course and the methods for bioanalysis and metabolism course.

This latter course was first run in 1992 and has run every other year since then, proving popular and successful. The aims of the course are to provide attendees with sufficient basic knowledge and background of bioanalysis to enable them to tackle the difficult analytical problems they face in a drug metabolism department such as pharmacokinetic determinations and metabolite identification.

Individual lecturers and tutors on the course are drawn from drug metabolism departments throughout the pharmaceutical industry and are usually specialists in their fields. Each lecture and tutorial was devised in consultation with all the other lecturers and the course is regularly updated as a result of the feedback obtained from delegates at each course. In this way we have tried to keep the course fresh, stimulating and up to date; we hope that the book reflects this, being, in essence, the course manual of the methods for bioanalysis and metabolism course. The notes from each lecture have been overhauled and edited to make a coherent whole and the order of the chapters is very similar to the order in which the lectures and tutorials on the course are given. We start with an explanation of the importance of the physico-chemical properties of various analytes. This is then related to extraction and analysis. From there the principles of the major analytical technique of liquid chromatography are dealt with, focusing on optimisation and detection methods.

Richard F. Venn

November 1998
Sandwich, UK

1

PHYSICO-CHEMICAL PROPERTIES OF DRUGS AND METABOLITES AND THEIR EXTRACTION FROM BIOLOGICAL MATERIAL

Hugh Wiltshire

1.1 Introduction

1.1.1 Metabolite isolation

In a 'typical' metabolic problem, rats weighing a total of 1kg were given a single dose of a ^{14}C-labelled drug at 20mg/kg. Liquid scintillation counting showed that 0–48 hour urine (80g) contained 70% (14mg-equivalents) of the dose. Reverse-phase high-performance liquid chromatography (HPLC) with radiochemical detection (Figure 1.1) suggested that the urine contained at least 15 components, of which four accounted for 71% (10mg-equivalents) of the dose recovered in the 0–48 hour urine sample. After being freeze-dried, the urine was found to contain 6g of solid.

Hence the concentration of drug-related material in the freeze-dried residue was 0.2%, the concentration of the major metabolite perhaps 0.07% and the concentration of a metabolite accounting for 2% of the total drug-related material less than 0.005% or 50p.p.m.

1.1.2 Bioanalysis

The ultraviolet (UV) chromophore of a drug suggests that a quantifiable peak would be obtained if 10ng were separated from endogenous material by reverse-phase HPLC. Assuming that 250µl of plasma are available and that 50% of the plasma extract can be analysed, the quantification limit of the assay will be 80ng/ml. As 1ml of plasma contains approximately 80mg of solid, the minimum detectable concentration of the drug in the plasma solids will be 1p.p.m.

1.1.3 Enrichment of drugs and metabolites

It is not feasible to apply 250µl of plasma to an analytical HPLC column or to try to isolate individual metabolites directly from 80g of urine. First, the drug and metabolites must be *enriched*, which means they must be separated from as much endogenous material as possible without loss of drug-related material.

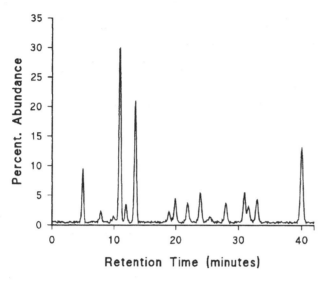

Figure 1.1 Reverse-phase HPLC analysis of the metabolites of Ru-1992 using radiochemical detection.

1.1.4 Differences between metabolite isolation and drug analysis

There are two fundamental differences between metabolite isolation and bioanalysis. In the first case, all the drug-related material must be extracted with minimal losses of any component (particularly the minor constituents). However, the isolation will, hopefully, be carried out only once or twice and so a complex process is acceptable. Excellent nuclear magnetic resonance (NMR) spectra should be obtained from 100µg of pure material and the metabolist will be trying to isolate at least this amount of as many metabolites as possible. It follows that it is essential to use volatile buffers during the purification stages or more impurities may be added than taken away.

On the other hand, in bioanalysis a good recovery of the drug, and possibly one or two major metabolites, is all that is required. However, thousands of analyses will be performed and so the enrichment and separation steps must be relatively simple. As there is no need to isolate the pure drug, the only criteria for the buffers are that they do not interfere with the chromatography or detection of the compound; non-volatile alkali metal phosphate buffers are perfectly acceptable for HPLC/UV methods but volatile buffers will be needed for HPLC/MS analyses.

1.2 Physico-chemical properties of drugs and solvents

1.2.1 Energy changes on solution

The Gibbs free energy of a system is defined as the maximum amount of useful work that it can provide and, by convention, is negative if the reaction is favourable. Hence, a compound will dissolve in a solvent, or two solvents will be miscible, provided that there is a decrease (ΔG) in the Gibbs free energy of the system (i.e. ΔG is negative). The free energy is made up of two components, enthalpy or heat of reaction (ΔH) and entropy (ΔS), a measure of the disorder of the system. The three terms are related by the equation:

$\Delta G = \Delta H - T\Delta S$ (where T is the temperature in kelvin).

If equal amounts of methanol and water are mixed, the temperature rises as heat is produced. This is an exothermic reaction (i.e. the enthalpy is negative; again by convention) and the two solvents are completely miscible. On the other hand, if equal amounts of acetonitrile and water are mixed, the temperature drops; the reaction is endothermic and the enthalpy is positive – and yet the two solvents are miscible.

As entropy is a measure of disorder, it will always increase as two compounds are mixed together, since a mixture will always be a more random state than two pure components. The result of mixing two solvents will therefore be a temperature-dependent decrease in ΔG ($\Delta G = \Delta H - T\Delta S$). In the case of the acetonitrile–water mixture, the increase in entropy as the two are combined more than offsets the increase in enthalpy, and a solution results.

1.2.2 Molecular phenomena behind solubility/miscibility

In order to dissolve a drug, a solvent must break the bonds that link the compound to its neighbours. At the same time, the dissolving solute must not break substantial numbers of intermolecular bonds of the solvent without replacing them with drug–solvent interactions. This is because the breaking of bonds is an endothermic process, requiring energy and causing an increase in enthalpy. Similarly, if the gain in entropy from the dissolution of two solvents in each other is insufficient to counteract any reduced amount of intermolecular bonding, they will not be completely miscible.

For example, the intermolecular forces linking water molecules are strong and cannot be broken by an organic solvent such as hexane which does not form intermolecular bonds of significant strength with water. On the other hand, water–methanol bonds are relatively strong and so methanol is miscible with water but hexane is not. The miscibility of methanol and hexane is a result of weak, but appreciable, bonding between the two solvents. Several types of intermolecular forces need to be considered.

Ionic bonds and 'salting-out'

The attractive force between two oppositely charged ions is given by the expression:

$F = (Q_1 \times Q_2)/(\varepsilon \times r^2)$

and the energy of interaction by:

$E = (Q_1 \times Q_2)/(\varepsilon \times r)$

where Q is the charge on the ions; ε, the dielectric constant of the medium and r is the separation of the ions. (The dielectric constant is a property of a substance whose

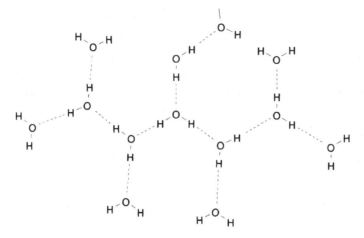

Figure 1.2   The molecular structure of water.

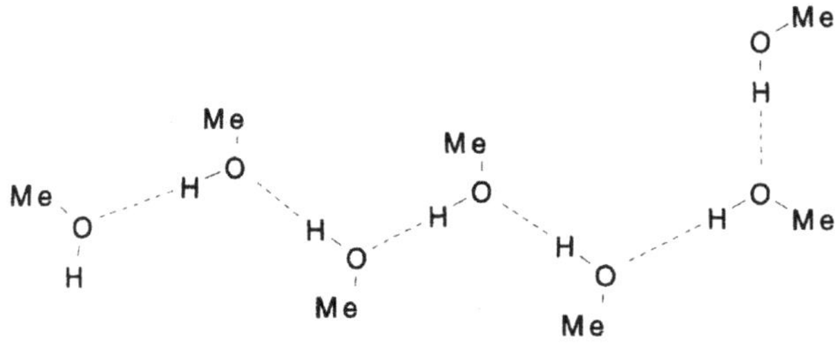

Figure 1.3   The molecular structure of methanol.

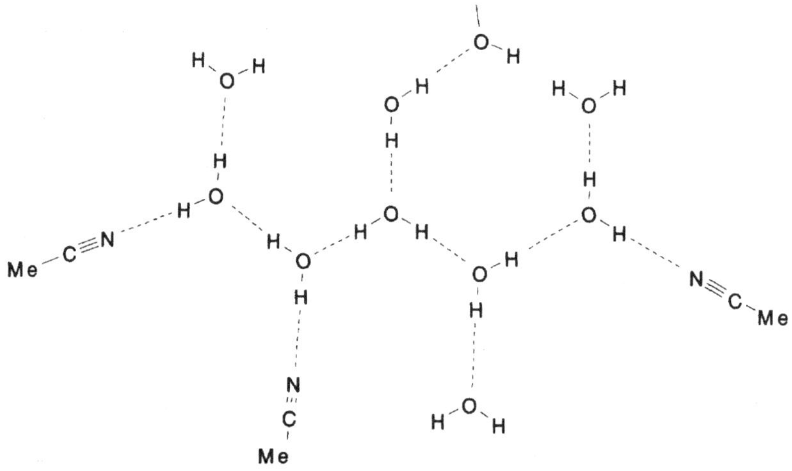

Figure 1.4   The molecular structure of aqueous acetonitrile.

dipoles can be aligned by an electric field. The dipole is a characteristic of a molecule which has a separation of charge, either permanent or induced by an electric field. Solvents, such as water, with strong intermolecular forces that can solvate ionic compounds have high dielectric constants.)

The interaction energies between ionic compounds, such as salts, are strong (40–400kJ/mol) and so these compounds tend to be high-melting and are only dissolved by solvents with high dielectric constants (e.g. water) which reduce the interionic coulombic attraction. Water has the ability to solvate (or surround) both anions and cations and this results in the formation of many solvent–ion bonds and an increased heat of solution. The energy of interaction between solvent and ion is given by:

$$E = (Q \times \mu)/(\varepsilon \times r^2)$$

where Q is the charge on the ion; μ, the dipole moment of the solvent; ε, the dielectric constant of the solvent and r is the separation of the ion and solvation shell (the sphere of solvent molecules around the ion). Ion–dipole forces are weaker than ion–ion forces, being in the range of 4–40kJ/mol.

The sodium and chloride ions of a saturated aqueous solution of common salt will be surrounded by many molecules of water. The result of the addition of a salt to water will be to reduce the number of water molecules available for interaction with other solutes and it will therefore be easier to extract these solutes into immiscible solvents. Similarly, the miscibility of hydrophilic solvents such as acetonitrile will be reduced. Addition of acetonitrile to saturated aqueous sodium chloride causes precipitation of some of the salt and the formation of two distinct layers. The precipitation of some of the sodium chloride is a result of water molecules now interacting with acetonitrile and so being unavailable for solvation of the sodium and chloride ions.

Hydrogen bonds

The hydrogen atom is unique in that it possesses a single electron and, when combined with an electronegative atom such as oxygen, fluorine, nitrogen or chlorine, its nucleus becomes partly deshielded. The result is that the hydrogen nucleus can interact directly with a second electronegative atom, often on another molecule. Hydrogen bonds are relatively weak, ranging from 4 to 40kJ/mol, but are responsible for the abnormally high boiling point of water and the secondary structures of proteins and nucleic acids. The hydrogen bonds in liquid water form a loose three-dimensional network (Figure 1.2).

Methanol can form hydrogen bonds in a similar way to water, except that it can only make chains as it lacks the second proton linked to an oxygen atom (Figure 1.3). Mixing the two solvents can increase the total number of hydrogen bonds formed and this explains the exothermic nature of the methanol–water interaction.

Acetonitrile, on the other hand, cannot form hydrogen bonds with itself. It can, however, accept a proton from water to form a weak hydrogen bond, but cannot donate one (Figure 1.4). Hence the endothermic reaction when the two solvents are mixed, as the total number of hydrogen bonds is substantially reduced.

Other organic solvents can also form hydrogen bonds; for example ethyl acetate is a good solvent for acids and chloroform for amines (Figure 1.5).

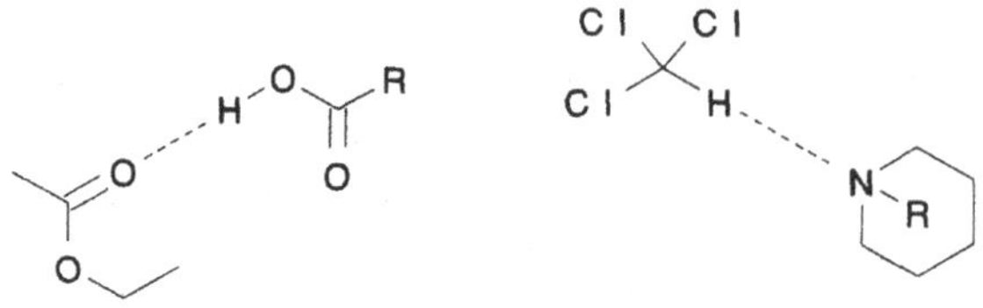

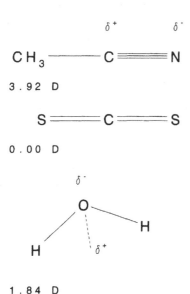

Figure 1.5 Hydrogen bonding in ethyl acetate and chloroform.

$$\delta^+ \qquad \delta^-$$

$$CH_3 \relbar\joinrel\relbar C \equiv\joinrel\equiv N$$

3.92 D

$$S =\joinrel=\joinrel= C =\joinrel=\joinrel= S$$

0.00 D

1.84 D

Figure 1.6 Dipole moments for acetonitrile, carbon disulphide and water.

Van der Waals forces

Many organic solvents lack the ability to form hydrogen bonds and most have too low a dielectric constant to dissolve salts – and yet they can be good solvents. This is the result of a combination of interactions known collectively as van der Waals forces.

Solvents and other compounds in which the centres of positive and negative charge do not coincide have a permanent dipole moment. Acetonitrile, for example, is a linear molecule with a particularly large dipole moment (3.92 debyes, where 1 debye $= 3.34 \times 10^{-30}$ coulomb metres) with the centre of negative charge near the nitrogen atom (Figure 1.6). Carbon disulphide is also linear but has no dipole moment as the centres of positive and negative charge coincide at the carbon atom (Figure 1.6). The three atoms of the water molecule do not lie in a straight line and its dipole moment has a centre of negative charge near the oxygen atom and a centre of positive charge approximately midway between the hydrogen atoms (Figure 1.6).

If two polar molecules are in close proximity the dipole–dipole energy of attraction is given by:

$$E = (2 \times \mu_1^2 \times \mu_2^2)/(3 \times \varepsilon \times r^6 \times k \times T)$$

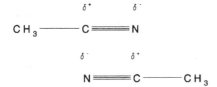

Figure 1.7 Dipolar interaction in acetonitrile.

where μ_1 and μ_2 are the dipole moments of the molecules; ε, the dielectric constant of the solvent; r, the separation of the molecules; k, Boltzmann constant and T is the absolute temperature.

The energy of this interaction is generally weak, in the range of 0.4–4kJ/mol. Liquids such as acetonitrile and nitromethane which have large dipole moments are good solvents mainly as a result of dipole–dipole interactions. The most favourable conformation for this interaction between two molecules of acetonitrile is when the positive and negative regions of the molecules are closest together (Figure 1.7).

A compound with a permanent dipole can induce a dipole in a non-polar molecule and the dipole–induced dipole energy is given by the equation:

$$E = (2 \times \mu^2 \times \alpha)/(\varepsilon \times r^6)$$

where μ is the dipole moment of the molecule; ε, the dielectric constant of the solvent; r, the separation of the molecules and α is the polarisability of the non-polar molecule.

Again the energy of interaction is relatively weak, comparable to that of the dipole–dipole. When carbon tetrachloride (dipole moment = 0D) is added to ethyl acetate (dipole moment = 1.78D) the temperature drops slightly as the number of dipole–dipole interactions in the ester is reduced (no hydrogen bonding is possible with these solvents). Liquids with large dipole moments, such as dimethyl sulphoxide, dimethylformamide, acetonitrile and nitromethane, are good solvents for a range of organic compounds as a result of their ability to cause these two kinds of polar interactions.

Benzene, carbon disulphide and carbon tetrachloride are non-polar, neutral molecules with no permanent dipoles and yet they are liquids at room temperature (and therefore must exert intermolecular interactions) and they can also be useful solvents. Although they have no permanent dipole, at any given time there is an instantaneous dipole caused by random movement of the electron charge density around the molecule. Dipoles in other molecules are momentarily induced and an attractive force results. This energy of interaction is given by the equation:

$$E = (3 \times I \times \alpha^2)/(4 \times r^6)$$

where α is the polarisability; I, the first ionisation potential of the molecule and r is the separation of the molecules. As with all van der Waals forces the strength of attraction is inversely proportional to the sixth power of the distance between the molecules and considerably stronger forces result when contact is maximised.

Table 1.1 Summary of intra- and intermolecular forces

Bond type	Examples	Bond strength (kJ/mol)
Covalent	Diamond, paraffins	200–800
Ionic	Salt	40–400
Hydrogen	Water/sugar	4–40
Dipole–dipole	Acetonitrile/peptide	0.4–4
Dipole–induced dipole	Acetonitrile/naphthalene	0.4–4
Ion–dipole	Water/salt	4–40
Dispersive	Benzene/naphthalene	4–40

The energy of interaction can be quite large (4–40kJ/mol) in readily polarisable compounds (e.g. benzene and other aromatic hydrocarbons). The importance of these 'dispersive' forces is demonstrated by the considerably higher boiling point of 'floppy' n-pentane (all of whose carbon atoms can interact with other molecules) than the relatively rigid neopentane (36°C as opposed to 10°C). These dispersive forces are also responsible for the ability of chains of polyethylene to form sheets.

The intermolecular forces discussed above are summarized in Table 1.1.

1.2.3 Water miscibility and water immiscibility

Alcohols can hydrogen bond with water and can also form dipole–dipole interactions and these intermolecular forces will aid miscibility. The presence of alkyl groups, on the other hand, will reduce solubility in water as they interact by means of dispersive forces which are not possible with water. Thus short-chain alcohols (methanol, ethanol, 1- and 2-propanols) are miscible with water, butanols are partially miscible and octanol is immiscible. Acids behave similarly. Nitriles and ketones and, to a lesser extent, esters and ethers can act as hydrogen bond acceptors and are also polar. Again, only the short-chain homologues are miscible with, or soluble in, water.

Hydrocarbons, such as toluene and the alkanes, are hydrophobic and dissolve compounds by means of dispersive forces. Halogenated hydrocarbons, such as dichloromethane, are more polar and their powers as solvents depend on both dispersive forces and dipolar interactions. These two classes of solvent form negligible intermolecular interactions with water and are therefore immiscible with it.

Hydrophilic groups such as hydroxyl, carbonyl, nitrile and amino (Table 1.2), which are both polar and able to act as donors or acceptors for hydrogen bonding, will encourage solubility in water. Carbon–carbon (particularly aromatic systems), carbon–hydrogen and carbon–halogen bonds are hydrophobic and will promote solubility in organic solvents. Ionic groups are particularly hydrophilic and ionisation of acids or protonation of bases will almost certainly greatly enhance water solubility. Drugs and metabolites which are charged at all values of pH, such as amino acids and N-oxides, will always be relatively hydrophilic.

Drugs with several aromatic rings will have poor solubility in water (unless they also contain hydrophilic groups) as a result of the lack of hydrogen bonding with water and the strong intermolecular dispersive forces of the solid drug. These compounds might be expected to be readily soluble in typical organic solvents. However,

Table 1.2 Hydrophobic and hydrophilic functional groups

Groups	Structures, examples
Hydrophilic groups	
Hydroxyl	–OH
Amino	–NH$_2$
Carboxyl	–COOH
Amido	–CONH$_2$
Guanidino	–NH(C=NH)NH$_3^+$
Tetra-alkylamino	–NR$_3^+$
Sulphate	–OSO$_3^-$
Hydrophobic groups	
Carbon–carbon	C—C
Carbon–hydrogen	C—H
Carbon–halogen	C—Cl
Olefinic	C=C
Aromatic	Benzene
'Neutral groups'	
Carbonyl	>C=O
Ether	–O—R
Nitrile	–C≡N

in some cases the intermolecular dispersive forces of the drug are so strong that solubility in all solvents is poor. Particularly intractable compounds are those that combine strong intermolecular dispersive forces (e.g. aromatic systems) and polar groups (e.g. amides).

1.3   Partition

When a drug is shaken with two immiscible solvents, it distributes between them such that the free energy of the entire system is at a minimum. The partition coefficient (P) for a compound undergoing an aqueous/organic separation is defined as the ratio of the concentration in the organic solution divided by the concentration in the aqueous layer. As this ratio can vary by many orders of magnitude, the base-10 logarithm (log$P$) is usually quoted. If log$P$ is significantly less than zero, the compound is hydrophilic, and if it significantly greater, it is lipophilic. Some examples of neutral drugs with differing values of log$P$ (distributed between an aqueous phosphate buffer at pH 7.4 and octanol) are given in Figure 1.8. The presence of groups that can form hydrogen bonds increases the hydrophilicity (log$P$ smaller) whereas carbocyclic aromatic systems exhibit a large log$P$ and are lipophilic.

1.3.1   Extraction efficiency

If the partition coefficient (P) of a drug between water and an organic solvent is equal to 2, the amount of compound extracted from an aqueous solution (initially equal to 100%) by an equal volume of solvent can be calculated using the equations:

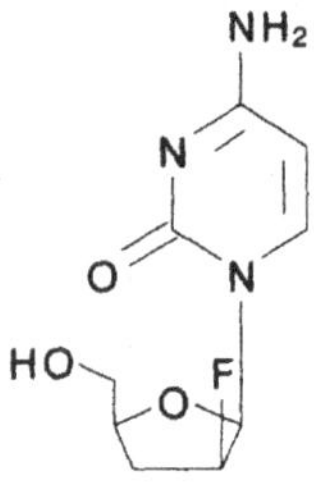

LogP = -1.1

Ro 31-6840/000

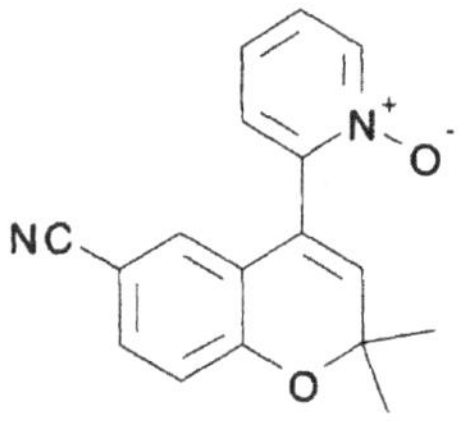

LogP = +1.2

Ro 31-6930/000

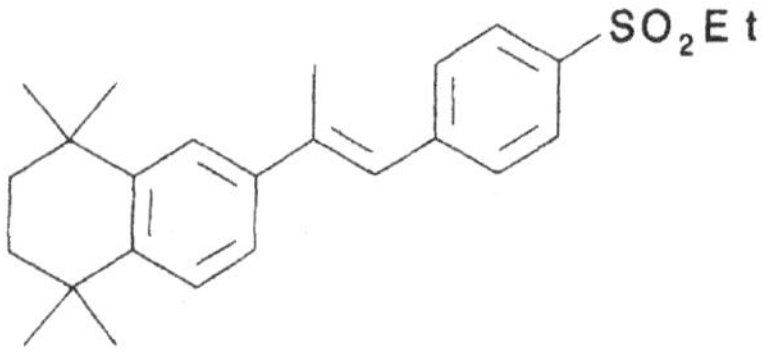

LogP = +7.4

Ro 15-1570/000

Figure 1.8 Some partition coefficients.

$$\text{conc(org)} = P \times \text{conc(aq)}$$

and

$$\text{amount(org)} + \text{amount(aq)} = 100\%.$$

Since the volumes are the same:

$$\text{amount(org)} = P \times \text{amount(aq)}$$
$$= P \times \{100 - \text{amount(org)}\}$$

and as $P = 2$

10

amount(org) = 2 × {100 − amount(org)} = 66.67%.

If the aqueous solution (now containing 33.33% of the drug) is re-extracted with an equal volume of solvent, two-thirds of the drug (22.22%) will be removed, so that 88.89% has been extracted.

If a single extraction had been carried out with 2 volumes of solvent, the amount removed would have been given by:

amount(org) = 2 × P × amount(aq) = 4 × {100 − amount(org)} = 80%.

Thus two extractions remove significantly more than does one with twice the volume. The improvement in yield is unlikely to outweigh the increased effort in a bioanalysis but will certainly be worthwhile for metabolite isolations.

1.4 Ionisation and its effect on the extraction of drugs

1.4.1 Ionisation, pH and pK

Pure water dissociates to a small extent into hydrated protons and hydroxide ions:

$$H_2O \rightarrow H^+_{aq} + OH^-_{aq}$$

and at equilibrium, by the law of mass action,

$$[H^+_{aq}][OH^-_{aq}]/[H_2O] = \text{a constant.}$$

As the concentration of water molecules is effectively constant at 55M, the product of the molar concentrations of the two ions, $[H^+_{aq}][OH^-_{aq}]$, is also constant. The ionisation constant for water, K_w is defined as:

$$K_w = [H^+_{aq}][OH^-_{aq}] = 10^{-14}.$$

The energy produced by the formation of the hydrated ions overcomes the covalent O–H bond energy of water. If the second proton of water is replaced by an electron-withdrawing group, such as phenyl or acetyl, the O–H bond strength is reduced and ionisation is facilitated. Thus with phenol:

$$PhOH + H_2O \rightarrow H^+_{aq} + PhO^-_{aq}.$$

The ionisation constant (K_a) is given by:

$$K_a = [H^+_{aq}][PhO^-_{aq}]/[PhOH] = 1.3 \times 10^{-10}.$$

In the case of acetic acid dissociation is even easier:

$$K_a = [H^+_{aq}][AcO^-_{aq}]/[AcOH] = 1.75 \times 10^{-5}.$$

11

The greater the ionisation, the more protons are produced and the stronger the acid; a strong acid such as sulphuric or hydrochloric acid or a sulphate conjugate will be completely ionised.

The concentration of both hydrogen and hydroxyl ions in pure water, or a neutral solution, equals 10^{-7} moles/litre. If the concentration of protons is raised, the concentration of hydroxide ions is correspondingly reduced, since the product of the concentrations of protons and hydroxide ions is constant at 10^{-14}M, and the solution becomes acidic. In order to avoid using small numbers, hydrogen ion concentration is quoted on a negative logarithmic scale, such that: $pH = -\log_{10}[H^+]$ and, as $[H^+] \times [OH^-] = 10^{-14}$, $pH + pOH = 14$. Similarly, the pK_as for phenol and acetic acid are defined as the negative logarithms of their respective equilibrium constants and are equal to 9.89 and 4.76, respectively.

A base has a lone pair of electrons and these have the ability to abstract a proton from water to form a cation and a hydroxide ion, e.g. with ammonia:

$$NH_3 + H_2O \rightarrow NH^+_{4\ aq} + OH^-_{aq}.$$

The excess of hydroxide ions will make the solution alkaline. As with the dissociation of acids, this reaction is governed by an equilibrium constant:

$$K_b = [NH^+_{4\ aq}][OH^-_{aq}]/[NH_3] = 1.8 \times 10^{-5}$$

and

$$pK_b = -\log(K_b) = 4.75.$$

The corresponding pK_a for the dissociation of the ammonium ion:

$$NH^+_{4\ aq} \rightarrow NH_3 + H^+_{aq}$$

is more usually quoted and equals $14 - 4.75 = 9.25$.

The substitution of one of the protons of ammonia by an electron-donating group increases the ability of the molecule to abstract a proton from water and makes it more basic. Thus the pK_a for dissociation of the methylammonium ion is 10.64. Quaternary ammonium salts and guanidines (e.g. that of arginine) are completely ionised and are strong bases.

Basic drugs (usually amines such as β-blockers) are commonly prepared as salts, often of mineral acids (e.g. hydrochloric), in order to improve the physical properties of the drug substance (solubility, melting point, etc.). As amines are weak bases and mineral acids are strong acids, the pH of an aqueous solution will be acidic. Conversely, many acidic drugs (e.g. penicillins) are formulated as sodium or calcium salts and aqueous solutions of these will be basic.

1.4.2 Titration curves

If a solution of an ionisable drug in a suitable aqueous solvent is titrated with an acid or base, the pH changes in a sigmoidal fashion approximately symmetrical about the point when the drug is half neutralised (i.e. half ionised). At this moment the concentrations of the acidic and basic forms are equal and the pK_a (or pK_b) of the compound

12

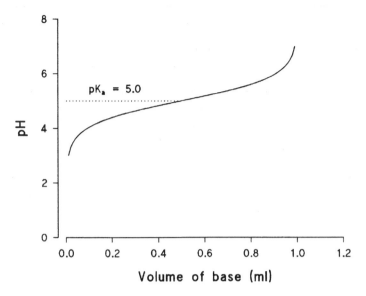

Figure 1.9   Titration curve for romazarit.

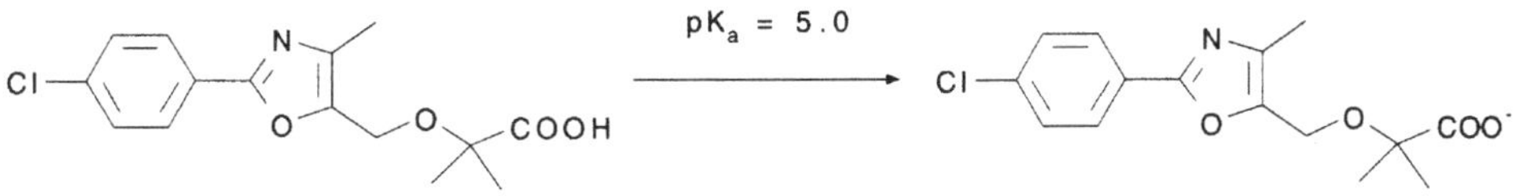

Figure 1.10   Ionisation of romazarit.

being investigated is equal to the pH of the solution (see Section 1.4.1). A titration curve for romazarit, a typical acidic drug, is shown in Figure 1.9 and its ionisation as a solution of aqueous base is added is shown in Figure 1.10.

The pK_a of salts of basic drugs (i.e. $14 - pK_b$) can be determined similarly, as shown in Figure 1.11 for a dibasic calcium-channel antagonist, mibefradil (Figure 1.12).

Drugs that contain both basic and acidic groups are a special case as they tend to form internal salts or zwitterions. The example in Figure 1.13 of the ACE inhibitor cilazapril (Figure 1.14) has two pK_a values of 2.35 and 10.9 units, corresponding to ionisation of the carboxyl group and the loss of one of the protons of the secondary amino group, respectively.

1.4.3   Henderson–Hasselbalch equation

The titration curve for a single ionisation can be described with reasonable accuracy by the Henderson–Hasselbalch equation:

$$pH = pK_a + \log([base]/[acid]).$$

Hence $pK_a = pH$ when the compound is half neutralised, i.e. [acid] = [base].

13

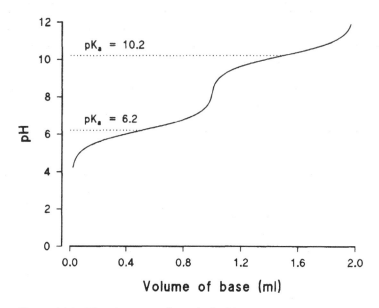

Figure 1.11 Titration curve for mibefradil.

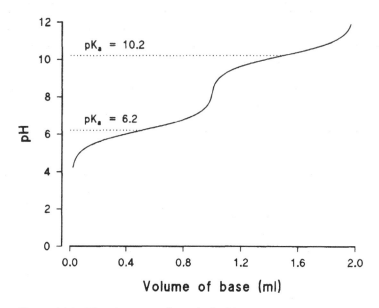

Figure 1.12 Ionisation of mibefradil.

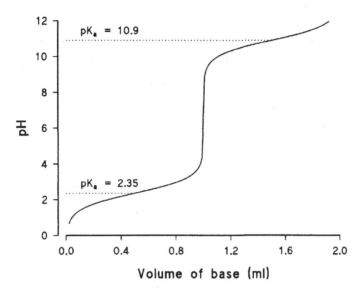

Figure 1.13   Titration curve for cilazapril.

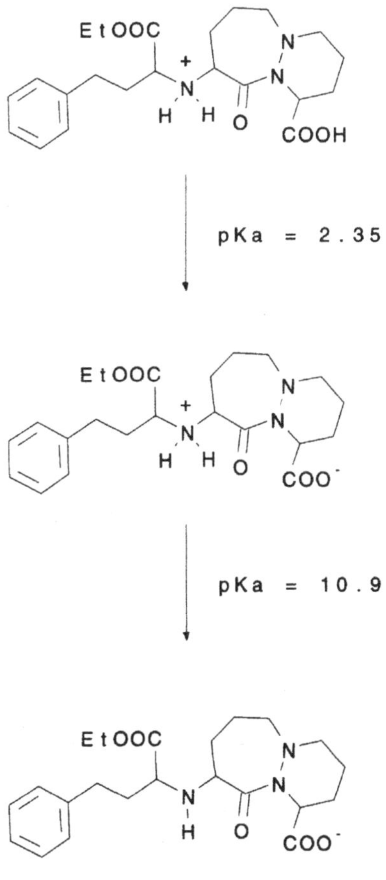

Figure 1.14   Ionisation of cilazapril.

If the pK_a is known, the Henderson–Hasselbalch equation can be used to calculate the amounts of ionised and unionised drug present at any pH. Thus for the acidic drug, romazarit, with a pK_a of 5.0: in plasma, pH = 7.4, log([base]/[acid]) = 7.4 – 5.0 = 2.4; and hence [base]/[acid] = 251 – the drug is almost entirely ionised.

Under acidic conditions, e.g. pH = 3, log([base]/[acid]) = 3.0 – 5.0 = – 2.0 and hence [base]/[acid] = 0.01 – the drug is almost entirely protonated.

1.4.4 Buffers

The relatively slow change in pH as base is added to an aqueous solution of a weak acid (e.g. Figure 1.9) is known as the 'buffering' effect of the acid. If 0.1mmole of alkali were added to a litre of pure water, the pH would theoretically rise from 7 to 10. The addition of the same amount of alkali to a 1.0mM solution of the acid of Figure 1.9 at pH 5.0 (50% ionised) removes hydrogen ions. As the acid re-ionises, the pH does not increase by as much as would be expected:

$$pH = pK_a + \log([base]/[acid])$$

and pK_a = 5.0, [base] and [acid] both = 0.5mM before the addition of 0.1mmole of alkali and 0.6mM and 0.4mM, respectively, afterwards. Hence, after the addition of the alkali:

$$pH = 5.0 + \log(0.0006/0.0004) = 5.18.$$

The buffering effect is much more pronounced near the pK. If the addition of alkali had been carried out at pH 6 with a 1.0mM solution of the same drug, the initial concentrations of acid and base would have been 0.1mM and 0.9mM, and the addition of the alkali would have completely ionised the acid to give a 1.0mM solution of the conjugate base. However, the anion will be protonated to a very small extent:

$$R\text{-}COO^- + H_2O \rightarrow R\text{-}COOH + OH^-; \quad pK_b = pK_w - pK_a = 14.0 - 5.0 = 9.0.$$

As all of the added hydroxyl ions have been used up in neutralising the acid, the only hydroxyl ions present will be produced by the reaction of the carboxylate with water, and therefore:

$$[R\text{-}COOH] = [OH^-]$$

and, as the concentration of water is effectively constant:

$$[R\text{-}COO^-] = (0.001 - [OH^-])$$
$$K_b = [R\text{-}COOH] \times [OH^-]/[R\text{-}COO^-]$$
$$\text{and } K_b = [OH^-] \times [OH^-]/(0.001 - [OH^-]).$$

Since $[OH^-] \ll 1mM$,

$[OH^-] = \sqrt{(0.001 \times K_b)} = 0.000001$ and $pOH = 6$

and $pH = 14 - 6 = 8$.

If 0.1mM of alkali had been added at pH 7 when the drug had been essentially completely ionised:

$[OH^-] = 0.0001M$

and as $K_b = [R\text{-}COOH] \times [OH^-]/[R\text{-}COO^-] = 10^{-9}$ ($pK_b = 9$),

$[R\text{-}COO^-]/[R\text{-}COOH] = 100\,000$

and $pH = pK_a + \log(100\,000) = 10$. No buffering occurs.

1.4.5 *Distribution coefficient*

The ionised forms of drugs will be much more hydrophilic than the unionised because of the hydration of the ions, and will therefore be extracted with difficulty into organic solvents. If it is assumed that only the unionised species will dissolve in the organic solvent, there will be two equilibria. In the case of an acidic compound:

$$R\text{-}COO^-_{aq} \xrightleftharpoons{\quad K_a \quad} R\text{-}COOH_{aq} \xrightleftharpoons{\quad P \quad} R\text{-}COOH_{org}$$

and the overall distribution coefficient 'D' is given by:

$$D = [R\text{-}COOH_{org}]/([R\text{-}COOH_{aq}] + [R\text{-}COO^-_{aq}]).$$

By applying the Henderson–Hasselbalch equation:

$[R\text{-}COO^-_{aq}] = \text{antilog}(pH - pK_a) \times [R\text{-}COOH_{aq}]$
and so $D = [R\text{-}COOH_{org}]/\{[R\text{-}COOH_{aq}] \times (1 + \text{antilog}(pH - pK_a))\}$.

But the partition coefficient 'P' = $[R\text{-}COOH_{org}]/([R\text{-}COOH_{aq}])$

and so $D = P/\{1 + \text{antilog}(pH - pK_a)\}$.

As with the partition coefficient, the distribution coefficient is usually expressed logarithmically, i.e. $\log D = \log P - \log\{1 + \text{antilog}(pH - pK_a)\}$.

Figure 1.15 shows how $\log D$ varies with pH for a typical acidic drug with a pK_a of 5.0 and a value of $\log P$ of 2.8. The drug is almost completely (99%) ionised at pH 7, 2 pH units above the pK_a, and only 1% ionised at pH 3. However, the compound is lipophilic and essentially quantitative extraction into organic solvent ($\log D = 2$) should be possible at pH 5.7. Conversely, the drug could be re-extracted efficiently into water ($\log D = -2.0$) from the organic solvent at pH 9.8.

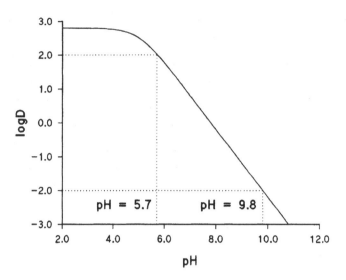

Figure 1.15 Relationship between log*D* and pH for an acid with log*P* = 2.8 and pK_a = 5.0.

1.5   Solvent extraction

The major application of solvent extraction is the separation of drugs and metabolites from polar endogenous material such as salts, sugars and peptides. It can, however, also be useful for separating metabolites from each other. For example, the basic *β*-blocker Ro 31-1411 is metabolised to basic, acidic, neutral and amphoteric (i.e. zwitterionic) derivatives (Figure 1.16).

In addition it is possible to distinguish between, and separate, the glucuronide (extracted) and sulphate (not extracted) conjugates of lipophilic metabolites by their partition between dilute aqueous acid and ethyl acetate (Figure 1.17). At pH 3 the sulphate, the salt of a strong acid, will be completely ionised and therefore insoluble in most organic solvents. The glucuronide, on the other hand, is a weak acid with a pK_a of between 4 and 5 and so will be largely protonated at pH 3 and extractable into organic solvent – provided that the aglycone is reasonably lipophilic.

1.5.1   *Choice of solvent*

Several factors need to be considered when choosing a solvent to extract a drug from a biological matrix in addition to its power to dissolve the required compounds. These include selectivity, density, toxicity, volatility, reactivity, physical hazards and miscibility with aqueous media.

• Ethyl acetate is a powerful solvent for many organic compounds and will therefore extract a considerable amount of endogenous material with the required drug. If the drug is relatively non-polar (e.g. Ro 15-1570, Figure 1.8), a more *selective* extraction could be obtained by using a hydrocarbon solvent.

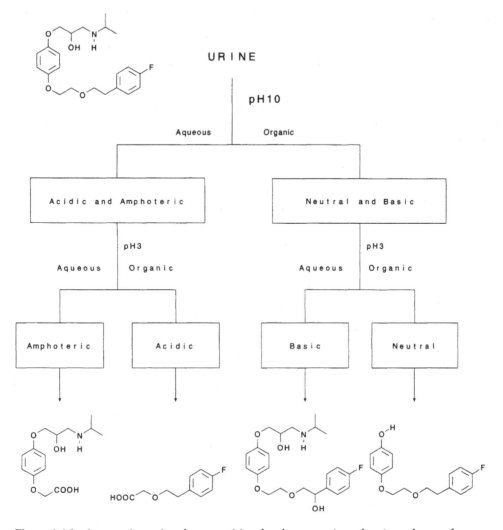

Figure 1.16 Aqueous/organic solvent partition for the separation of various classes of metabolite of Ro 31-1411 from urine.

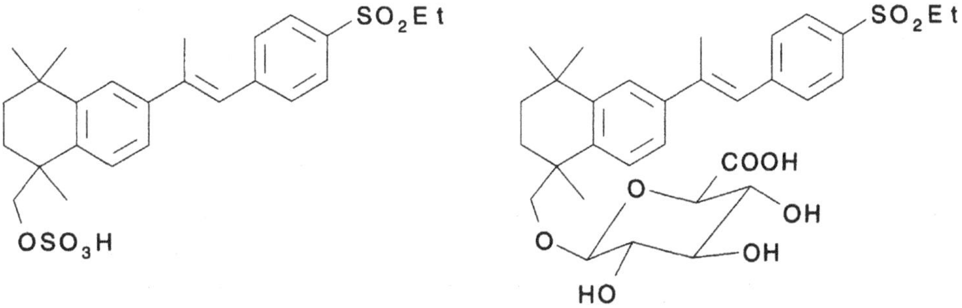

Figure 1.17 Sulphate and glucuronide conjugates of a metabolite of Ro 15-1570.

Table 1.3 First-choice solvents for extraction of drugs from aquous media

Solvent	Mwt	Bpt (°C)	Density (g/ml)	Dielectric constant	Dipole moment (D)
Dichloromethane	84.9	40	1.33	8.9	1.60
Ethyl acetate	88.1	77.1	0.90	6.0	1.78
Butyl chloride	92.6	77	0.89		
Diethyl ether	74.1	34.5	0.71	4.3	1.15
Pentane	72.2	36.1	0.63	1.8	0
Cyclohexane	84.2	80.7	0.78	2.0	0
n-Hexane	86.2	69	0.66	1.9	0.08
Toluene	92.2	110.6	0.87	2.4	0.36
Isopropanol*	60.1	82.4	0.79	18.3	1.66
Tetrahydrofuran*	72.1	66	0.89	7.3	1.63
Acetonitrile*	41.1	81.6	0.78	36.2	3.92
n-Butanol	74.1	117	1.00	7.3	1.5
[Water*	18.0	100	1.00	78.2	1.84]

* Solvents miscible with water.

- Halogenated hydrocarbons like chloroform and dichloromethane are excellent, volatile solvents. However, they are *denser* than water which makes them difficult to use for analysis when many small-scale separations have to be performed (it is trickier to remove the lower layer with a pipette). Replacement by butyl chloride, which is less dense than water, will often simplify the assay.
- Benzene is a useful solvent, reasonably volatile, inert and immiscible with water. However, its *toxicity* precludes its use.
- Toluene has similar properties as a solvent to benzene, but is not particularly toxic. However, its boiling point is 111°C and it is not really sufficiently *volatile* for use as a solvent in bioanalysis.
- Chloroform is an excellent solvent but *reactivity* with bases reduces its uses with basic drugs that need to be extracted at high pH.
- Di-isopropyl ether is less miscible with water than is di-ethyl ether but is much more likely to form *explosive* peroxides and is best avoided.
- Diethyl ether is a good, volatile solvent. However, it is quite soluble (~4%) in water and this *miscibility* means that extracts are wet and can be difficult to blow to complete dryness.

Suggested 'first-choice' solvents for extraction of drugs from aqueous media are given in Table 1.3.

1.5.2 Mixed solvents

In some cases pure solvents will not be satisfactory for the extraction of the compound of interest. Alcohols are excellent solvents but those with lower boiling points are too soluble in water to be useful for extraction and the less miscible ones have high boiling points. However, the use of mixed solvents containing alcohols can solve some problems. For example, the addition of a small amount of amyl alcohol (~1%)

to hexane has been shown to reduce losses of drugs that adhere to the glass of test tubes. Similarly, the addition of 20% of butanol to halogenated solvents can improve the extraction of polar drugs from water. Conversely, the addition of hexane to ethyl acetate might improve selectivity by reducing the extraction of endogenous materials. Although acetonitrile is completely miscible with water, it has more affinity for many organic solvents and when added to toluene is largely immiscible with water and can be used for the extraction of organic molecules. A 1 : 1 mixture of tetrahydrofuran (which is miscible with water) and dichloromethane (which is not) is a powerful solvent for the extraction of polar compounds from aqueous solutions.

1.5.3 Dealing with plasma proteins and emulsions

The presence of proteins can cause difficulties in extracting drugs from plasma. Emulsions are often formed (although these can sometimes be removed by the addition of small amounts of octanol and vigorous centrifugation) and partial precipitation can obscure the interface between the two layers. The proteins can be precipitated in a number of ways, such as the addition of 10–20% trichloroacetic acid or about five volumes of a water-miscible solvent such as acetonitrile. Both methods can cause problems. Strong acid may degrade the drug or metabolites and may well be extracted with the compounds of interest. The addition of large amounts of organic solvent adds to the volume that will need to be evaporated, and will prolong the assay time. If the compound is water-soluble, however, it may be possible to remove the acetonitrile by extraction with toluene (see example below, Figure 1.18).

1.5.4 Choice of pH for solvent extraction

Organic acids and bases are usually much less soluble in water than are the corresponding salts. As a general rule, extractions of bases into an organic solvent should be carried out at high pH (about 2 pH units above the pK_b) and extraction of acids at low pH. If the drug of interest is a reasonably non-polar base, it could be back-extracted from the organic solvent into acid, basified and re-extracted into the organic solvent. This process will ensure that the drug is only contaminated by other basic compounds. A similar, but opposite, procedure could be used to separate an acidic drug from endogenous basic and neutral components. For example, the extraction of a series of acidic drugs related to romazarit from plasma involved precipitation of proteins with acetonitrile, removal of the acetonitrile and neutral compounds by extraction with toluene, acidification to pH 3 and extraction of the drugs with ethyl acetate (Figure 1.18). Neutral drugs can be extracted into an organic solvent and purified by washing with both aqueous base and acid.

1.5.5 Artefacts arising during the extraction of drugs and metabolites

Care must always be exercised when using extreme values of pH that drugs and metabolites are not degraded. For example, esters (e.g. mibefradil, Figure 1.19) are usually labile under basic conditions and if the drug of interest is an ester, the alcohol is a likely metabolite. The use of too high a pH will cause some hydrolysis of the drug (e.g. Figure 1.19), produce a 'metabolite', and the performance of the assay will be

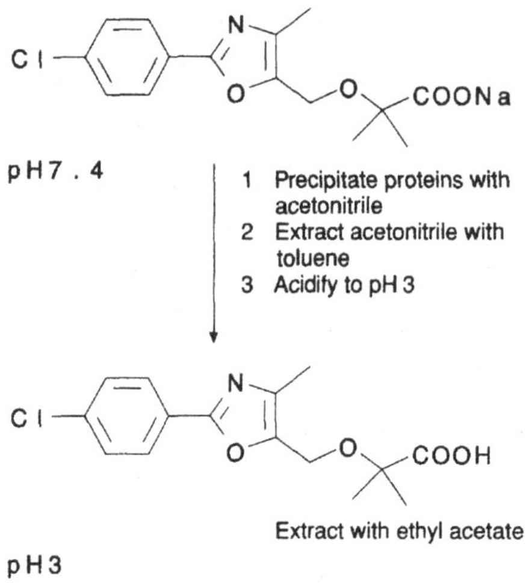

Cl
N
O
O
COONa
pH7.4
1 Precipitate proteins with acetonitrile
2 Extract acetonitrile with toluene
3 Acidify to pH 3
Cl
N
O
O
COOH
Extract with ethyl acetate
pH3

Figure 1.18 General method for the extraction of acid drugs from plasma.

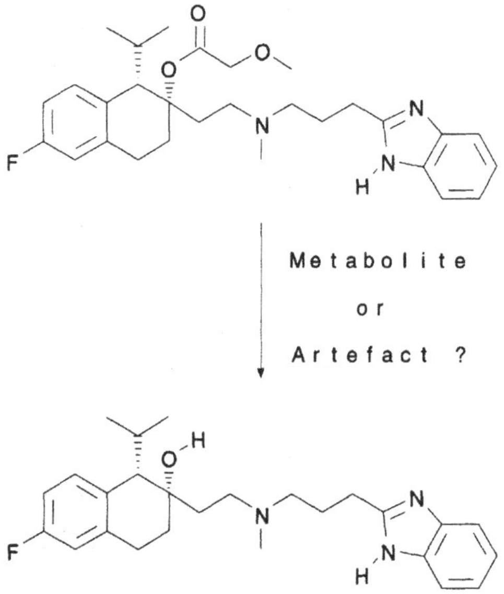

O
O
O
N
N
N
H
F
Metabolite
or
Artefact ?
O
H
N
N
N
H
F

Figure 1.19 Mibefradil and its 'metabolite'.

reduced. It would probably be better to use a lower pH buffer and to accept a poorer recovery of the drug.

This sort of quandary can be overcome in bioanalysis to some extent by the use of appropriate standards and controls. However, when isolating metabolites this is not possible, and the use of strong bases should be avoided. For example, acyl glucuronides

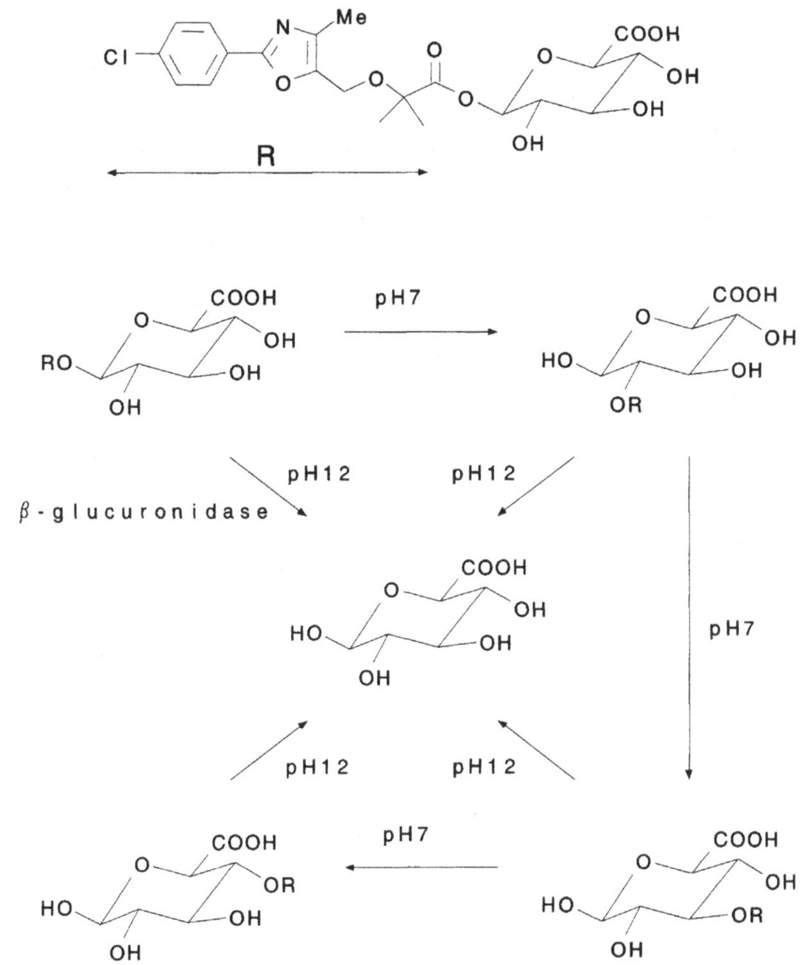

Figure 1.20 Artefacts arising from the acyl glucuronide of romazarit.

are common metabolites of organic acids (e.g. romazarit, Figure 1.20) and are unstable under basic conditions, rearranging at pH values between 7 and about 9 and being hydrolysed under more drastic conditions (e.g. Figure 1.20). Reducing the pH to 4–5 will obviate this problem.

Another potential pitfall is the use of alcohols under acidic conditions, whether for the precipitation of proteins, extraction of metabolites or chromatography. Acids are readily esterified and esters undergo even more facile trans-esterification. Acids are common metabolites and a metabolic study will be made more complex by the artefactual formation of a number of methyl esters by the careless use of acidic methanol. Figure 1.21 shows some of the artefacts and 'real' metabolites identified during metabolic studies with the β-blocker, Ro 31-1411.

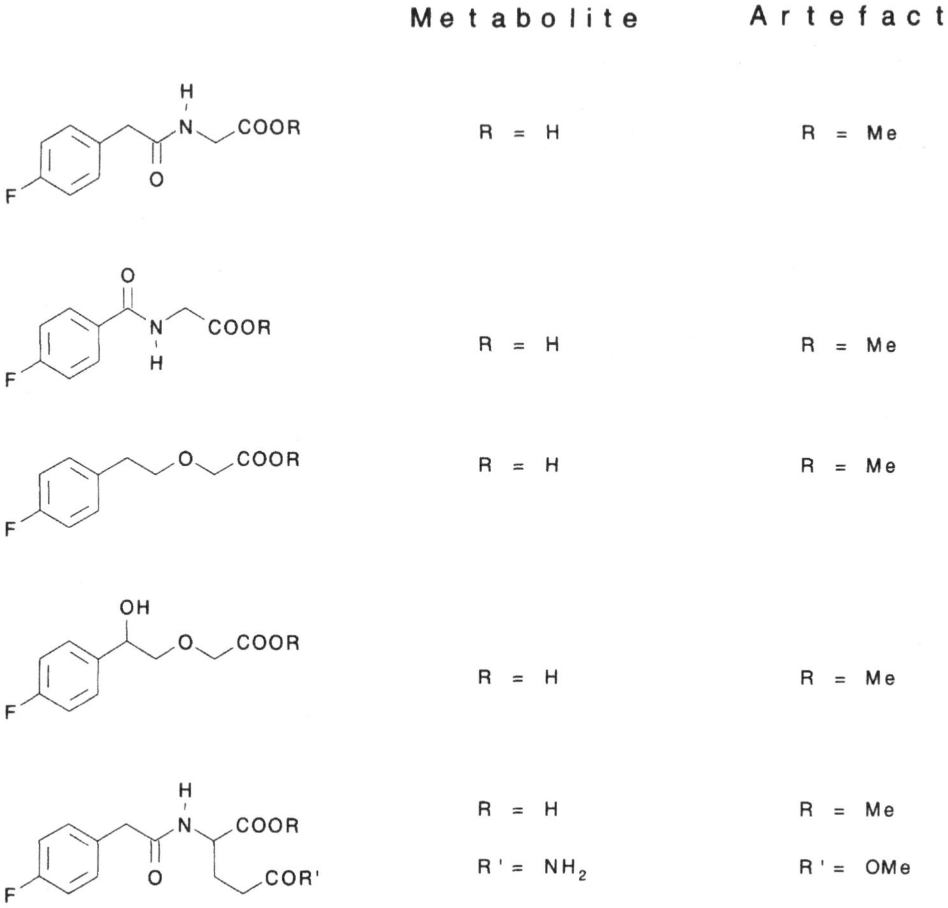

Metabolite   Artefact

R = H   R = Me

R = H   R = Me

R = H   R = Me

R = H   R = Me

R = H   R = Me

R' = NH$_2$ R' = OMe

Figure 1.21 Some artefacts arising from the isolation of the acidic metabolites of Ro 31-1411.

1.5.6 Modification and derivatisation of drugs and metabolites

In some cases it will be desirable to modify a drug or metabolite in order to assist analysis or isolation. If gas chromatography is the preferred method of analysis, it might be necessary to derivatise the molecule in order to improve volatility, or sensitivity to electron capture detection. With HPLC, on the other hand, it might be necessary to add a UV chromophore. Advantage should be taken of the derivatisation process to improve the selectivity of the extraction process whenever possible (e.g. Figure 1.22).

Polar metabolites such as conjugates are relatively difficult to isolate owing to the fact that there are larger amounts of polar than non-polar endogenous compounds in biological materials. Hydrolysis of partially purified metabolites by specific enzymes followed by isolation and identification of the less polar aglycones may well be the best method for characterisation of the conjugate.

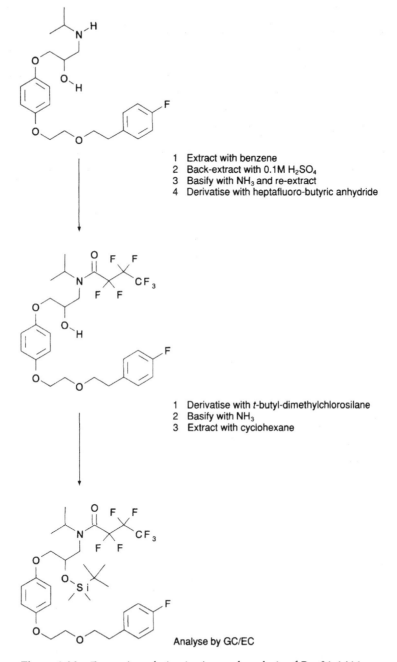

1 Extract with benzene
2 Back-extract with 0.1M H_2SO_4
3 Basify with NH_3 and re-extract
4 Derivatise with heptafluoro-butyric anhydride

1 Derivatise with *t*-butyl-dimethylchlorosilane
2 Basify with NH_3
3 Extract with cyclohexane

Analyse by GC/EC

Figure 1.22 Extraction, derivatisation and analysis of Ro 31-1411.

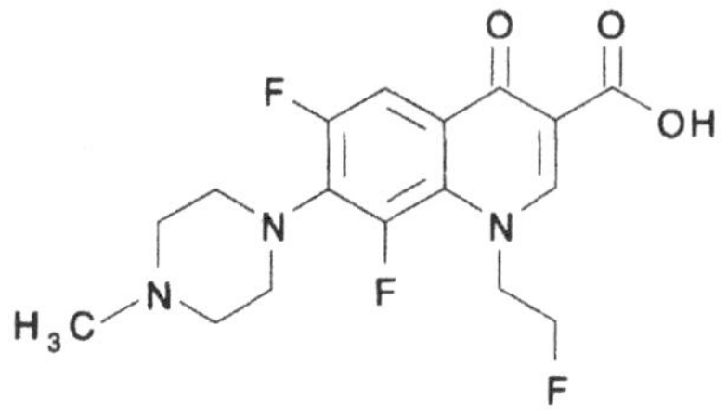

Figure 1.23 The structure of fleroxacin.

1.5.7 Ion-pair extraction

Amino acids and other zwitterions present a particular problem for solvent extraction as they are ionised whatever the pH. One answer is the use of an ion-pairing agent which forms a salt with one of the ionic centres of the drug. As the ion-pairing reagent has a lipophilic 'tail' it will dissolve in organic solvents and bring the drug with it. Figure 1.23 shows the structure of the zwitterionic fluoroquinolone antibiotic fleroxacin. A successful reverse-phase HPLC analytical method employs initial extraction from acidified plasma containing 3–8mM sodium dodecyl sulphate with a 7 : 3 mixture of dichloromethane and isopropanol.

1.5.8 Recoveries

Poor recoveries lead to poor reproducibility of assays. They also make the isolation of reasonable amounts of metabolites difficult. It follows that the recovery of any extraction process should be determined and optimised. Normally, recoveries of metabolites at each stage of an isolation study will be monitored by liquid scintillation counting of aliquots. Radiochemical methods are also the simplest way for determining recoveries for bioanalytical methods. Robust assays can be developed, even when recoveries are low, as long as they are reproducible. However, a 50% recovery will double the theoretical quantification limit, if the sensitivity is governed by the signal-to-noise ratio.

1.6 The 'first law of drug metabolism'

The hardest part of metabolic and analytical studies is the development of suitable methods for the separation of the drugs and metabolites of interest from endogenous material. The large number of endogenous compounds in plasma and excreta means that interference is extremely likely. Nevertheless, the many kinds of intermolecular forces which are responsible for solubility and for solid-phase interactions mean that it should not be difficult to separate two dissimilar compounds by one of the available methods.

The 'first law of drug metabolism' requires that the scientist should use a variety of methods for analytical and metabolic studies. For example, solvent or ion-exchange extraction (at different values of pH) followed by reverse-phase HPLC chromatography might be suitable for an analytical method. Alternatively, before a gas chromatography

(GC) analysis, reverse solid-phase extraction, derivatisation and solvent extraction might be an efficient procedure. When devising a strategy for the isolation of metabolites, it should be possible to employ solvent extraction, ion-exchange methods, normal-phase thin-layer chromatography (TLC) and reverse-phase HPLC (with a variety of solid and mobile phases).

1.7 Bibliography

Barrow, G.M. (1996) *Physical Chemistry*, 6th edn, New York: McGraw-Hill (ISE).

Chang, R. (1977) *Physical Chemistry with Applications to Biological Systems*, New York: MacMillan Publishing.

Moore, W.J. (1972) *Physical Chemistry*, 5th edn, Harlow: Longman.

2

SOLID-PHASE EXTRACTION

Chris James

2.1 Introduction

Over the past decade solid-phase extraction (SPE) techniques have largely replaced liquid/liquid methods as the preferred technique to extract drugs from biological fluids prior to quantitative analysis. As the name suggests, extraction is performed by absorbing the analyte(s) from the matrix onto a solid support (sorbent). SPE methods typically have four main steps: a priming step to ensure conditions are optimal for retention of the analyte, a retention step where the matrix is applied to the cartridge and the analyte is retained, a rinsing step where potential interferences are eluted while the compound of interest remains bound to the sorbent, and an elution step where the purified analyte is collected for analysis (Figure 2.1).

SPE has significant advantages compared with traditional liquid/liquid (L/L) extraction methods. The wide range of sorbents available (Table 2.1) make the technique

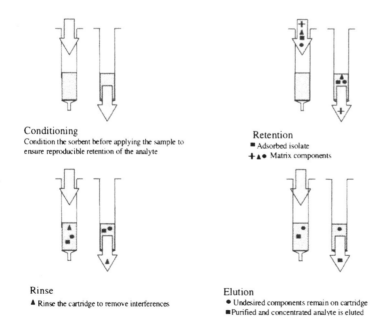

Conditioning
Condition the sorbent before applying the sample to ensure reproducible retention of the analyte

Retention
■ Adsorbed isolate
+▲● Matrix components

Rinse
▲ Rinse the cartridge to remove interferences

Elution
● Undesired components remain on cartridge
■ Purified and concentrated analyte is eluted

Figure 2.1 Principle of solid-phase extraction.

Table 2.1 List of sorbents

Sorbent description	Structure	Endcapped*	Capacity	pKa	Mode
C18 Octadecyl	$-Si-C_{18}H_{37}$	–			Strong non-polar
C8 Octyl	$-Si-C_8H_{17}$	–			Moderate non-polar
C2 Ethyl	$-Si-C_2H_5$	–			Weak non-polar
C1 Methyl	$-Si-C_1H_3$	–			Weak non-polar
Ph Phenyl	$-Si-\bigcirc$	–			Moderate non-polar
CH Cyclohexyl	$-Si-\bigcirc$	–			Moderate non-polar
CN Cyanopropyl	$-Si-CH_2CH_2CH_2CN$	–			Moderate non-polar/polar
2OH Diol	$-Si-CH_2CH_2CH_2OCH_2CH-CH_2$ $OH\ \ OH$				Polar
Si Silica	$-Si-OH$				Polar
CBA Carboxylic acid	$-Si-CH_2COO^-$		0.3meq/g	4.8	Weak cation exchange
PRS Propylsulphonic acid	$-Si-CH_2CH_2CH_2SO_3^-$		0.8meq/g		Strong cation exchange
SCX Benzenesulphonic acid	$-Si-CH_2CH\bigcirc SO_3^-H^+$		0.6meq/g		Strong cation exchange
NH₂ Aminopropyl	$-Si-CH_2CH_2CH_2NH_2$			9.8	Weak anion exchange/polar
PSA Primary-secondary amine	$-Si-CH_2CH_2CH_2NCH_2CH_2NH_2$ H			10.1 10.9	Weak anion exchange/polar
DEA Diethylaminopropyl	$-Si-CH_2CH_2CH_2N(CH_2CH_3)_2$		1.0meq/g	10.7	Weak anion exchange/polar
SAX Quaternary amine	$-Si-CH_2CH_2^+CH_2N(CH_3)_3$		0.7meq/g		Strong anion exchange
PBA Phenylboronic acid	$-Si-CH_2CH_2CH_2NH\bigcirc$ $B(OH)_2$				Covalent
Polymeric sorbents, e.g. Waters Oasis™ HLB	Poly(divinylbenzene-co-N-vinylpyrrolidone)				Non-polar/polar

* Many sorbents are now available in both endcapped and non-endcapped forms.

applicable to most classes of drugs (and metabolites), and SPE can be used for highly polar and ionic compounds that are difficult to extract by liquid/liquid methods. Careful selection of the sorbent and refinement of the extraction conditions will normally allow a highly selective extraction to be developed. SPE is relatively easy and quick to perform, particularly when extracting large numbers of samples, and methods are simpler to automate than liquid/liquid procedures.

The recent introduction of SPE materials in 96-well microtitre plate format further aids processing large numbers of samples, whether conducted manually or via automated instruments. SPE methods also minimise the use of toxic solvents.

2.2 General properties of bonded silica sorbents

Although other materials are available (e.g. polymeric resins, XAD-2, alumina, florisil), the vast majority of SPE extractions are carried out using bonded silica materials similar to those used in HPLC columns (except the particle size is larger, typically 40μm diameter instead of 5μm). More recently 'disc'-based cartridges have been introduced (Section 2.7), but again the functional chemistry of these is based on bonded silica sorbent incorporated into the disc material.

Bonded silica phases are produced by the reaction of organosilanes with activated silica, allowing the introduction of a wide range of functional groups (Table 2.1). These are typically categorised into those containing non-polar, polar, ion-exchange or covalent functionalities. The terminology used for SPE differs slightly from that used for HPLC, and non-polar extractions are essentially analogous to reverse-phase HPLC separations, and polar SPE to normal-phase chromatography.

Bonded silica materials are rigid and do not shrink or swell like polystyrene-based resins in different solvents. They equilibrate rapidly and flow through a 40μm sorbent bed can be achieved by application of moderate pressure (10–15 psi). Processing cartridges involves drawing various solutions and the sample through the sorbent, which can be achieved by applying vacuum, by positive pressure (e.g. from a syringe) or by centrifugation. Vacuum manifolds are most commonly used for manual processing of samples, whereas automated instruments (Chapter 10) use either vacuum or positive pressure applied via a syringe pump.

Use of too high a flow rate when processing cartridges may affect retention of analytes, particularly during the sample loading and elution steps. For a manual extraction with a 1 ml cartridge (100 mg sorbent), minimal vacuum should be applied to draw a 1 ml loading solution or elution solution through the sorbent in 30–45 seconds. For automated instruments, flow rates of 1–2 ml/min are generally recommended for these critical steps in the assay.

The capacity of most sorbents is approximately 5% of sorbent mass, and the bed volume is about 120μl/100mg of sorbent. Minimum elution volume for a cartridge is about two bed volumes of elution solvent; however, to ensure robustness larger volumes are usually advisable. Potentially, the capacity of the cartridge will be affected by all components from the sample, not only the analytes of interest. For this reason, if testing the selectivity of a method against a co-administered drug, one should check both for co-elution in the chromatographic system, and that high concentrations (equivalent to C_{max}) of the co-administered drug do not affect recovery of the analyte during the SPE process.

Cartridges or 96-well plates are available commercially containing from 10 mg to 10 g of bonded silica packing, and materials are also available in the form of sheets or discs. The smaller sizes (10–500 mg) are normally used for bioanalytical and metabolism applications.

2.3 Sorbent/analyte interactions

2.3.1 Solvation

Solvation of the sorbent is essential, particularly for non-polar phases, to achieve reliable retention of analytes. This is performed by passing a small volume of a water-

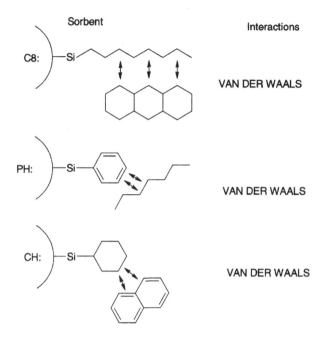

Figure 2.2 Non-polar interactions.

Table 2.2 Bond energies of various interactions

Interaction	Energy (kJ/mol)
Covalent	200–800
Ionic	40–400
Hydrogen bond	4–40
Dipole–dipole	0.4–4
Induced dipole–dipole	0.4–4
Dispersion	4–40

miscible organic solvent (typically methanol or acetonitrile) through the sorbent. This 'wets' the phase to create a suitable environment for adsorption of the analyte from an aqueous medium. Excess solvent is removed from the cartridge with a water or aqueous buffer wash prior to loading the sample. Drying of the sorbent by excessive application of vacuum during extraction can result in a loss of this solvation and variable retention of the analyte.

2.3.2 Non-polar

Retention occurs via non-polar interactions between carbon–hydrogen bonds of the analyte and carbon–hydrogen bonds of the sorbent functional groups ('Van der Waals' or 'dispersion' forces; Figure 2.2). Bond energies of various types of interaction are shown in Table 2.2. In theory, high bond energies are advantageous since strong

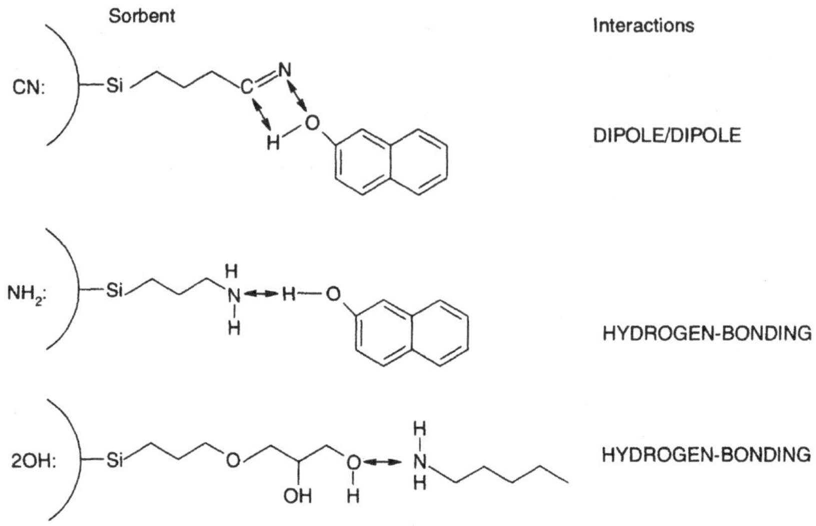

Figure 2.3 Polar interactions.

wash solvents can be used to remove interferences. In practice, non-polar phases are most commonly used as, although bond energy is weak, there are many possible sites of interaction of the sorbent with organic compounds and good retention is often obtained. Retention of analytes will occur from polar (i.e. aqueous) environments and consequently most biofluids can be applied directly to the cartridge following dilution with water or buffer. Analytes are eluted with a less polar solvent (e.g. acetonitrile, methanol, acidified methanol). Mixtures of acetonitrile or methanol and water are commonly used as wash solvents.

2.3.3 Polar

Retention occurs via polar interactions, e.g. hydrogen bonding, dipole–dipole, induced dipole–dipole and pi–pi (Figure 2.3). These phases can offer a highly selective extraction procedure capable of separating molecules with very similar structures. Their main drawback is that the analyte must be loaded onto the sorbent in a relatively non-polar organic solvent such as hexane. This requires biofluid samples to be first subjected to liquid/liquid extraction or SPE to transfer the analyte into a suitable solvent. Elution is accomplished with more polar solvents such as ethyl acetate or dichloromethane. Typically, mixtures of solvents (e.g. ethyl acetate : hexane) are used to provide wash solvents of intermediate polarity.

2.3.4 Ion exchange

Ionic interaction occurs between an analyte carrying a charge (positive or negative) and a sorbent carrying the opposite charge (Figure 2.4). The pH of the matrix must be adjusted so that both the analyte and the sorbent are ionised for retention to

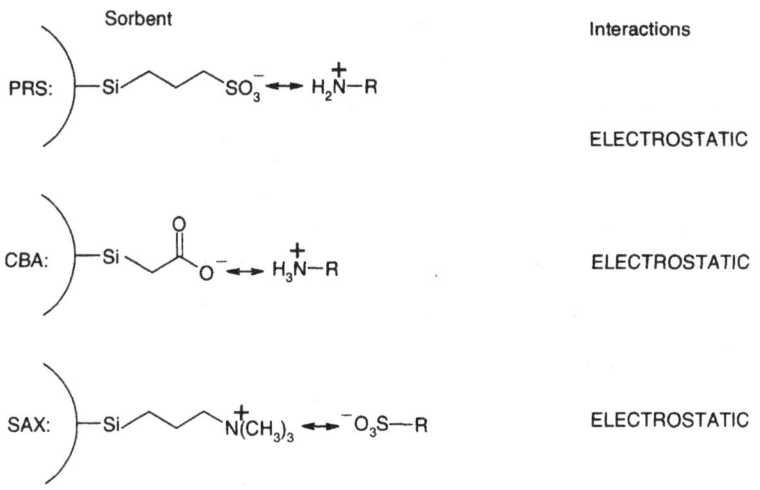

Figure 2.4 Ion-exchange interactions.

occur. For example, to retain a protonated basic compound with a pK_a of 10.5 on a CBA cartridge (pK_a 4.8), the pH would have to be a least two pH units below the pK_a of the analyte (pH < 8.5), but at least two pH units above the pK_a of the CBA phase (pH > 6.8). Elution from an ion-exchange sorbent is achieved by adjusting the pH to suppress the charge on either the analyte or sorbent (i.e. in the example above, a pH above 12.5 or below 2.8 would be needed for elution), or by the introduction of a counter-ion at a high ionic strength (i.e. buffer > 0.1M). With strong anion- or cation-exchange cartridges (e.g. SAX, SCX) the charge on the phase cannot easily be suppressed, so elution must be performed by suppressing the charge on the analyte, or use of a high ionic strength buffer. As SPE cartridges are used only once, the pH limits normally applied for HPLC columns are not important and basic or highly acidic solutions can be used during the extraction, if applicable. An organic modifier may also be required to overcome secondary non-polar interactions.

Biofluids can usually be applied directly to ion-exchange sorbents following dilution of the sample with water or a buffer, and possibly adjustment of pH. However, elution from strong ion-exchange sorbents can be a problem as high ionic strength, or extremes of pH, may be required, which may affect analyte stability or further processing of the sample (e.g. the sample may still contain buffer salts). Use of ion-exchange SPE can often improve the overall selectivity of a method, as it provides a contrast to the retention mechanism of a typical reverse-phase HPLC separation interaction.

2.3.5 Covalent

Reversible covalent interactions can be achieved between vicinal diols (e.g. catecholamines, *cis*-diols) and phenylboronic acid (PBA) sorbents. An alkaline pH is required to condition the phase prior to retention, and elution is effected under acid

Table 2.3 Type of interactions on different sorbents

	Non-Polar	Polar	Anion exchange	Cation exchange
C18	●	○		✤
C8	●	○		✤
C2	●	●		✤
CH	●	○		✤
Ph	●	○		✤
CN	●	●		✤
2OH	●	●		✤
Si		●		✤
NH₂	○	●	●	✤
PSA	○	●	●	✤
DEA	○	●	●	✤
SAX	○	○	●	✤
CBA	○	○		●
PRS	○	○		●
SCX	●	○		●

●, Primary; ○, secondary; ✤, silanol activity.

conditions. Highly selective extractions can be developed due to the high bond energy of the reversible covalent link and the specificity of the reaction for these functional groups. Strong wash solvents (e.g. 50% methanol) can be used provided a high pH is maintained, eluting contaminates that are retained by non-specific, rather than the covalent mechanism.

2.3.6 Mixed-mode interactions

Although SPE phases are conveniently grouped into non-polar, polar, ion exchange, etc., virtually all sorbents provide the possibility of more than one mode of interaction occurring between the sorbent and analyte (Table 2.3). These secondary interactions can be exploited, or they can be a potential pitfall. Residual silanol groups in particular can be a problem as virtually all silica-based SPE sorbents, even endcapped phases, contain unbound acidic silanol groups. These can retain basic compounds by cation exchange. An extraction method that utilises such secondary interactions can fail in routine application as the percentage of free silanol groups can vary between batches of sorbent. Investigation of retention, wash and elution solvents can often indicate that such secondary interactions are significant, e.g. a basic compound may not elute completely with pure acetonitrile but will elute completely with acidified methanol.

A number of 'mixed-mode' phases (Figure 2.5) have been introduced to exploit multiple mode interactions, and to make them more reliable. Such phases were originally designed to screen for drugs of abuse but they have found wider use in bioanalytical applications and metabolite isolation (Bogusz et al. 1996; Ingwersen 1993).

Combinations of functionality can be used to retain mixtures of analytes with different properties, e.g. a lipophilic parent drug and polar metabolites. Alternatively, the multiple interactions can be exploited to give a cleaner extract for a compound that will be retained by multiple interactions (e.g. ion exchange and non-polar);

34

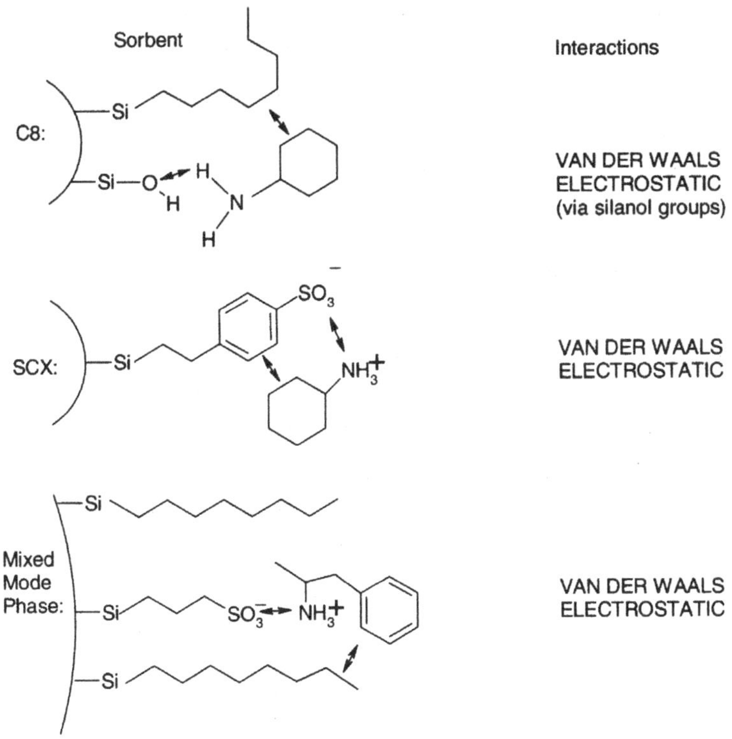

Figure 2.5 Mixed-mode interactions.

strong wash solvents for each 'mode' can be applied sequentially to elute contaminants retained by one mode while the analyte is retained throughout. It is also possible to fractionate complex mixtures by adjusting elution solvents. As such phases are manufactured to provide both ionic and hydrophobic characteristics, there is little risk of the method failing due to batch-to-batch variation, which would be a significant risk, if, for example, secondary interactions with silanol groups on a non-polar phase were required for analyte retention. Examples of these phases are Varian's Certify I™ and Isolute-Confirm HCX™, which combine strong cation exchange with non-polar functional groups, and Varian Certify II™ and Isolute-Confirm HAX™, which combine strong anion exchange and non-polar functionality. There is even a cartridge (Isolute-Multimode™) which combines cationic, anionic and non-polar functionalities. Not all manufacturers will reveal the precise chemistry of these phases, but some cartridges are produced by simply packing a mixture of two different sorbents with the required functionalities.

2.3.7 Polymeric sorbents

Polymeric materials (e.g. styrenedivinylbenzene) have been available for a number of years and, in terms of functionality, are generally comparable with the non-polar silica phases (C8, C18, etc.). Potentially, they do not have the problems associated with the silanol groups of silica-based materials, but slight swelling of the gel leads to higher back-pressure and consequently a significant practical disadvantage when processing large numbers of samples. Recently a new polymeric material, Waters Oasis™ (poly(divinylbenzene-co-N-vinylpyrrolidone)) has been introduced. Although this material still needs slightly greater vacuum to process samples, it is reported as having a mixture of lipophilic and hydrophilic retention characteristics, which enhance its ability to retain polar materials. In addition, the phase can be wetted by aqueous solutions and consequently it is very resistant to drying and loss of solvation during the sample extraction process; with less concern about drying of cartridges, processing of samples should be both easier and more robust.

2.3.8 Miscellaneous

A number of other SPE sorbents are available. For example, Chemelut™ and Toxelut™ cartridges (Varian) contain a diatomaceous earth and allow liquid/liquid extraction to be performed on a solid support. Cartridges are also available for very specific assays, e.g. THC cannabinoids. Highly selective extraction cartridges produced from antibodies (immunoaffinity extraction) or 'molecular imprinting' techniques are discussed in Chapter 9.

The vast majority of SPE phases are packed into cartridges having polypropylene barrels and polyethylene frits. For most assays these materials do not cause any major problems. However, if leaching of interferences from the frits is suspected, or other problems occur, cartridges can also be obtained with glass barrels and/or stainless-steel frits.

2.4 Sample pretreatment of different biological matrices

2.4.1 Liquid samples

The only pretreatment normally required for liquid samples such as plasma, urine and serum is dilution with water or buffer. Retention is generally superior from dilute samples but the pH must be appropriate to achieve the required interaction between the sorbent and analyte. Any pH adjustment is therefore best performed with relatively large volumes of dilute buffer (< 0.1M), ensuring that sufficient buffering capacity exists to make the intended pH change. Urine may contain high salt (and possibly high drug and/or metabolite concentrations) and care should be taken not to exceed the capacity of the cartridge. As a rule of thumb, up to 1ml of plasma or up to 100μl of urine can be extracted onto a 100mg cartridge.

2.4.2 Protein binding

Problems due to plasma protein binding of drugs are normally indicated by a high recovery when the analyte is loaded onto the cartridge in aqueous solution, but low recovery when the analyte is loaded in spiked plasma. Drug–protein interactions may be overcome by adjustment of pH or, more commonly, by precipitating plasma proteins with acetonitrile, perchloric acid, trichloroacetic acid, etc., prior to applying the sample onto the cartridge. However, many highly protein-bound drugs can be extracted successfully without special pretreatment as passage through the sorbent is sufficient to disrupt the protein binding.

2.4.3 Solid samples

Solid samples (e.g. tissues, faeces) are normally homogenised with a buffer or an organic solvent, the remaining solids removed by centrifugation, and the diluted sample applied to the cartridge.

2.5 Developing SPE methods

In developing a method, the properties of the isolate, the sorbent and the matrix should be considered. Although one should try to understand the mechanisms of retention, it should also be remembered that various secondary interactions may occur, and that a selective extraction from biofluid also requires the sorbent *not* to extract a large number of unknown compounds from the matrix. The best conditions are not easily predicted and the method needs to be developed experimentally in combination with the analytical method being used.

The following stages are recommended for method development:

1 Consider physico-chemical properties of analyte, nature of matrix and known chromatographic properties of analyte. Consider likely interactions with SPE sorbents.
2 Screen a range of cartridges (e.g. C18, C8, C2, Ph, CBA, SCX, SAX, PRS, NH$_2$, DEA, CH, Si, Certify I and II, polymeric sorbent – including the ones that are not expected to work) – under simple conditions (e.g. from aqueous buffer solutions) looking for good retention of the analyte. If radiolabelled analyte is available, this can conveniently be used to track analyte in such screening experiments. This experiment should not only identify likely cartridges for use in the assay, but should be used to try to confirm possible mechanisms of interaction between the analyte and sorbent.
3 Select a more limited number of sorbents (2–3) and examine conditions for loading/wash/elution (consider if pH control is needed, possible strength (% organic solvent) of wash solvents). Try extraction from biofluid. Select sorbent for final development.
4 Final development of extraction conditions and chromatographic analysis. Consider the robustness of the assay when finalising extraction conditions; for example, do not select a wash solvent where a small change in composition could lead to elution of analyte from the cartridge.

The choice of different cartridges, manufactures and formats is becoming so extensive that it can appear almost overwhelming. It is generally better to build experience with a more limited set of sorbents, perhaps concentrating on cartridges from only one manufacturer. Also concentrate on investigating the use of cartridges with different functional groups (i.e. test C18, C8, C2, Phenyl, CBA, SCX, etc.), and those that use a contrasting mechanism to the analytical separation (Simmonds *et al.* 1994).

Table 2.4 Example of a SPE Assay for PNU-78875; comments are given explaining reason for various steps in the extraction

Step	Comment
Add 50µl internal standard to 0.5ml plasma	A 'good' internal standard (close structural analogue for HPLC; stable label for LC/MS) is a useful, but not essential for SPE extractions
Dilute to 1ml with 0.05M phosphate buffer pH 7.4 (PB)	pH control with dilute buffer used to minimise co-extracted plasma components. For high sensitivity assays dilution with high-purity water sometimes minimises interferences, even if pH control with a buffer would, in theory, be more correct
Prime cartridges (C2, 100mg/1ml size) with: 2×1ml CH_3CN	Use of two 1ml aliquots to ensure cartridge is fully solvated
1×1ml H_2O	Excess acetonitrile is removed with water wash to avoid possible precipitation of buffer
1×1ml PB	Adjust pH in cartridge as part of priming step
Load cartridges with diluted plasma	Retention is normally better from dilute sample
Wash cartridge with 2×1ml H_2O and 2×1ml 20% methanol	Wash steps do not use buffer; can give cleaner extract and avoids risk of buffer salts precipitating in final extract and blocking column
Elute cartridges with 0.6mL 60% CH_3CN, 0.1% trifluoroacetic acid	As 100% CH_3CN not required for elution, use lower conc. for improved selectivity. TFA helps overcome secondary (e.g. silanol) interactions, and is also volatile, so can be evaporated
Partially evaporate extracts to approx. 200µl	Partial evaporation is quicker; CH_3CN is evaporated more easily so extract solution is mostly aqueous, allowing injection of large volume into RP-HPLC. Partial evaporation is also useful if poor recovery after complete evaporation step is noted, e.g. due to binding to the tube wall after complete evaporation
Inject 150µl into HPLC	

2.6 Example of an SPE method

Table 2.4 shows the SPE method to extract a potential anxiolytic drug, Pharmacia and Upjohn compound PNU-78875 from human plasma; comments are provided on choice of various conditions for the extraction. Following SPE, extracts were analysed by HPLC with fluorescence detection.

2.7 Disc cartridges

Over the past few years SPE materials in a disc cartridge format (Lensmeyer *et al.* 1991; Blevins and Schultheis 1994) have become available (principally from 3M

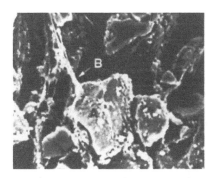

Figure 2.6 Structure of Empore™ disc SPE material showing silica particles (A) embedded in a matrix of PTFE fibrils (B).

(Varian) and Anasys). These sorbents are manufactured in continuous sheets and discs (typically 4 or 7mm diameter) of the material are incorporated into the extraction cartridges. Both types of disc incorporate bonded silica sorbents with similar functional groups to those available with standard cartridges, although fewer phases are currently available. The two types of disc are:

1 3M Empore™ (Varian). Bonded silica particles (7μm) are incorporated into a matrix of polytetrafluorethylene (PTFE) fibrils (Figure 2.6). The ratio of sorbent to PTFE is 90 : 10. In addition to Empore™ disc cartridges with 1, 3 and 6ml reservoirs incorporating 4, 7 and 10mm diameter discs, the Empore™ material is available as 25, 47 and 90mm diameter discs. These can be used to extract large volumes (e.g. 500ml) and have applications such as the extraction and concentration of metabolites from relatively large volumes of urine.
2 SPEC™ (Anasys, available from Shandon HPLC in the UK). Bonded silica phase (30–40μm) is embedded in glass fibre discs. Discs incorporating 15mg and 1.5mg sorbent are available and each has a capacity of 5–10% of the sorbent mass.

2.7.1 Potential advantages

The products should be more reliable as the materials are mass produced in continuous sheets and it is not necessary to pack individual columns with loose, particulate materials. No frits are required, so samples are not contacting frits with a large surface area that may adsorb analytes, or from which contaminates might be leached (frits used in SPE are typically made from polyethylene). The bed volume is much reduced and volumes of wash solvents, elution solvents and possibly sample can be reduced proportionately. With careful adjustment of conditions, the final eluent can often be injected directly in the HPLC or LC/MS system, hence avoiding an evaporation step which is often needed with large cartridge formats. With the increasing batch size of many analyses, and particularly the use of SPE materials in 96-well format, the avoidance of an evaporation step can give a big improvement in assay efficiency.

2.7.2 Disadvantages

The flow characteristics of discs can be very different from those of conventional cartridges, and methods may not transfer directly to disc cartridges without modification. Resistance to flow is fairly high with Empore™ cartridges compared with standard cartridges, requiring greater vacuum to be applied. Both types of disc are prone to blocking when biological samples are extracted, and both Varian and Anasys now have disc cartridges available with a coarse prefilter to reduce this problem. It is far easier to dry out the disc cartridges and for solvation of the phase to be lost.

2.8 96-Well format (e.g. Porvair Microsep™ system)

The physical 'packaging' of the extraction materials has developed considerably over the past few years, with disc materials and different sizes and formats of cartridge. A particularly significant development is the availability of SPE materials in the 96-well microtitre plate format (Allanson *et al.* 1996; Kaye *et al.* 1996; Plumb *et al.* 1997). This format is already used extensively for many other types of analysis, chemistry and sample handling. Its use in bioanalysis can be expected to increase greatly, not only as an SPE format, but also for general liquid handling and to replace individual autosampler vials. With increasing automation and LC/MS analysis, batches sizes often exceed 100 or even 200 samples, and it is no longer efficient to handle 100 or more individual tubes, SPE cartridges, etc. All major sorbent suppliers are now offering SPE materials in the 96-well block format, and disc cartridges are also available (Figure 2.7). These plates can be processed manually using a special vacuum manifold, or automated instruments can be used (Chapter 10). Even with manual processing, the 96-well plates are significantly easier and quicker to process than a similar number of individual cartridges.

Figure 2.7 96-Well extraction plate and vacuum manifold.

2.9   Direct injection of plasma

There have been a number of attempts to produce HPLC columns which would allow the direct analysis of plasma without any extraction. These have been referred to as internal surface reverse-phase (ISRP) supports, restricted access materials or Pinkerton columns (Pinkerton *et al.* 1986). The principle of all these columns is that proteins are excluded from the pores of the material which contain lipophilic groups suitable for retention of small analyte molecules. The outer surface of the silica particles is modified to be more polar and non-adsorptive to proteins. In practice, these columns seem to be prone to blockages, and provide less efficient chromatography compared with standard reverse-phase HPLC columns. The restricted access materials have not been widely used, and may have greater utility as pre-columns in column-switching systems (Chapter 10).

2.10   Other new developments

2.10.1   Fines

Users of SPE cartridges will be familiar with the problem of variations in resistance to flow between individual cartridges, or different wells in a 96-well plate. When processing a large number of samples on a vacuum manifold, many users have a syringe (and rubber bung!) to apply positive pressure to the few cartridges that are more resistant to solutions passing through them. Although centrifugation of plasma samples can reduce such blockages, this problem also relates to fines within the cartridges. These are silica particles with a diameter significantly less than the nominal size range of the material (normally 30–40μm) that have been generated in the manufacturing process. Certain suppliers have made efforts to greatly reduce the presence of fines (e.g. Isolut™ materials, IST). This not only provides more consistent flow between cartridges, but reduces the resistance to flow in general. With certain formats (e.g. 3ml cartridges containing 50 or 100mg sorbent), most priming, loading, wash and elution solvents will flow through the cartridges under gravity, or with the minimum of applied vacuum. This can make the processing of a large number of samples much easier. Similarly, the trend with automated systems using 96-well plates is to reduce the amount of sorbent (to as little as 10mg), to both reduce elution volume and minimise resistance to flow.

2.10.2   A cartridge in a pipette tip?

Pipette tips containing disc SPE material (SPEC PLUS PT™, Anasys) should be available shortly. These will provide two interesting possibilities. First, for immediate processing of single, or small numbers of samples, using a simple hand-held pipette away from the laboratory. For bioanalysis, this is most likely to be useful for samples containing unstable analytes, which could be extracted rapidly after collection from a patient to prevent degradation. Secondly, a number of standard liquid-handling laboratory robots (e.g. Tecan Genesis, Packard, Gilson 215) can already pick up and discard standard pipette tips. Such robots would therefore be able to perform SPE extraction with no additional equipment or special adaptation.

2.11   Conclusions and future perspectives

1   SPE is now the method of choice for extracting drugs from biological fluids for quantitative analysis, as it combines efficient extraction with good sample clean-up and ease of use.

2   As long as future analytical systems require significant sample clean-up before injection, SPE looks set to continue as the preferred extraction method.

3   SPE is a mature technology and the functional chemistry of the cartridges has not changed much over the past decade, though multimode phases are a useful addition to the analyst's options. The wide range of chemistries available ensures that SPE is applicable to most classes of drug compounds.

4   Recent advances have focused mostly on differing physical formats to 'package' the SPE phases, to improve reliability and consistency of cartridges (e.g. Empore™ discs), make extraction easier to perform and allow ever greater numbers of samples to be extracted (e.g. 96-well format).

5   Although it is important to understand the chemistry of the sorbent/analyte interaction, choice of the physical aspects of the method (cartridge size, format, volumes and flow rates of solution, etc.) is becoming increasingly important to allow practical processing of large numbers of samples.

6   Highly selective analytical techniques, such as LC–MS–MS, have reduced the need for very selective methods of extraction, and consequently the need to optimise the selectivity of the SPE of each individual assay. We can therefore expect greater interest in 'generic' SPE methods, which are applicable to a range of compounds, and require little addition method development for each new compound.

7   Where highly selective extraction is required, immunoaffinity or materials generated by 'molecular imprinting' (Chapter 9) have some promise, although it seems unlikely that such methods will be needed for the majority of assays.

2.12   Bibliography

Allanson, J.P., Biddlecombe, R.A., Jones, A.E. and Pleasance, S. (1996) The use of automated solid phase extraction in the '96 well' format for high throughput bioanalysis using liquid chromatography coupled to tandem mass spectroscopy. *Rapid Communications in Mass Spectroscopy*, **10**, 811–16.

Analytichem International (1985) *Proceedings of the Second Annual International Symposium on 'Sample Preparation and Isolation Using Bonded Silicas'*, 14–15 January, Philadelphia USA. Analytichem International Inc, 24201 Frampton Avenue, Harbor City, CA 90710 USA.

Blevins, D.D. and Schultheis, S.K. (1994) Comparison of extraction disk and packed-bed cartridge technology in SPE. *LC–GC*, **12**, (1), 12–16.

Bogusz, M.J., Maier, R.D., Schiwy-Bochat, K.H. and Kohls, U. (1996) Applicability of various brands of mixed-phase extraction columns for opiate extraction from blood and serum. *Journal of Chromatography B: Biomedical Applications*, **683**, (2), 177–88.

Horne, K.C. (ed.) (1985) *Analytichem International Sorbent Extraction Technology Handbook*, Analytichem International Inc. 24201 Frampton Avenue, Harbor City, CA 90710 USA.

Ingwersen, S.H. (1993) Combined reversed phase and anion exchange solid phase extraction for the assay of the neuroprotectant NBQX in human plasma and urine. *Biomedical Chromatography*, **7**, (3), 166–71.

Kaye, B., Herron, W.J., Macrae, P.V., Robinson, S., Stopher, D.A., Venn, R.F. and Wild, W. (1996) Rapid, solid phase extraction technique for the high-throughput assay of darifenacin in human plasma. *Analytical Chemistry*, **68**, (9), 1658–60.

Krishnan, T.R. and Ibraham, I. (1994) Solid phase extraction technique for the analysis of biological samples. *Journal of Pharmaceutical and Biomedical Analysis*, **12**, 287–94.

Lensmeyer, G.L., Wiebe, D.A. and Darcey, B.A. (1991) Application of a novel form of solid-phase sorbent (Empore Membrane) to the isolation of tricyclic antidepressant drugs from blood. *Journal of Chromatographic Science*, **29**, 444–9.

Moors, M., Steenssens, B., Tielemans, I. and Massart, D.L. (1994) Solid-phase extraction of small drugs on apolar and ion-exchanging silica bonded phases: towards the development of a general strategy. *Journal of Pharmaceutical and Biomedical Analysis*, **12**, 463–81.

Pinkerton, T.C., Perry, J.A. and Rateike, J.D. (1986) Separation of furosemide, phenylbutazone and oxyphenbutazone in plasma by direct injection onto internal surface reversed-phase columns with systematic optimisation of selectivity. *Journal of Chromatography*, **367**, (2), 412–18.

Plumb, R.S., Gray, R.D. and Jones, C.M. (1997) Use of reduced sorbent bed and disk membrane solid-phase extraction for the analysis of pharmaceutical compounds in biological fluids, with applications in the 96-well format. *Journal of Chromatography B: Biomedical Applications*, **694**, (1), 123–33.

Simmonds, R.J., James, C.A. and Wood, S.A. (1994) A rational approach to the development of solid phase extraction methods for drugs in biological matrices, in: Stevenson, D. and Wilson, I.D. (eds) *Sample Preparation for Biomedical and Environmental Analysis*, New York: Plenum Press.

3

BASIC HPLC THEORY AND PRACTICE

Andy Gray

3.1 Origins

The current favourite expansion of the abbreviation HPLC is high-performance liquid chromatography although alternatives of high pressure and highly priced may be equally appropriate! The technique is relatively recent and has rapidly become probably the most frequently used (or abused) analytical method.

HPLC arose in the early 1960s by a combination of the experiences gained with both gas chromatography and ordinary column chromatography. There seems some debate about who should be credited with the discovery of HPLC. However, the major experimental advances that led to practical HPLC systems were made by Horvath *et al.* (1967), Huber and Hulsman (1967) and Kirkland (1969), although many of the theoretical advantages were first described by Giddings (1965). The first commercially available system was marketed by Dupont, and led to the rapid development of HPLC as a routine analytical tool.

HPLC complements gas chromatography (GC). In HPLC the mobile phase is liquid rather than a gas. In contrast to GC, HPLC is a suitable technique for the analysis of compounds with a wide range of polarities, high molecular weights, and those that are thermally unstable or have a tendency to ionise in solution.

The popularity of HPLC has resulted from its flexibility, both in terms of the diverse nature of the components that it can be used to separate and the different modes that can be used (reverse phase, ion exchange, analytical, preparative, etc.), and that it can be used as a reliable and reproducible quantitative analytical method. Although HPLC is easy to use, there are many variables that determine the overall chromatographic separation (solvents, stationary phases, flow rates, gradients, temperature, etc.). In this chapter I hope to provide the reader with a basic understanding of how these various parameters influence the chromatographic process and provide a basis for optimisation of the separation, as discussed in Chapter 4.

3.2 Applications

Although thin-layer chromatography (TLC) and column chromatography are of proven value in the analysis of compounds and immunoassay techniques have a role to play, in the majority of cases the analytical method of choice is either GC or HPLC. GC is generally selected where the compounds to be analysed are volatile, of low molecular weight, thermally stable and non-polar. In most other circumstances HPLC would be chosen.

In HPLC the analyst has a wide choice of chromatographic separation methodologies, ranging from normal to reverse phase, affinity to ion exchange, and a whole range of mobile phases using isocratic or gradient elution techniques. Various detectors are also available for HPLC, including electrochemical (ECD), refractive index (RI), fluorescence, radiochemical and mass-sensitive detectors, although by far the most popular remains the UV detector. This diversity of modes of separation and detection makes HPLC suitable for the analysis of a wide range of compounds in the pharmaceutical, agrochemical, food, industrial chemical and biochemical industries.

One major advantage of HPLC over GC is the ease of recovery of the sample from the eluate by simple evaporation of the solvents. In addition, HPLC methods are easily scaled up. Separation methods developed for analytical work with picograms, nanograms or possibly micrograms of compound are simply transferred to larger columns at higher flow rates for preparative chromatography for the isolation of milligrams or grams of material. Similar-scale preparative GC would be impractical, if not impossible, in the laboratory. This ease of scale-up has resulted in the extensive use of preparative HPLC in the clean-up and isolation of compounds from synthetic brews and biological samples.

3.3 Apparatus

A stylised HPLC system is illustrated in Figure 3.1.

3.3.1 Column

In order to achieve rapid analyses with a separating power approaching that of GC, it is necessary to obtain rapid equilibration between the mobile and stationary phases and a rapid flow rate. To achieve this, columns with small-particle solid phases (typically 3–10mm diameter) and pressurised solvent delivery systems have been developed. In order to avoid excessively high pressures, analytical HPLC columns tend to be short (generally only 10–25cm long) and have a bore of 2–5mm to maintain a high **linear flow velocity** (i.e. a rapid rate of progress along the column) for a given flow rate. This latter property is important in order to keep analysis times short.

Most suppliers supply their versions of the main types of column support. However, the experienced chromatographer will know that a C18 column from one manufacturer may produce a different separation from another. Indeed, occasionally batch variations may lead to columns from the same source having different selectivities. Such variations may be partly a result of different carbon loadings. Monomeric phases have a reasonably homogeneous covering of the silica with the bonded ligand, which may result in some free silanol groups (Section 3.6.2). Polymeric phases usually have a relatively higher carbon loading. Monomeric phases are generally more efficient in the separation of polar molecules, whereas polymeric phases are recommended for solutes of a intermediate or non-polar nature. The sample capacity of the polymeric phases is greater than that of monomeric phases and may be more useful in preparative chromatography.

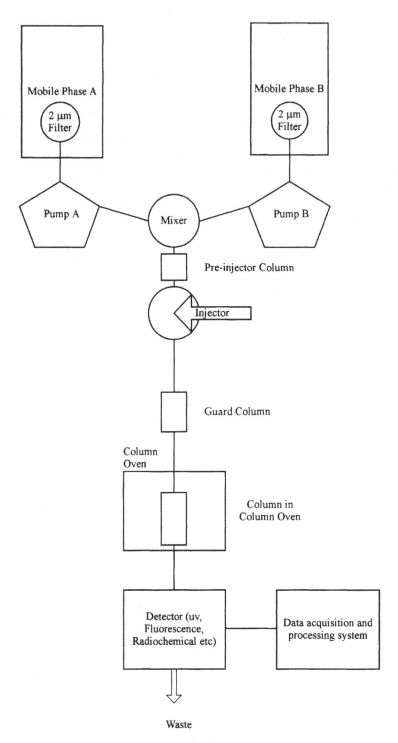

Figure 3.1 Diagrammatic representation of an ideal HPLC system, consisting of two solvent reservoirs with inline solvent filters; two high-performance pumps and solvent mixer; a pre-injector column to pre-saturate the solvents with stationary phase; an injector or autosampler to introduce analyte; a guard column, containing the same packing material as the analytical column; an analytical column; a suitable detector to allow measurement of solutes as they are eluted; and a data collection system, generally computer based.

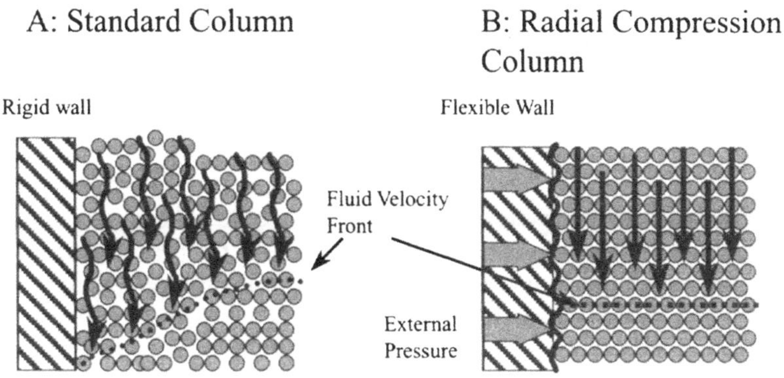

A: Standard Column

B: Radial Compression Column

Rigid wall

Flexible Wall

Fluid Velocity Front

External Pressure

Figure 3.2 Comparison of conventional and radial compression cartridges. In (a), flow through a standard steel column, the solid-phase particles are irregularly arranged, leaving voids, particularly at the edges of the column, resulting in an uneven flow profile and peak broadening. Radial compression of the cartridge (b) forces the solid phase to conform to a more regular arrangement, minimising voids, especially at the walls, hence maintaining a uniform solvent front and reducing band spread.

The most common analytical column is a stainless-steel tube, generally of 2–5mm internal diameter. The main drawback to such columns is their expense, since the whole column and fittings are disposed of once the stationary phase has expired. However, such systems do permit the chromatographer maximum choice in the type, size and source of stationary phase.

Cartridge columns are cheaper and consist of a stainless-steel tube similar to a column, but without end fittings. These cartridges are clamped into a column holder which includes the fittings.

Radial compression cartridges dispense with the stainless steel altogether; the stationary phase is held in a plastic tube. The cartridge is held in a special device that compresses the cartridge radially. The main advantage of the radial compression systems is that the wall of the cartridge is compressed to conform with the shape of the stationary phase (Figure 3.2), thereby reducing voids and decreasing peak broadening (Section 3.5.3). In addition, these cartridges are relatively cheap to replace. However, the radial compression system requires some capital outlay and, like the stainless-steel cartridges, ties the user to one supplier.

3.3.2 Plumbing

Although the degree of peak separation (**resolution**) achieved depends largely on the column and mobile phase, many other aspects of the HPLC system do influence this. For analytical work it is important to ensure that the tubing used between the column and detectors is of a suitably narrow bore (0.12 or 0.17mm) and that the length of tubing and number of fittings, even if of the 'low dead volume' variety, are kept to a minimum. Poor column fittings will result in voids that will cause eddies and **peak broadening** (Section 3.5.3) which will be particularly detrimental to the resolution of early eluting peaks. Further peak broadening can occur in the detector cell and its volume should be minimised while maintaining maximum sensitivity (i.e. maximising pathlength and minimising volume). In practice, for the majority of analytical HPLC applications, a UV flow cell of 8ml capacity is used, although both smaller and larger cells are available (Chapter 5).

3.3.3 *Pumps*

The pumps used in analytical HPLC systems are generally capable of flow rates in the 0.1–10ml/min range. Indeed, the majority of analyses are conducted between 0.5 and 2ml/min although, for preparative work, flow rates of 50ml/min are not uncommon. The solvent delivery rate must be closely controlled over a wide range of solvent viscosities. The pumps are generally capable of operating up to 6000psi, although the vast majority of HPLC analyses are now conducted at less than 2500psi (hence the change of name from high pressure to high performance). Since many detectors (ECD, RI and to a lesser extent UV) are very sensitive to fluctuations in pressure, the solvent delivery must be pulseless to avoid artefactual peaks or baseline variations.

For an **isocratic separation** (i.e. one in which the mobile phase remains the same throughout the separation), a simple system consisting of a solvent reservoir, pump, injector, column and detector is all that is required. However, for metabolite profiling, where several components with diverse physico-chemical properties (ionisation, lipophilicity, etc.) are to be separated, gradient elution is required. Reverse-phase gradient HPLC is generally the method of choice for metabolite profiling and isolation studies.

Gradient elution involves changing the relative amounts of usually two, but occasionally three or four, mobile phases during a chromatographic separation. This may be achieved by two methods. The more traditional method requires one pump for each mobile phase and a gradient controller which changes the relative amounts of mobile phase delivered by each pump according to a predefined programme. The mobile phases are then passed through a mixing chamber at a relatively high pressure prior to passing through the injector and the column (Figure 3.3).

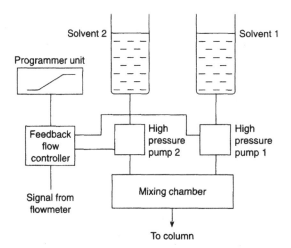

Figure 3.3 Schematic of a high-pressure mixing solvent delivery system consisting of two separate high-pressure pumps.

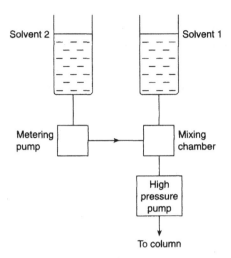

Figure 3.4 Schematic of a low-pressure mixing solvent delivery system consisting of a single high-pressure pump.

A second gradient elution system is gaining favour in many analytical laboratories that utilises a low-pressure mixing system, reducing the expense by requiring only one high-pressure pump (Figure 3.4). These systems are also popular for preparative chromatography, as higher flow-rate pumps (e.g. 50ml/min) are readily available. These systems may be less precise in their control of the relative proportions of the mobile phases due to variations in solvent viscosity.

Whenever possible (for analytical assays) an isocratic separation should be favoured over a gradient elution method since, as no re-equilibration is required between each run, it allows much more rapid sample analysis. In addition, the added complexity of the equipment and solvent systems used in gradient elution can result in greater potential for day-to-day variation and reduced robustness. However, for metabolite profiling and isolation work it is often necessary to use gradient elution in order to separate all the components.

3.3.4 Injectors

The most popular manual injection system, and one that is utilised by several auto-injectors, is the syringe-loop injector of the Rheodyne™ type (Figure 3.5). The sample to be analysed is injected into the sample loop while the mobile phase is running at its initial settings. To inject the sample onto the column the valve is rotated and the mobile phase flushes the sample loop contents onto the column. Such injectors may be used to inject variable volumes, by only partly filling the sample loop, or as a fixed-volume injector if the sample loop is overfilled.

The latter method is more reproducible provided that the loop is genuinely overfilled (i.e. injection volume at least 10% greater than loop volume), since it ensures that exactly the same volume of sample is injected each time. Most laboratories utilise fully automated HPLC systems for the analysis of large numbers of samples. In these

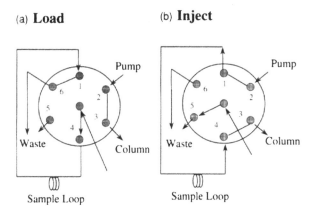

Figure 3.5 Diagrammatic representation of a Rheodyne (Model 7125) type injector. (a) In the load conformation, sample introduced into the needle port via a syringe or autosampler is transferred into the sample loop and displaced solvent in the loop goes to waste. During loading, the HPLC mobile phase is fed directly from the pumps to the column. (b) Once the injector valve is rotated to the inject position, the mobile phase back-flushes the content of the sample loop onto the analytical column and the needle port can be flushed to waste to ensure no sample carryover.

circumstances an autoinjector (or autosampler) can be programmed to inject set volumes of samples and to include quality control (to check accuracy and precision of the assay) and calibration standards (to allow generation of a calibration curve) at predetermined intervals. Such systems permit unattended repetitive analyses of samples, thereby allowing greater throughput and more efficient use of such an expensive resource.

3.3.5   Column ovens

It is becoming increasingly more common to heat the HPLC column to temperatures of up to 60°C to overcome retention variations due to poor control of environmental temperature and to improve peak shape and resolution (Section 3.5.4). Due to the use of flammable solvents, it is essential that the ovens used are designed for use with HPLC systems. Since the retention of a solute can be dramatically affected by temperature, very precise control of column temperature must be maintained.

3.3.6   Detectors

Types of detector and the selection of the most appropriate detector are the subject of Chapter 5.

3.4   The chromatographic process

3.4.1   Basic principles

Essentially, HPLC is a technique that enables the separation of the components of a mixture by virtue of their differential distribution between the mobile (liquid) phase and the stationary (solid) phase. Migration of a solute component can only occur while it is dissolved in the mobile phase. Thus, solutes that have a high distribution into the stationary phase will elute more slowly than those that distribute more readily into the mobile phase and the two will therefore undergo chromatographic separation.

3.4.2 Molecular forces

There are five elementary types of forces that individually, or in combination, result in retention of the solutes by the stationary phase:

1 Van der Waals (non-polar) forces between molecules causing momentary distortion of their electrostatic configuration.

$$-CH_3\cdots\cdots H_3C-$$

2 Dipole–dipole interactions arise between molecules when the electron cloud is temporarily distorted, resulting in electrostatic attraction between molecules.

$$^{\delta-}O=^{\delta+}C<$$
$$\vdots \quad \vdots$$
$$>C^{\delta+}=O^{\delta-}$$

3 Hydrogen bonding interactions between proton donors and proton acceptors.

$$-O^{\delta-}\cdots\cdots^{\delta+}H-$$

4 Dielectric interactions resulting from electrostatic attraction between the solute molecules and a solvent of high dielectric constant.

5 Electrostatic attractions, e.g. ion–ion and ion–dipole.

$$-N^+H\cdots\cdots^-O_2C-$$

3.4.3 Distribution

The separation of components by chromatography depends on the differences in their **equilibrium distribution coefficients** (K) between the stationary and mobile phases:

$$K = C_s/C_m$$

where C_s and C_m are the concentrations of the solute in the stationary and mobile phases, respectively.

Since migration of solutes through the column only occurs when they are in the mobile phase, solutes with high K values elute more slowly than those with low K values.

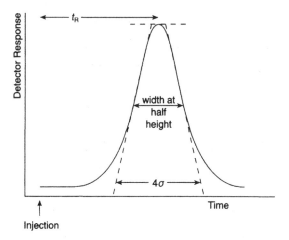

Figure 3.6 A Gaussian peak, with lines constructed at tangents to the points of inflection providing the measurement of 4σ at their intercept with the baseline. The retention time (t_R) is indicated from the time of injection of the sample until the peak maximum.

3.4.4 Theoretical plates

The efficiency of a column is measured by the number of **theoretical plates**, represented as N, to which the column is equivalent. This term was originally derived to describe the process of fractional distillation. The value of N can be calculated from the equation:

$$N = (t_R/\sigma)^2$$

where σ represents the standard deviation of a Gaussian peak and t_R is the retention time of the peak (usually the last peak of interest).

Provided that the peak is Gaussian, lines constructed at a tangent to the points of inflection of the peak would intercept the baseline at 4σ (Figure 3.6).

$$N = 16 \times (t_R/4\sigma)^2$$

where 4σ is the distance between the tangential lines at the intercept with the baseline measured in units of time (minutes or seconds).

However, as it is difficult to precisely determine the point of inflection of the peak, it is more accurate to measure the peak width at half height (Figure 3.7) and use the equation:

$$N = 5.54 \times (t_R(\text{time})/\text{peak width at half height in units of time})^2.$$

The value of N is an indication of the column performance: the larger the number, the better the column. In general, N increases with a reduction in stationary-phase particle size, flow rate and mobile phase viscosity and with increases in temperature and column length. A typical 10cm × 4.6mm column with 5mm particle size ODS (octadecylsilane) stationary phase and a 1ml/min flow rate would usually have an N value of around 7000.

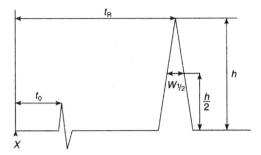

Figure 3.7 Measurement of peak retention time (t_R), retention time for an unretained peak (t_0) and peak width at half height ($W_{1/2}$) from a typical chromatogram.

The value of N is determined without making any allowance for the retention time of unretained peaks (t_0) and this will play a significant role in determining the apparent efficiency of the column for poorly retained solutes. This deficiency may be overcome by the use of the **effective plate number** (N_{eff}):

$N_{eff} = 5.54 \times [(t_R - t_0)/\text{peak width at half height in seconds}]^2$.

In order to make comparisons between columns of different length, the term **plate height** is determined, using the equation:

$H = L/N$

where L is the length of the column (in millimetres). Since better columns have higher N values, it follows that columns with *low* H values are best. The H value for the typical column given above would be 14μm.

The dispersion of a solute, usually referred to as **band broadening** (Section 3.5.3), at the end of a column is given by:

$4\sqrt{(H \times L)}$

or, substituting for H,

$4L/\sqrt{N}$.

Thus, columns should be kept short to minimise band spread.

The plate height is a function of the particle size of the stationary phase and, to permit comparisons between columns containing different-sized particles, it is useful to include this parameter in the equation to obtain the **reduced plate height** (h):

$h = H/d_p = L/(N \times d_p)$

where d_p is the particle diameter in millimetres (and therefore L must also be in millimetres). The value of h is dimensionless and the lower the better. A value of 3 for the example above indicates that this is a reasonably good column.

3.5 The chromatogram

3.5.1 Retention

As a result of retention on the stationary phase, solutes move through the column more slowly than the mobile phase. The time taken to elute a component, retention time (t_R), is measured, in units of time, from the point of injection to its peak maximum (Figure 3.7). Alternatively, retention may be quoted in terms of the volume required to elute a component (**retention volume**, V_R). The values t_R and V_R are related by the equation:

$$V_R = t_R \times f$$

where f is the flow rate of the mobile phase in ml/unit time.

A component that is not retained will elute at t_0 and the volume of mobile phase that has passed through the system since the injection is called the **void volume** (V_0).

The retention volume (V_R) is also related to the equilibrium distribution coefficient (K) of the solute between the stationary and mobile phases:

$$V_R = V_m + (K \times V_s)$$

where V_m is the volume of the mobile phase in the column and V_s the volume of the stationary phase. In adsorption chromatography, V_s is replaced by A_s, the surface area of the adsorbent.

The **retention** or **capacity factor** (k') is a measure of the degree of retention of a solute, which can be calculated from the following equations provided the flow rate remains constant:

$$k' = (V_R - V_0)/V_0 \text{ or } (t_R - t_0)/t_0.$$

k' is related to K by the equation:

$$k' = K \times (V_s/V_m).$$

3.5.2 Resolution

For two components to be separated they must have different retentions. The separation of solutes A and B (where B is the later-eluting peak) can be expressed as a **selectivity factor** (α) which is the ratio of their capacity factors:

$$\alpha = k'(B)/k'(A).$$

The resolution (R_s) of two components is a measure of how well they are separated and is defined by the following equation:

$$R_s = 2(t_R \text{ peak A} - t_R \text{ peak B})/\text{sum of the peak widths at base in units of time.}$$

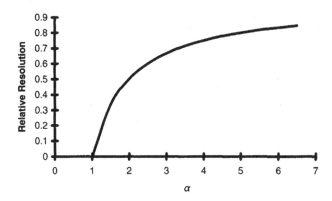

Figure 3.8 The relationship between selectivity factor (α) and the relative resolution of peaks in a chromatography system. As α is increased from 1 the selectivity term in the resolution equation initially increases dramatically but after α values of 5 are achieved, further improvements in resolution are small.

However, as stated previously, it is often difficult to measure the peak width accurately at the base. The resolution factor can be calculated more easily using the following expression:

$$R_s = 0.25 \times [(\alpha - 1)/\alpha] \times [k_2'/(1 + k_2')] \times \sqrt{N}$$

where k_2' and N are the capacity factor and theoretical plates for the last-eluting peak of interest.

Since the selectivity factor (α) is the ratio of the capacity factors of the two solutes, the greater the differences in these values the better the resolution; i.e. resolution increases with increasing α. Changing the selectivity (i.e. the relative distributions of the solutes between stationary and mobile phases) will have the most influence on the resolution. This requires modification of the relative capacity factors by changing the stationary or mobile phases. It is worth noting that sometimes large and unpredictable changes in resolution may be achieved in reverse-phase HPLC by replacing methanol with acetonitrile in the mobile phase, even to the extent of reversing the elution order. Variation of pH or changing the counter-ion can also modify selectivity in ion-exchange chromatography (Section 3.6.5).

Figure 3.8 shows the relationship between resolution and the selectivity factor. When $\alpha = 1$ the capacity factors for both solutes are equal and no separation will occur. Marked increases in resolution are obtained by increasing selectivity from 1 to 3. Increasing the selectivity factor above five only results in minor improvements in the resolution. For example, if α is 1.1, the selectivity term in the resolution equation will be 0.09, but if this is increased by only approximately 30% to 1.4, this term increases threefold to 0.29, indicating a dramatic improvement in resolution.

The example in Figure 3.9 illustrates how, by changing the mobile-phase constituents from acetonitrile/water to tetrahydrofuran/water and thereby affecting the selectivity of the components, a marked improvement in resolution of the three steroids is obtained.

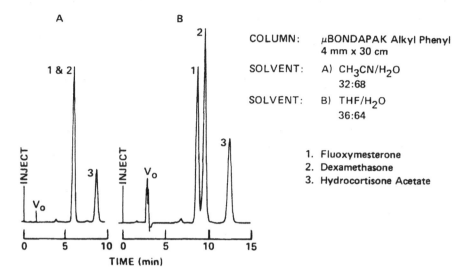

Figure 3.9 An example of the improvements in resolution that may be achieved by changing the selectivity of the chromatographic system. Both systems use the same μBondapak™ alkyl phenyl column (4mm × 300mm) for separation of a steroid mixture. System A which has a mobile phase of acetonitrile : water (32 : 68) fails to resolve components 1 and 2. In contrast, system B with tetrahydrofuran : water (36 : 64) as mobile phase provides almost baseline separation of components 1 and 2 while maintaining the resolution of component 3, although at a cost of a 30% increase in run time.

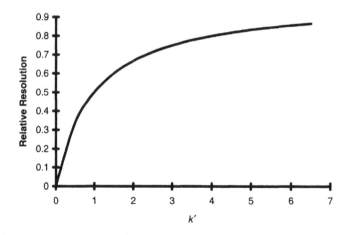

Figure 3.10 The relationship between capacity factor (k') and the relative resolution of peaks in a chromatography system. As k' is increased from zero the capacity term in the resolution equation initially increases dramatically, but after k' values of 5 are achieved, further improvements in resolution are small.

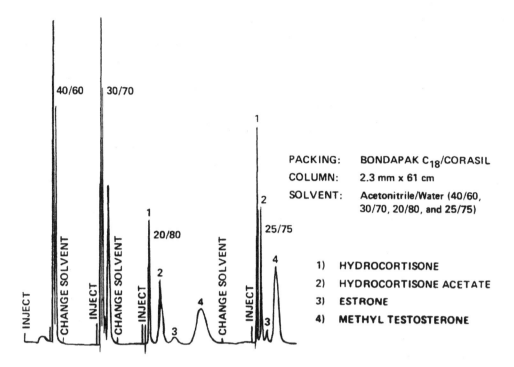

Figure 3.11 An example of the improvements in resolution that may be achieved by changing the capacity factor term of a chromatographic system. All four chromatograms were generated using the same μBondapak C18/Corasil column (2.3mm × 610mm) with a mobile phase of acetonitrile : water in varying ratios of 40/60, 30/70, 20/80 and 25/75. Decreasing the solvent strength by reducing the acetonitrile content of the mobile phase does not affect the selectivity of the separation, but increases the retention time and hence the capacity factor. In the third chromatogram, good resolution of the four components is achieved. However, the run time is unnecessarily long and the fourth chromatogram is optimised using an acetonitrile concentration which is between those of the second and third chromatograms.

The capacity factor is particularly important in determining resolution when k' is small (less than 5; Figure 3.10), i.e. when the peaks elute within five times the void volume. The example shown in Figure 3.11 illustrates how, by increasing the capacity factor by decreasing the solvent strength from 40% acetonitrile to 25% acetonitrile (fourth chromatogram), a marked improvement in resolution is obtained. This is because in the first trace, k' is too small with little retention of the steroids on the column. In the third chromatogram, although resolution is achieved, the run time is unnecessarily long.

Resolution is related to $\sqrt{N}$ and therefore although improvements in resolution can be achieved by increasing N, by increasing column length or decreasing particle size, it is impractical to increase resolution significantly by changing N alone. For example, a doubling of column length would, at best, double N and only increase resolution by $\sqrt{2} = 1.4$ fold. In practice, since peak width also increases with column length, the actual improvements in resolution would be less than this. In addition, exceptionally long columns will result in higher back pressures and longer analysis times.

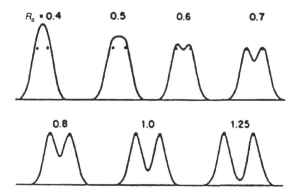

Figure 3.12 Schematic representation of the chromatographic separation of two components that would be associated with different resolution (R_s) values. R_s values of 0.5 and below appear as a single peak, baseline separation is not achieved until an R_s value of 1.5 is achieved.

In practice, a resolution factor of 1.5 is considered the lowest practical value for quantitative analysis (Figure 3.12) and, provided both peaks are Gaussian, implies baseline separation.

3.5.3 Peak shape

Skewed peaks

Peak symmetry (A_{10}) may be calculated by determining the ratio of the distances between the peak maximum and the leading and trailing edges at 10% of peak height (Figure 3.13):

$$A_{10} = b/a.$$

Values of A_{10} of between 0.85 and 1.3 are generally acceptable.

The shape of the peak depends on the relationship between concentration of the solute and its distribution coefficient, K, between the stationary and mobile phases. The peak is Gaussian when there is a linear relationship between the concentrations in the stationary and mobile phases, called a linear **isotherm** (Figure 3.14). Skewed peaks result when the isotherms are non-linear. As a result, the retention time will change with changes in sample concentration. In practical terms, for the small sample concentrations used in most analyses (i.e. nanogram quantities injected), symmetrical Gaussian peaks are obtained even when the isotherms are not linear. However, there may be implications for preparative HPLC and care should be taken not to overload the column. It should be remembered that all of the components in the sample can contribute to the column loading, not just the few micrograms of metabolite that is being isolated.

Peak tailing is usually the result of a void forming in the column packing, often at the top of the column. This may be overcome by the addition of extra packing or fine glass beads to the top of the column to fill this void. Tailing may also be caused by secondary interactions (Section 3.6.2) or by the overloading of the column (possibly

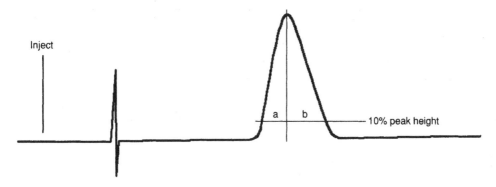

Figure 3.13 Calculation of peak symmetry. The ratio of the peak widths at 10% of the peak height either side of a vertical line through the peak maximum provides a measure of peak symmetry. Values of between 0.85 and 1.3 are generally considered acceptable.

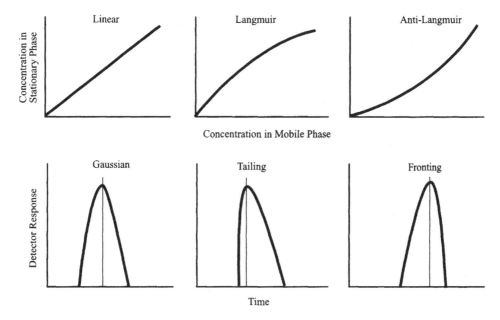

Figure 3.14 Three basic isotherm shapes and their effect on peak shape and retention time. Skewed peaks occur when the isotherms are non-linear (Langmuir and anti-Langmuir), resulting in asymmetrical peak shapes.

by extra-sample material). The chromatograms in Figure 3.15 illustrate that, at low loadings, there is a minimal effect of sample size on column efficiency. However, as the injection mass is increased still further, resolution of the two components declines and peak tailing is evident. Peak fronting (Figure 3.14) is usually due to voids that have formed along the wall of the column (edge effect) and usually indicates the demise of the column.

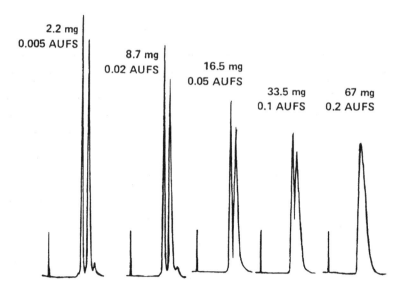

Figure 3.15 Illustration of the effect of sample loading on column efficiency. As sample load is increased from 2.2mg to 67mg peaks are broadened and resolution is lost.

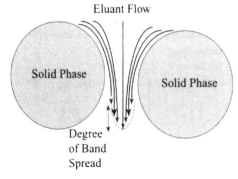

Figure 3.16 Band spread resulting from eddy diffusion occurs due to the flow of mobile phase in the centre of the voids between the stationary phase being faster than the flow near the stationary-phase surface. Hence the smaller the particle size, the smaller the voids and the less eddy diffusion occurs, giving higher-efficiency columns.

Band broadening

As a solute passes through a column, the band (peak) width increases (broadens) due to molecular diffusion, eddy diffusion and mass transfer. Although important in GC, molecular (or longitudinal) diffusion is not a major contributor to peak broadening in HPLC, as diffusion in liquids is considerably slower than in gases.

Eddy diffusion is caused by channels of variable sizes between stationary-phase particles. The solvent (and solutes) flow faster through wider channels and in the centre of the channel, and slower in narrow channels and close to the stationary phase (Figure 3.16).

60

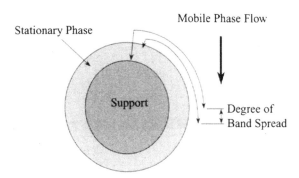

Figure 3.17 Band spread resulting from mass transfer effects occurs due to differing rates of movement of the solute into the stationary phase and back into the mobile phase. Some solute may diffuse deep into the stationary phase, whereas some may interact with stationary phase at the surface and will rediffuse into the mobile phase more rapidly.

Mass transfer of the solute between the mobile and stationary phases may contribute to peak broadening when the stationary phase coating is thick and solutes can diffuse further into it. Differences in the depth of diffusion of solutes into the stationary phase (Figure 3.17) result in a variable delay in the diffusion back into the mobile phase. Mass transfer is increased with temperature, generally resulting in improved peak shape.

It is also worth noting that extra-column factors can produce peak broadening, for example excessive diameter or lengths of tubing, large volume detector cell or fittings.

Artefact peaks

The most common causes of artefact peaks are electronic spikes and bubbles in the eluate. If the former is a problem, it is often easily resolved by the installation of a voltage stabiliser plug. The latter problem may have several causes, including inadequate degassing of solvents (particularly water with aqueous/methanol gradients) or to pressure changes, often in the flow cell. Degassing by vacuum filtration, sonication, bubbling the solvent with helium gas or the addition of a flow restrictor on the detector cell outlet are usually effective in resolving these problems.

3.5.4 Effect of temperature

One of the major advantages in the maintenance of column temperature above ambient is that it minimises variations in chromatography resulting from fluctuations in environmental temperature. However, raising temperature also tends to increase the solubility of solutes in the mobile phase and increase mass transfer, thereby increasing the efficiency of the column, and reduces analysis time. Elevation of temperature also decreases solvent viscosity, resulting in reduced inlet pressures and contributing to improved mass transfer. Elevated temperature is generally to be recommended in any application in which such a rise would not cause decomposition of the sample. Temperature is usually increased to between 40°C and 60°C and must be closely controlled (variation less than 0.2°C).

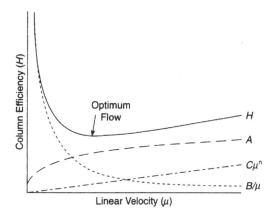

Figure 3.18 Relationship between linear velocity (μ) and column efficiency (H) and the relative contributions of eddy diffusion (A), longitudinal diffusion (B/μ) and mass transfer ($C\mu^n$) to column efficiency.

In reverse-phase chromatography, as a general rule, a 10°C increase in temperature will result in a halving of the capacity factor (k') and a reduction in analysis time. However, in some circumstances retention may increase with temperature as a result of temperature-dependent alterations in molecular configuration or ionisation of the solute or buffers. These latter factors can be especially important in reverse-phase ion-pair chromatography. In normal-phase chromatography, elevating temperature can have variable effects on retention; retention is generally decreased with temperature rises up to 40°C but solutes are retained longer at temperatures greater than 40°C.

Temperature programming is not generally used in HPLC, mainly because of the large volumes of mobile phase and bulky columns take too long to achieve equilibrium.

3.5.5 Effect of flow rate and linear velocity

The speed of elution of a solute from a column is dependent on its capacity factor and the **linear velocity** (μ) of the mobile phase. The linear velocity is the rate at which the mobile phase moves along the column and is determined by the flow rate and column dimensions. Band spreading or peak broadening (Section 3.5.3) is a function of three factors: longitudinal (molecular) diffusion, eddy diffusion and mass transfer, and these are influenced to different degrees by the linear velocity (Figure 3.18). The contribution of longitudinal diffusion to column efficiency is proportional to the inverse of the linear velocity, i.e. as flow rate is increased, band spreading due to longitudinal diffusion decreases. This factor is considerably less important in HPLC than in GC. Eddy diffusion is independent of flow rate, but is increased proportionately with the particle size of the stationary phase. Band spreading due to mass transfer effects is increased with linear velocity.

The complex inter-relationship between these factors and linear velocity is described by the Van Deemter equation:

$$H = A + B/\mu + C\mu^n$$

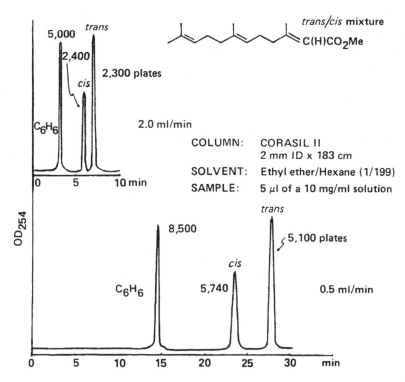

Figure 3.19 Example of how column efficiency may be improved by decreasing flow rate. In the top chromatogram, at a flow rate of 2ml/min the last peak has a much lower *N* value (2300 plates) than in the lower chromatogram (5100 plates) in which the flow rate has been reduced to 0.5ml/min.

where *A* is the eddy diffusion factor, B/μ the linear diffusion factor, $C\mu^n$ the mass transfer factor with $n < 1$ and μ is the linear velocity.

The optimum flow rate for maximum column efficiency (Figure 3.18) is generally much lower than is practically useful, since the speed of the analysis is an important factor in any routine analysis. At the flow rates used for most separations mass transfer is the only significant factor affecting band spread and this increases linearly with the linear velocity of the mobile phase. It is, therefore, often possible to improve the efficiency of a column by decreasing flow rate (Figure 3.19). This example depicts how column efficiency can be improved by decreasing flow rate. In the top example at a flow rate of 2ml/min, the last peak has an N of 2300 plates and the first peak a value of 5000 plates. This illustrates how N tends to decrease with *k'* (due to increased band spread from longitudinal diffusion and mass transfer effects). In the lower trace, the flow rate has been decreased to 0.5ml/min and the N value for all three components has increased considerably, resulting in improved resolution. It is notable, however, that adequate resolution was probably achieved in the first example with a more acceptable run time.

The optimum flow rate for a column is also dependent on the mobile phase, due to the contribution of viscosity to the diffusion and mass transfer factors (Figure 3.20).

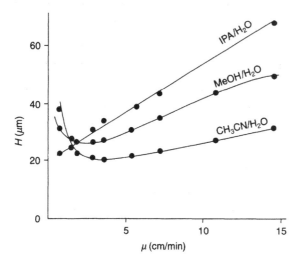

Figure 3.20 The influence of mobile phase on the optimal flow rate for column efficiency for three mobile-phase systems. The isopropyl alcohol mobile phase, which has the highest viscosity, gives the highest efficiency at the lowest flow rates whereas the methanol and acetonitrile mobile phases, which have much lower viscosity, give greatest column efficiencies at approximately 3cm/min.

3.5.6 Effect of sample volume

In general for isocratic separations, the smaller the sample volume the better the column efficiency. The sample is not concentrated on the top of the column and will elute in at least as large a volume as the injection volume. This does not apply to gradient elution provided the sample is prepared in solvents of similar, or lower, strength to the initial column conditions, since this will result in concentration of the sample at the top of the column.

The sample capacity of a column depends largely on its size and loading. Analytical columns will generally have a maximum optimum loading of a few milligrams. Increasing the loading will result in decreased column efficiency (Figure 3.15). It is worth noting that all the sample constituents contribute to the column loading and that column capacities may be severely reduced if biological extracts are analysed without some sample clean-up.

3.6 Separation mode

3.6.1 Normal phase

Normal-phase chromatography is carried out on polar stationary phases. The chromatographic process is generally referred to as **adsorption** chromatography. The traditional stationary phases are silica and alumina, in which hydroxyl groups are involved in the interactions with solvent and solute molecules. However, these phases are prone to changes in selectivity and retention unless the mobile phase is carefully maintained. The cyano, diol and amino **bonded** phases are preferable due to their greater chromatographic stability. The mobile phase consists of a non-polar solvent, usually hexane or heptane, and a polar modifier such as a short-chain alcohol, ethyl acetate or dichloromethane. The solute is retained by two possible interactions with the mobile and stationary phases:

1. **displacement** of *solvent* molecules, which are adsorbed on the stationary phase;
2. **sorption** of *solute* molecules into a layer of *solvent* molecules without their displacement from the stationary phase.

In practice a combination of both of these processes probably occurs since neither theory is able to account for all separations.

The retention of solutes on normal-phase columns decreases with polarity. Thus polar compounds with ionised groups (e.g. carboxylic acids and amines) are best retained, followed by polarisable groups (e.g. aldehydes and ketones), and least retained are the non-polar hydrocarbon compounds. Increases in the polarity (or solvent strength) of the mobile phase will decrease retention (k'). Thus, decreasing the proportion of dichloromethane or replacing it with a similar proportion of acetonitrile would result in a decrease in the retention of a solute.

In the example shown in Figure 3.21, in which a mixture of phthalates are analysed on a μPorasil column as a normal-phase separation, the diethyl and diphenyl phthalates are poorly resolved with the ethyl acetate mobile phase. Changing to butyl acetate decreases the solvent polarity and increases retention (k'). This also results in a reversal of the order of elution of the diphenyl and diethyl phthalates due to a change in the selectivity (α) of the two components for the new mobile phase.

3.6.2   Reverse phase

As implied by the name, the order of elution of solutes from a reverse-phase column would be the reverse of that of a normal-phase column. However, the interactions involved in reverse-phase chromatography are quite different from those of normal phase, although similar to those of GC. The predominant process is one of **partition** chromatography in which components are separated according to their relative partition coefficients between the mobile and stationary phases. Highly lipophilic (high log P), non-polar solutes will be better retained by the stationary phase and hence have longer retention times (higher k'). In contrast, polar molecules will be less well retained.

In reverse-phase liquid chromatography the stationary phase is prepared by **chemically bonding** a relatively non-polar group onto the stationary phase support (Figure 3.22). Traditionally the support has been spherical silica but, due to problems in the dissolution of silica at basic pH, an inert polymer support has been developed. However, to date, the use of such polymer-based supports has resulted in columns of lower efficiency than for equivalent silica-based columns. The most frequent non-polar group bonded to the stationary support is octadecylsilane (ODS or C18) which gives a highly lipophilic stationary phase. Less lipophilic stationary phases are produced when octylsilane (C8), C2, phenyl or cyanopropyl-bonded phases are used. In the last case, some separations may occur due to dipole–dipole interactions, but only when the mobile phase is predominately non-polar. With the phenyl-bonded phases π–π interactions may contribute to solute retention, whereas ion-exchange processes can occur with the amino-bonded phases. The shorter silanyl chain bonded phases

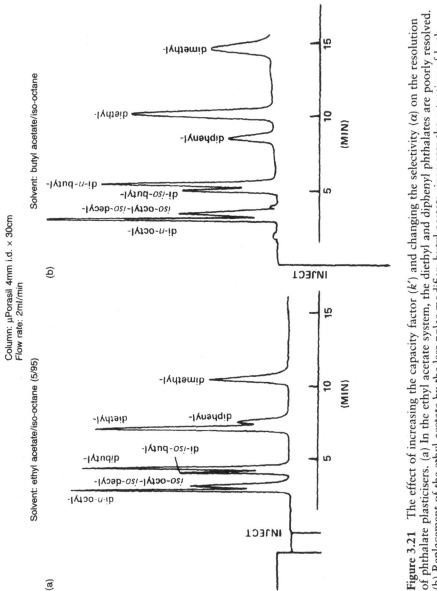

Column: µPorasil 4mm i.d. × 30cm
Flow rate: 2ml/min

Solvent: ethyl acetate/iso-octane (5/95)

Solvent: butyl acetate/iso-octane

Figure 3.21 The effect of increasing the capacity factor (*k'*) and changing the selectivity (α) on the resolution of phthalate plasticisers. (a) In the ethyl acetate system, the diethyl and diphenyl phthalates are poorly resolved. (b) Replacement of the ethyl acetate by the less polar modifier, butyl acetate, increases the retention of both analytes, but also changes their selectivity, resulting in a reversal of their elution order.

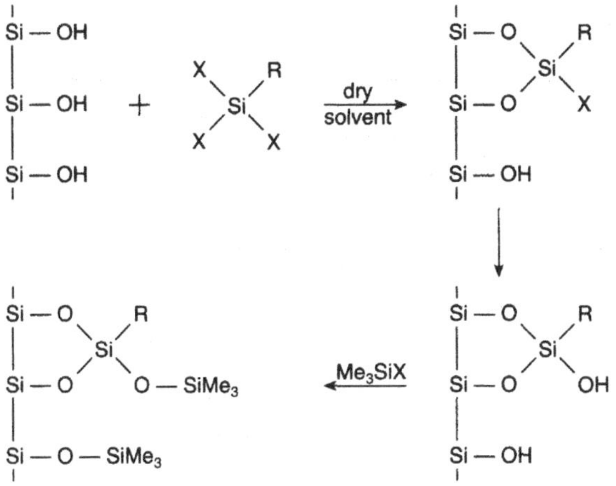

Figure 3.22 Steps in the chemical derivatisation of silica in the preparation of bonded reverse phases. X = Cl or OR′ and R is one of the commonly applied bonded phases such as C18, C8, C2, cyano, amino, phenyl.

(C2 and C8) are often most appropriate when highly lipophilic solutes are to be separated which would be highly retained on C18 columns (the equivalent of highly ionised polar solutes on normal-phase columns).

The mobile phase in reverse-phase HPLC is polar, generally consisting of water and a water-miscible organic solvent, such as methanol or acetonitrile. Although methanol and acetonitrile are frequently interchanged as they are regarded as having similar solvent properties (UV cutoff, refractive index and strength), they have different selectivities. Methanol is electron withdrawing and has a greater potential for hydrogen bonding, whereas acetonitrile is electron donating. In addition, acetonitrile has a lower viscosity than methanol and greater flow rates are possible for any given column inlet pressure. Tetrahydrofuran (THF) may also be added to increase solvent strength and aid elution of retained solutes (see Figure 3.9).

In bonded phases the silanol groups of the silica support are reacted (Figure 3.22) with alkyl groups. The alkyl-phase loading varies (expressed as percentage carbon by weight) for different types of column, from 5% to more than 20%. With low loading the alkyl groups probably form a monolayer over the surface of the silica. Such monomeric phases should provide a high separation efficiency since mass transfer between the solid and mobile phases should be rapid. However, there may be some non-alkylated silanol groups that may act as adsorption sites for the solutes and hence may cause peak tailing. Such unreacted silanol groups can be blocked with trimethylsilanyl groups by treatment of the phase with trimethylchlorosilane. This is known as **endcapping**.

Due to the complex nature of the stationary phases, several potential secondary interactions might take place. Examples include: adsorption of polar solutes with free silanol groups on the reverse-phase stationary support or interactions with the buffer to form ion-pairs or ion-exchange process. Although occasionally these secondary interactions will contribute to the overall separation, more often they result in peak tailing and a reduction in resolution.

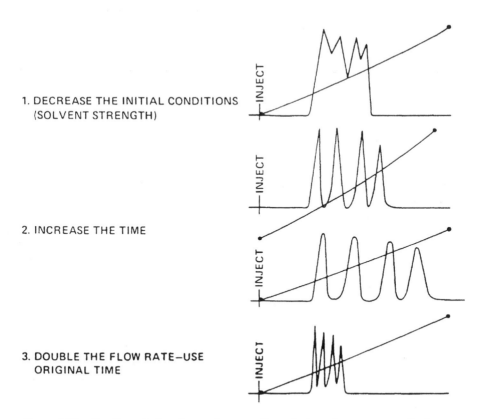

1. DECREASE THE INITIAL CONDITIONS
   (SOLVENT STRENGTH)

2. INCREASE THE TIME

3. DOUBLE THE FLOW RATE—USE
   ORIGINAL TIME

Figure 3.23 Gradient elution: improving overall resolution of components. Each chromatogram represents the separation of four components and the gradient profile is illustrated by a single line. In the second chromatogram resolution has been improved by decreasing the initial solvent strength (reduced organic), in the third chromatogram a slower gradient has been used and in the fourth chromatogram the same gradient conditions have been used but at an increased flow rate.

3.6.3 Gradient reverse phase

When a number of solutes with widely differing properties need to be resolved in a single chromatographic separation, it is often necessary to modify the solvent constituents during the analysis. This is usually the case when attempting to separate a drug and its more polar metabolites. Reverse-phase chromatography is the method of choice, with the initial conditions selected to retain the most polar material in the sample on the column and obtain a k' value of 3–5, and then increasing the organic solvent content until the least polar solute is eluted, usually the parent drug. The simplest gradients have a linear change with time and this is often the best first choice. However, if resolution of all the peaks of interest is still not achieved, or there are large barren areas in the chromatogram, more complex non-linear or multigradient curves can be used. Examples of the approach are illustrated in Figures 3.23–3.25.

In Figure 3.23 resolution is poor throughout the chromatogram. Initially, the solvent strength should be decreased (increase the aqueous content) and also the gradient should be slowed (more gradual increase in organic solvent proportion).

1. DELAY THE RATE AT THE START AND
   INCREASE THE RATE TOWARD THE END

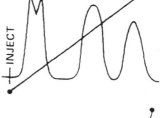

2. AS ABOVE AND VARY DEGREE

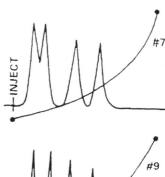

3. DECREASE INITIAL CONDITIONS

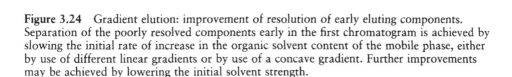

Figure 3.24 Gradient elution: improvement of resolution of early eluting components. Separation of the poorly resolved components early in the first chromatogram is achieved by slowing the initial rate of increase in the organic solvent content of the mobile phase, either by use of different linear gradients or by use of a concave gradient. Further improvements may be achieved by lowering the initial solvent strength.

Increasing the flow rate (yes, *increasing flow rate*) may also improve the separation as more mobile phase at any given strength will pass through the column and the components will be separated on the basis of differential solubilities and k'. In Figure 3.24, resolution is poor at the start of the chromatogram. This may be improved by decreasing the initial organic content, delaying the start of the gradient or changing the gradient profile to a concave curve. In Figure 3.25, adequate resolution has been achieved early in the chromatogram but not at the end. In this case the rate of increase in solvent strength towards the end of the chromatogram should be decreased, possibly even maintained isocratic, or a convex gradient curve used.

3.6.4 Ion suppression and ion pairing

Weak acids or bases often give poor peak shapes, variable retentions or even multiple peaks when chromatographed in unbuffered or non-pH-controlled solvent systems. The simplest solution is to suppress ionisation by the addition of acid (phosphoric, sulphuric or perchloric acid) to suppress ionisation of weak acids, or ammonia or ammonium carbonate for weak bases.

1. INCREASE THE RATE AT THE BEGINNING AND
   DECREASE THE RATE TOWARD THE END.

2. AS ABOVE AND VARY DEGREE.

3. DECREASE FINAL CONDITION.

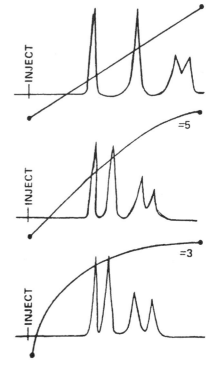

Figure 3.25 Gradient elution: improvement of resolution of late eluting components. Separation of the poorly resolved components late in the first chromatogram is achieved by slowing the rate of increase in the organic solvent content of the mobile phase during the terminal phase of the gradient, either by use of different linear gradients or by use of a convex gradient. Further improvements may be achieved by lowering the final solvent strength.

However, since a sample may contain weak acids, bases and neutral components, it may be more efficient to control ionisation by buffering the mobile phase. The ideal buffer would:

1 be chemically stable;
2 have uniform buffering capacity over a wide pH range (2–7);
3 be transparent to light down to at least 200nm;
4 be highly soluble in the organic solvents used in reverse-phase chromatography, especially methanol and acetonitrile;
5 have a rapid rate of proton transfer to reduce kinetic contributions which could contribute to band spread;
6 have the potential to mask free silanol groups on the stationary phase.

The last ability is expressed by buffers that contain an amine or ammonium component, for example ammonium acetate or phosphates of vicinal diamines. This results in a reduction of peak tailing caused by secondary interactions with the stationary phase, particularly with positively charged solutes. Ammonium acetate is probably

Table 3.1 Ion-pairing reagents

Ion-pairing agent	Application
Quaternary amines, e.g. $(CH_3)_4N^+$ (tetrabutylammonium, TBA)	Strong or weak acids, e.g. carboxylic acids
Tertiary amines, e.g. trioctylamine	Sulphonic acids
Alkyl sulphonic acids, e.g. camphor sulphonic acid, heptane sulphonic acid	Strong and weak bases, e.g. catecholamines
Alkyl sulphates, e.g. lauryl sulphate	Strong and weak bases, e.g. catecholamines

the most popular buffer for reverse-phase LC. The main disadvantage is that it has a UV cutoff of 220nm, although this is not a problem for isocratic separations. Ammonium acetate may also form ion pairs with solutes and the NH_4^+ ions may associate with the stationary phase to give some ion-exchange properties.

Ion-pair chromatography is also known by a number of alternative names, solvophobic-ion chromatography, counter-ion chromatography, surfactant chromatography and ion-association chromatography. This technique is generally used in reverse-phase chromatography with a C18 stationary phase and a mobile phase of methanol/water or acetonitrile/water. Ion pairing is often useful if the components are highly basic or acidic, providing better retention and higher selectivity than that afforded by the column and organic solvents alone.

The ion-pairing reagent is usually a large ionic or ionisable organic molecule (Table 3.1). Although perchloric acid ($HClO_4$) may also be used, its corrosive properties often make it impractical (dissolves metal tubing, valves and pumps).

The actual mechanisms for the retention of solutes in ion-pair chromatography are uncertain; several models have been proposed:

1   The free lipophilic counter-ions are adsorbed onto the stationary phase and then interact with the charged ions of the solute by an ion-exchange process.
2   The counter-ion forms an ion pair with the charged solute and the uncharged ion pair is then retained by the usual partition chromatography principles.
3   A combination of both the above mechanisms, possibly involving an exchange of the solute from an ion pair in the mobile phase to a counter-ion retained on the stationary phase.

If no ion-pair agent is present (Figure 3.26a), retention is controlled by reverse-phase partitioning. The acidic molecule, HA, will exhibit a maximum k' value (HA unionised) when the mobile phase pH is less than two units below the pK_a for the acid. At higher pH values the acidic molecule will be ionised, and retention decreased significantly. If a large concentration of ion-pair agent is present, the ion-pair–solute interaction dominates the retention (Figure 3.26b). At low pH, less than two units below the pK_a for the acid, the unionised acid will not interact with the ion pair but may still be retained due to reverse-phase interactions with the bonded phase. As pH is increased, HA ionises to form ion A^- which will form an ion pair and be retained by either ion-exchange or reverse-phase ion-pair retention mechanisms.

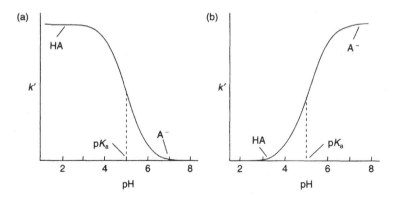

Figure 3.26 Chromatographic retention of an acidic sample HA as a function of pH using (a) reverse phase and (b) ion pair with a high concentration of ion-pair reagent.

3.6.5 Ion exchange

The ion exchangers in HPLC are generally silica stationary phases chemically bonded with anionic or cationic groups, usually aminopropyl, tetra-alkylammonium or sulphonic acid groups. Molecules are separated on the basis of their molecular charge on the principle of opposite charges attracting each other, i.e. cations (positive ions) would be retained on a cation exchange (negatively charged, e.g. SO_3H) bonded phase and vice versa. Resolution is influenced by pH and by the ionic strength of the buffer. Ionised compounds, particularly zwitterionic compounds (i.e. molecules containing both +ve and −ve charged groups, such as amino acids), are amenable to ion-exchange chromatography. Reverse-phase ion pair, or reverse-phase with pH-buffered mobile phases, is currently a more popular method for the separation of moderately polar compounds, probably as this procedure permits the concurrent separation of components with a wide range of polarities.

3.6.6 Others

Size exclusion chromatography separates on the basis of size or molecular weight. The stationary phase is generally a semi-rigid, porous material in which the pore size is selected to be similar to that of the solute particles. Traditional supports include cross-linked dextrans (Sephadex™) and polyacryamide gels (Biogel™) which are only suitable for aqueous column chromatography. For HPLC, semi-rigid cross-linked polystyrene (Styragel™) or controlled-pore-size rigid glasses or silicas are used. The mobile phase plays no role in the separation process, the only requirement being that the solutes are soluble in it.

As in to the development of GC, where packed columns were superseded by micro-bore columns, similar developments are taking place in HPLC. Microbore columns, whether of the packed or wall-coated open tubular (high-speed) variety (Chapter 6) display high resolving power ($N > 20\,000$–$100\,000$). The open tubular high-speed columns have an extremely low volume and, even at the low solvent flow rates (μl/min) used, have high linear velocities, resulting in very rapid analysis times and minimal solvent usage. The main disadvantage of these high-speed chromatography columns

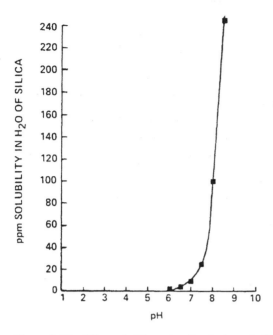

Figure 3.27 Effect of pH on the dissolution of silica. The silica matrix of columns will be gradually dissolved by mobile phases with pH values above 7, resulting in a deterioration in column performance.

is their limited sample capacity, and the need for extremely low-volume detector cells (0.01–1μl) to minimise peak broadening may limit sensitivity.

3.7 Column care

The voids that form at the top of columns are not due to the pressures causing compaction of the stationary phase but are due to dissolution of the silica which occurs at extremes of pH (> 7.3; Figure 3.27). This may be alleviated by including a small pre-column between the pumps and injector, containing a large particle-size silica to pre-saturate the mobile phase before it reaches the analytical column.

A second pre-column (guard column) should be used between the injector and the analytical column to remove any contaminants from the mobile phase or sample, and thereby extend the life of the analytical column. These columns should contain the same packing as the main column but should not be relied upon to pre-saturate the solvent with silica, since voids in the guard would also lead to peak broadening and reduced resolution.

Solvents (and samples, if appropriate) should be filtered to prevent microparticulates entering the system and blocking the column or frit at the top of the column. Generally a 2mm in-line filter placed on the end of the solvent lines is adequate. HPLC-grade solvents should be used to minimise the accumulation on the column of impurities that may be present in lower-grade solvents.

Contaminants from samples and solvents on the column can be removed using a strong solvent wash. For example, tetrahydrofuran (THF) in acetonitrile or methanol can be used to remove lipid materials, and a gradient of acetonitrile and propanol containing 0.1% trifluoroacetic acid may remove protein contaminants. Hydrophobic materials may be eluted with acetonitrile or methanol and repeat injections of THF (100μl–200μl). Occasionally the top of a column may need to be removed and the void replaced with fresh packing or fine glass beads. Care should be taken not to inhale the fine particulate stationary phase when carrying out this procedure.

Avoid extremes of pH as silica and ODS columns may be destroyed by high pH (> 7) as the silica dissolves. Reverse-phase columns should not be used below pH 2.5 as the bonded ligand may be hydrolysed.

Prior to storage, columns should always be thoroughly flushed to remove any acids, bases or salts that may be present. Reverse-phase columns should be flushed and kept filled with acetonitrile/water or methanol/water when not in use. Normal-phase columns should be stored filled with hexane. All columns must be firmly capped to prevent solvent loss and drying out of the stationary phase, as voids may occur resulting in reduced column efficiency.

3.8   Bibliography

Bidingmeyer, B.A. (1987) *Preparative Liquid Chromatography; Journal of Biochemistry Library*, Vol. 38, Oxford: Elsevier.

Fallon, A., Booth, R.F.G. and Bell, L.D. (1987) *Applications of HPLC in Biochemistry: Laboratory Techniques in Biochemistry and Molecular Biology*, Amsterdam: Elsevier.

Giddings, J.C. (1965) *Dynamics of Chromatography*, New York: Marcel Dekker.

Hamilton, R.J. and Sewell, P.A. (1982) *Introduction to High Performance Liquid Chromatography* (2nd edn), London: Chapman & Hall.

Horvath, C., Preiss, B. and Lipsky, S.R. (1967) Fast liquid chromatography. Investigation of operating parameters and the separation of nucleotides on pellicular ion exchangers. *Anal. Chem.*, 39, (12), 1422–8.

Huber, J.F.K. and Hulsman, Y.A.R.J. (1967) Liquid chromatography in columns. Time of separation. *Anal. Chim. Acta*, 38, (1–2), 305–13.

Kirkland, J.J. (1969) *J. Chromatogr. Sci.*, 7, 7.

Parris, N.A. (1984) *Instrumental Liquid Chromatography; Journal of Biochemistry Library*, Vol. 27, Oxford: Elsevier.

4

HPLC OPTIMISATION

David Bakes

4.1 Objective

The objective of this chapter is to provide the reader, who may already have some HPLC experience, with a logical approach to the optimisation of a chromatographic separation of compounds of pharmaceutical interest in a biological matrix, and covers reverse-phase, normal-phase, ion-pair, ion-exchange and chiral HPLC.

This chapter covers *only* the chromatographic separation on an HPLC system; no assumptions are made about sample preparation or about the choice of detectors. They are, however, closely related and it is suggested that Chapters 3–5, 10 and 16 be considered as a whole (Figure 4.1).

The nature of the sample preparation will affect the HPLC optimisation: a 100% specific preparation step greatly simplifies the analytical separation! However, since much of our work is concerned with biological matrices, it is often not possible to achieve such a clean sample; there are usually a large number of potential interferences injected onto the analytical column at the high sensitivities dictated by pharmacokinetic analysis.

Similarly, the choice of detector may aid specificity and reduce the need for a 100% selective chromatographic separation, as may be the case with LC–MS–MS detection where peaks co-eluting with the analytes are not detected. Conversely, fraction collection of the peaks of interest for structural identification, or the use of a universal detector, require that the peaks are as pure as possible.

4.2 System parameters

Optimisation in HPLC is the process of finding a set of conditions that adequately separates and enables the quantification of the analytes from the endogenous material

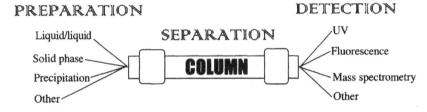

Figure 4.1 Overall concept of optimisation.

with acceptable accuracy, precision, sensitivity, specificity, cost, ease and speed. In order to optimise this process some measure of 'separation' is needed. This is given by the term 'resolution'. The higher the resolution of the system the more selective, sensitive and rugged the analytical method. Resolution (R) may be calculated thus:

$$R = \tfrac{1}{4}[(\alpha - 1)/\alpha]\sqrt{N}[k_b/(k_a + 1)].$$

For baseline separation an R value of at least 1.5 between the peaks is usually necessary. This is generally the minimum resolution required for a good analytical chromatographic method. In order to calculate R it is therefore necessary to define the other terms.

k is the retention factor and is a measure of the time the sample component resides in the stationary phase relative to the time it resides in the mobile phase: it expresses how much longer a sample component is retarded by the stationary phase than it would take to travel through the column with the velocity of the mobile phase. Mathematically, it is the ratio of the adjusted retention volume and the hold-up time. In a fully optimised system, all the peaks of interest will normally elute from the column between $k = 3$ and $k = 20$, i.e. between 3 and 20 column volumes.

α is the separation factor of a chromatographic system and is calculated using $\alpha = k_b/k_a$, where k_b represents the later-eluting peak. Obviously the greater the separation factor between any two peaks the easier it is to separate them.

N is a measure of the separating potential of the whole analytical system and is termed the plate number. There are a number of ways of calculating N, one of the easier and more popular is to measure the peak width at half height:

$$N = 5.54(t_R/W_h)^2$$

where t_R is the peak retention time and W_h the peak width at half peak height. N is expressed in plates per column or plates per metre. The higher the value of N, the sharper the chromatographic peaks.

To summarise the objectives: HPLC optimisation aims to provide a resolution of at least 1.5 between the analyte(s) and endogenous material, with all peaks of interest eluting from the column such that $3 < k < 20$.

Just how this can be achieved is the subject of the following sections, but first a word about quantification. The objective of HPLC optimisation is not just the separation of peaks of interest but their accurate and precise quantification. This is facilitated by the use of either internal or external standards. Although their use is discussed in Chapter 16, it is important to remember that it is the optimal chromatography of *all* peaks of interest that produces the best results in both qualitative and quantitative work.

4.3   Reverse-phase HPLC

In the pharmaceutical industry the great majority of all chromatographic analyses are by reverse-phase HPLC, thus illustrating the almost universal applicability of this technique. As a consequence its optimisation will be discussed in considerable detail.

> *In reverse-phase HPLC the main interaction is that between the solutes and the mobile phase. Retention is effected by adjusting the strength of this interaction as opposed to that of the analytes with the stationary phase.*

An appropriate point to start is to take note of the structure of the analyte(s), particularly with respect to reactive groups, namely acidic and basic groups such as carboxyls or sulphonyls and amines, respectively. All such groups can play a significant part in determining the peak shape, relative retention (k), separation factor (α) and hence resolution (R) in a reverse-phase chromatographic system.

It is preferable in reverse-phase HPLC for the analytes to migrate through the column in a neutral form. If they possess no ionisable groups, then mobile phases based on water and organic solvent may be sufficient. If, as is usually the case in pharmaceutical research, ionisable groups are present, then the mobile phase will require pH adjustment to control this. If the ionization can be removed from the analytes by pH control then the interaction with the non-polar stationary phase increases and the partitioning into the aqueous component of the mobile phase decreases. Unfortunately, many pharmaceutical analytes are mixed species possessing both acidic and basic groups. In order to suppress the ionization of the weakly acidic groups (pK_a 4–6), a mobile phase pH of say 3.5 would be suitable. However, basic groups generally have pK_b values greater than 7 and it is not possible to suppress their ionization and hence reactivity by raising the pH of the mobile phase significantly above pH 7.0 since silica-based stationary phases in HPLC are only tolerant of pH over the approximate range pH 2.5–7.0. Below pH 2.5 there may be hydrolysis of the bonded phase, and above pH 7.0 the silica support starts to dissolve. However, there are a number of modern silica-based columns that may be tolerant up to around pH 10.0, with polymeric columns stable up to pH 13.

Let us therefore consider the nature of the reaction of ionised basic species with the column support.

The great majority of reverse-phase HPLC separation is done on bonded silica columns, where alkyl silane hydrophobic groups are bonded to silanols on the silica surface. Due to the relative sizes of the hydrophobic groups, e.g. octadecylsilane (C18), it is not possible for all the silanols to be covered, there will always be some remaining. Manufacturers try to overcome this traditionally by producing endcapped columns which have residual silanols reacted with, for example, the smaller trimethylchlorosilane. Novel chemistries are always being produced to overcome this, but manufacturers are naturally reticent to divulge the details of their proprietary procedures.

There are three types of silanol group – free, geminal and vicinal – and all will be present to some extent on the silica, but it is the free silanols that are the most acidic, giving rise to secondary interactions with protonated bases in reverse-phase HPLC that are mainly responsible for the peak tailing observed with the analysis of many basic compounds. The other mechanism of peak tailing through secondary interaction is by hydrogen bonding between protonated species in the mobile phase and residual alkali metals in the silica remaining from the manufacture of the silica itself. No amount of endcapping will overcome this. It is only in recent years, as the chemistry of silica manufacture has advanced, that the purity of the water used to precipitate the silica from solution has been identified as a source of metal ion contamination. Manufacturers at the leading edge of quality column manufacture maintain the utmost purity of all their reagents, resulting in silicas that are more than 99.999% pure. In all applications these are the columns that will afford the best peak shape. They are, unfortunately, the more expensive and the more difficult to pack in narrow-bore columns.

The peak-tailing effects of ionised basic groups may be countered by using columns that demonstrate a higher suitability for such compounds. Examples at the time of writing are brand names such as Luna and Prodigy from Phenomenex, Inertsil from GL Sciences, Symmetry and Symmetry Shield from Waters and Genesis from Jones Chromatography, although this is by no means an exclusive list, with many new and excellent columns being produced continually. Alternatively, an amine modifier may be used to saturate the free silanol groups on the support, thus leaving little if any to interact with the protonated sample. The most commonly used such modifier in HPLC is probably triethylamine (TEA). This additive works very well in maintaining good peak shape for compounds containing amino groups analysed using acidic mobile phases. It is, however, not without its difficulties. It is not usually possible to use TEA for high-sensitivity analyses with UV detection at wavelengths much below ~230nm due to the strong absorbance of TEA at these wavelengths. Nor is its use feasible where electrochemical detection is indicated, since impurities in the TEA, brought about by oxidation in the atmosphere, will cause a very 'noisy' signal due to an increased background current. For mass spectrometry it is necessary to use a base-deactivated column since the non-volatile TEA will rapidly accumulate on and contaminate the ion source. The only modifiers acceptable in mass spectrometry are volatiles, such as ammonium acetate and formate and acetic and formic acids, for example.

A mobile phase containing 50mM ammonium acetate and up to 5mM triethylamine may be used to produce a 'generic' aqueous mobile-phase component, where the bonded phase alone is responsible for retention, and may be used in optical detectors above ~230nm. For LC–MS–MS applications a generic mobile phase of 2g/l ammonium acetate and up to 5ml/l formic acid in the aqueous/organic mixture is usually adequate.

After the column and ionic additives have been selected, the next step to consider in reverse-phase HPLC optimisation is the organic additive. The easiest way to do this is to use a four-way or quaternary pumping system:

Solvent A: Water, 50mM phosphoric acid, adjusted to pH 3.5 with sodium hydroxide (use 50mM ammonium acetate instead if necessary for LC–MS–MS applications)
Solvent B: Acetonitrile
Solvent C: Methanol
Solvent D: Tetrahydrofuran (THF)

As stated previously:

> *In reverse-phase HPLC the main interaction is that between the solutes and the mobile phase. Retention is effected by adjusting the strength of this interaction as opposed to that of the analytes with the stationary phase.*

It is primarily the interactions between sample, both analytes and endogenous material, and solvent molecules that will determine the relative retention and separation factor in reverse-phase HPLC. These differential sample/solvent interactions are mainly due to dipole attraction and hydrogen bonding. This means that solvent selectivity can be categorised by solvent dipole moment, relative solvent basicity (proton acceptor) and relative solvent acidity (proton donor).

It is obvious that the solvents to effect the greatest changes in selectivity should differ as much as possible in their polar interactions, this means the use of one proton donor, one proton acceptor and one with a strong dipole moment. Mixtures of these solvents, with water to control the retention range, can therefore approximate the selectivity of any solvent within the confines of the triangle. Obviously, in the case of reverse-phase HPLC these solvents must also be miscible with water, incorporating a weak buffer of appropriate pH. The three solvents that are most appropriate, bearing in mind the ideals of low viscosity and UV transparency are methanol, acetonitrile and tetrahydrofuran. Using this four-solvent mobile-phase optimisation provides effective control over the selectivity of the reverse-phase HPLC system. In LC–MS–MS applications the preferred solvents are methanol and acetonitrile with very weak, volatile acetate or formate salts or, for peptide work ~0.05% trifluoroacetic acid. For coulometry the preferred solvent is methanol with ultra-pure dibasic phosphate salts. Tetrahydrofuran may be used in conventional reverse-phase HPLC with caution. It is not recommended for mass spectrometry or coulometry. THF, as an ether, is prone to oxidation through contact with atmospheric oxygen to generate highly reactive peroxide free-radicals. These can oxidise system components *in situ*, causing poor quantification and reproducibility. THF also attacks PEEK tubing and the Tefzel Rheodyne rotor in autosamplers. If it is to be used, then stainless-steel tubing and a Vespel™ rotor should be employed. In spite of its obvious limitations, THF can have some useful selectivity applications as Table 4.1 shows.

As stated at the beginning of this chapter, the objective of this optimisation is to provide a k range of 3–20 for all peaks of interest, with R at least 1.5 for any peak to

Table 4.1 Comparative solvent effects

Effect compared to methanol	Tetrahydrofuran	Acetonitrile
Acceleration	Aliphatic and aromatic alkyls	Aliphatic OH
Weak acceleration		Aliphatic and aromatic alkyls, phenolic OH, chlorides
Retardation	Phenolic OH, nitrates	Nitrates, cyanates
Negligible	Ketones, esters, aliphatic OH, 2° and 3° amines	Ketones, esters

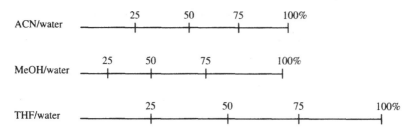

Figure 4.2 Nomograph of relative eluting strengths of different solvents.

be quantified. Initially a 90% acetonitrile mobile phase is used and an injection made, with successive runs gradually decreasing the amount of acetonitrile and increasing the aqueous content until the right k values are obtained. It is, of course, necessary to consider the precipitating effect of acetonitrile on buffer salts at high organic percentages. If the separation factor and resolution appear satisfactory, then the development is finished. It is quite probable though, especially with biological extracts, that some or all of the peaks of interest will be poorly resolved, either from each other or from endogenous material. With the aid of the nomograph (Figure 4.2) a methanol/H_2O mobile phase of the same eluting strength but differential selectivity may be chosen and evaluated.

Follow Figure 4.3 until a suitable separation has been achieved. If this procedure is followed, then the HPLC system should be optimised to run in the reverse-phase mode.

There are other ways to change the resolution in reverse-phase HPLC and they are given below in decreasing order of priority, i.e. the most predictable, effective and reliable first:

1   *pH*: the most effective variable if the analytes contain ionisable species.
2   *Alternative HPLC method*: ion-pair chromatography for ionisable analytes with pK_a values outside the range 3–7.
3   *Columns from different sources*: columns from different manufacturers have different coverage, surface area and relative suitability for acidic and basic samples.
4   *Column type*: in reverse-phase HPLC the different bonded phases have different relative retentive strengths. A few relative strengths of different column types relative to a C8 column (Snyder *et al.* 1988) are:

C18	C8	Phenyl	Cyano
1.26	1.0	0.46	−0.64

5   *Temperature*: as ionisation can be temperature dependent, it is possible to alter separation by changing the temperature over the range 4°C to ~70°C. Be aware of possible column damage and detector bubbles at this upper limit.

A 100% acetonitrile mobile phase may be thought to be the strongest, yet there may be times when it gives unacceptably high k values. This is probably because the

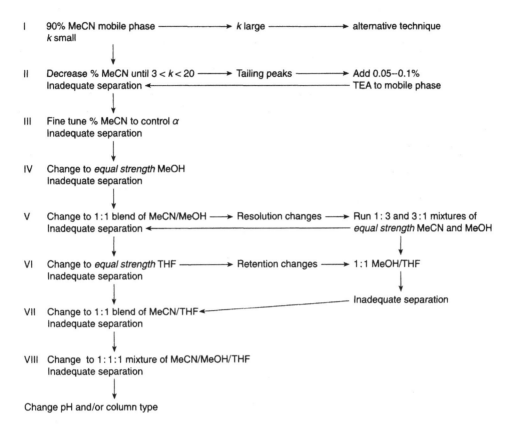

Figure 4.3   Flow sheet of reverse-phase HPLC optimisation, see chapter glossary for explanation of abbreviations. (Adapted from Snyder *et al.* 1988, by permission of John Wiley and Sons.)

dipolar moment alone is not sufficient to displace the hydrogen bonds or weak ionic interaction responsible for retention. In this case the addition of 0.1% aqueous ammonia or 10mM ammonium formate will probably dramatically reduce k and sharpen peak shape. It is as though the mobile phase requires an ionic or proton acceptor/donor component to function. If this is the case, it will be observed in acetonitrile/water mixtures that at 100% acetonitrile k is large, as water is introduced it decreases rapidly and on the further addition of water k again increases. If 0.1% aqueous ammonia in acetonitrile is used, the *apparent* pH will be much greater than pH 9. Normally there is the serious risk of silica dissolution in aqueous mobile phases at these pH values. However, since silica has such low solubility in acetonitrile, such high pHs may be used without significant damage to the column.

The optimisation procedure in Figure 4.3 is based on an optical detector. If mass spectrometry is to be used then there will be some differences. Notably that:

*It is not necessary to obtain time resolution between the analyte and all other material, unless they possess the same parent **and** product ions.*

The great selectivity of the mass spectrometer means that co-eluting material, providing it is not quenching the analyte's ionisation, is not a problem. In this case the reverse-phase optimisation for mass spectrometry will focus primarily on separating the analytes from V_0, to prevent ion-source contamination, and on peak shape and peak elution volumes, covered in Column conditions (Section 4.10).

4.4   Ion-pair HPLC

Reverse-phase HPLC is the most popular chromatographic method for samples that are soluble in water/organic mixtures, even with samples that contain ionic analytes. However, ion-pair HPLC may be a useful alternative for such samples, especially if it is neither practical nor possible to neutralise the charged groups on the analytes. Such compounds are usually poorly retained and/or exhibit poor peak shape. In this situation an aqueous/organic mixture is used, with a buffer and an ion-pairing agent to provide a higher selectivity than is afforded by the column and organic modifier functioning in a strict reverse-phase mode. It is unfortunate that the non-volatile nature of the reagents make this technique *unsuitable* for mass spectrometry.

Retention in ion-pair HPLC can be described as:

> 1   *The production of an ion pair comprising a sample ion and the oppositely charged ion-pair component of the mobile phase, followed by non-polar interaction of the ion pair with the stationary phase.*
> 2   *A twofold process in which the ion-pair agent first adsorbs to the stationary phase via non-polar interaction, and the sample ion is then retained in the column through an ion-exchange mechanism with the bound ion-pair reagent.*
> 3   *A combination of 1 and 2.*

The ion-pair retention of an acidic compound, HA, is illustrated in Figure 4.4. Note that the ion-pair diagram assumes a large concentration of the ion-pair agent. If no ion-pair agent is present, then the retention follows conventional reverse-phase theory. The molecule, HA, has a maximum k when the pH of the mobile phase is 2 units lower than the pK_a of the acid, i.e. more than 99% is in the unionised form. As the pH increases, the acidic molecule will become ionised, and its retention will be decreased as it shows a greater affinity for the polar mobile phase. If sufficient ion-pair agent is present, in this case a tetra-alkylammonium salt would be appropriate, then the stationary phase will be covered with the ion-pair agent through non-polar interaction. The ion-exchange process then controls retention, as shown in the second half of the figure, with k increasing as the analyte becomes more charged, reaching a maximum at, or near, 2 pH units above its pK_a. At a low pH the neutral molecule is not retained by ion exchange, but may be slightly retained due to reverse-phase interaction with the silane-bonded phase; indeed, presence of the ion-pair agent may reduce retention by this mechanism due to competition between the two for the stationary phase.

This example assumes either that there is no ion-pair agent or that there is enough to change the retention mechanism to ion exchange. Through the careful control of

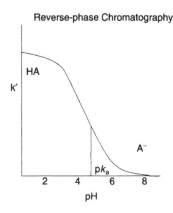
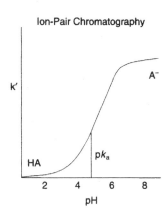

Figure 4.4 Retention of an acidic sample as a function of pH. (Adapted from Snyder *et al.* 1988, by permission of John Wiley and Sons.)

the ion-pair concentration it is possible to cause retention to be governed by both non-polar *and* ion-exchange principles, thus ion-pair concentration is a major selectivity factor in ion-pair chromatography. The other is a corollary of this, namely pH, since the strength of association within the ion pair is also a function of the degree of analyte ionisation.

The alkyl content of the ion-pair agent (e.g. pentyl versus heptyl sulphonate used for basic analytes and methyl versus butyl ammonium for acidic analytes) can also affect k through non-polar interaction, since, simply, the greater the carbon density in the ion-pair reagent, the greater its retention in a reverse-phase system. This reasoning identifies the main selectivity parameters in ion-pair HPLC as:

1 a change in retention mechanism from reversed phase to ion exchange (or vice versa) by varying the concentration of the ion-pair agent in the mobile phase;
2 a change in retention of compounds with different pK_a values, by varying mobile-phase pH (Snyder *et al.* 1988).

As reverse-phase and ion-exchange HPLC function through different mechanisms, the analytes will show changes in *separation factor* as the concentration of ion-pair agent and pH are changed, since ionic analytes have a variety of pK_as, and as this mode combines carbon density and ionisation to increase k, whereas in reverse-phase HPLC they are antagonistic.

Ion-pair concentration and mobile-phase pH are the two primary parameters to control selectivity in ion-pair HPLC.

The selectivity optimisation described for reverse-phase HPLC can be modified to apply to ion-pair HPLC, as shown in Figure 4.5. However, one organic solvent (usually methanol) is used in optimising ion-pair separations as it is less prone to precipitate buffer salts. The other three components of the mobile phase for *basic* analytes are:

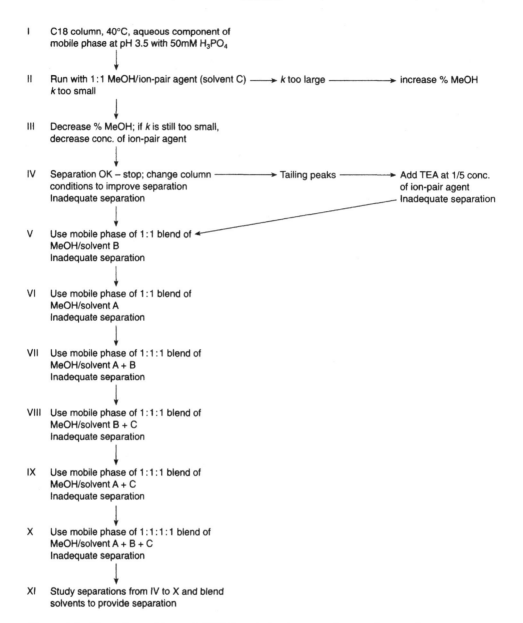

Figure 4.5 Flow sheet of ion-pair HPLC optimisation; see chapter glossary for explanation of abbreviations. (Adapted from Snyder *et al.* 1988, by permission of John Wiley and Sons.)

Solvent A: Buffer pH 3.5 : 50mM H_3PO_4 adjusted to pH 3.5 with NaOH
Solvent B: Buffer pH 7 : 50mM H_3PO_4 adjusted to pH 7 with NaOH
Solvent C: Ion-pairing agent. 0.02M heptane sulphonate in water adjusted to pH 5.0

This procedure applies to the ion-pair separation of a basic analyte, and that heptane sulphonate is the ion-pair agent. Of course, other ion-pair agents can be used. When separating acidic analytes it is necessary to use an anionic ion-pair agent such as a tetra-alkylammonium ion. There are many such reagents to choose from. It is important to remember that in ion-pair HPLC both k and α are affected primarily by the total quantity of ion-pair agent adsorbed onto the column, they are less affected by the carbon content of the chosen ion-pair agent. Usually, a higher molecular weight agent will be more strongly retained, so that lower concentrations of that agent in the mobile phase are required for the same effect. However, high molecular weight ion-pair agents are also more difficult to remove from the column when changing from ion-pair to reverse-phase HPLC. Indeed, it is good practice to devote a column to ion-pair work once it has been used for such as it is not always possible to remove all the bound ion-pair reagent and so return the column to a true reverse-phase mode.

Given below is a selection of other options, after pH and ion-pair concentration, that may be applied to optimise an ion-pair separation. As in the reverse-phase section they are listed in order of approximate decreasing priority:

1 *Temperature*: as ionisation can change with temperature, vary the temperature between ambient and ~70°C, but bear in mind potential problems with bubbles in the mobile phase at the detector and column stability.
2 *Percentage strong solvent*: change the ratio of methanol : water. Since ion-pair HPLC has a separation component derived from reverse-phase HPLC, changing this parameter can effect changes in α. For the same reason, the type of organic modifier may be changed within the constraints of buffer precipitation.
3 *Mobile-phase additives*: in the analysis of basic analytes using sulphonates as the ion pair, add triethylamine, up to ~10% of the concentration of the ion-pair agent, to improve peak symmetry. This will interact with the active silanols on the bonded phase and give more symmetrical later peaks.
4 *Ionic strength*: the buffer concentration may be changed or salt may be added to the mobile phase as in conventional ion-exchange HPLC.

4.5   Ion-exchange HPLC

Ion-exchange HPLC (IEC) separations are similar in concept to those carried out by ion-pair HPLC (IPC) and compounds that are ionic or ionisable can often be separated by either HPLC method. In IPC the surface of the column packing is coated by the adsorbed ion-pair agent, resulting in a charged stationary-phase surface. This acts very much like that in an IEC column, except that in the case of IEC, the charged groups are covalently bonded to the surface of the packing material. Ionised analytes are retained by the displacement of a counter-ion, typically hydrogen or chlorine, that is initially associated with the ionic group bound to the particle surface. In IEC, compounds having different charges, i.e. ±1, 2 or 3, will also show changes in separation factor, in a similar way that an increase in carbon density increases retention in a reverse-phase system.

Since the ionisation of acids and bases is pH dependent:

Table 4.2 Counter-ion selectivity in ion-exchange HPLC

Cations		Anions
Ba^{2+}	Highest counter-ion selectivity	Benzene sulphonate
Ag^+		Citrate
Pb^{2+}		I^-
Hg^{2+}		HSO_4^-
Cu^+		ClO_3^-
Sr^{2+}		NO_3^-
Ca^{2+}		Br^-
Ni^{2+}		CN^-
Cd^{2+}		HSO_3^-
Cu^{2+}		BrO_3^-
Co^{2+}		NO_2^-
Zn^{2+}		Cl^-
Cs^+		HCO_3^-
Fe^{2+}		IO_3^-
Mg^{2+}		HPO_4^-
K^+		Formate
Mn^{2+}		Acetate
NH_4^+		Propionate
Na^+		F^-
H^+		OH^-
Li^+	Lowest counter-ion selectivity	

Differences between successive listings may not be significant.

> *High pH values ionise and therefore increase retention of acids on strong anion exchangers, as do low pH values for bases on strong cation exchangers.*

A number of variables can be used to optimise an ion-exchange separation, they include:

1 *Choice of column*: for basic analytes that can be protonated, cation exchangers such as benzyl sulphonic acids (strong) and carboxylic acid (weak) bonded phases may be used. For acidic analytes, quaternary amine, aminopropyl and diethyl-aminopropyl phases are appropriate for anion exchange.

2 *pH*: the retention increases in cation exchange and decreases in anion exchange, with a decrease in mobile phase pH, and vice versa.

3 *Buffer salt*: the eluting strength and selectivity of the mobile phase are determined by the competing counter-ion; the relative strength of counter-ion to analyte affects retention. Table 4.2 shows counter-ion selectivity.

4 *Ionic strength*: mobile-phase strength generally increases with an increase in molarity. There is little change in selectivity except where analytes of different valency are to be separated.

5 *Organic modifier*: ion-exchange columns are usually bifunctional, with an organic ligand supporting the ion exchanger. With hydrophobic analytes these supports

Table 4.3 Summary of the main differences between ion-exchange and ion-pair HPLC (adapted from Snyder *et al.* 1988, by permission of John Wiley and Sons)

	IEC	IPC
Sample	Any ionic sample	Any ionic sample
Column	Anion or cation exchange	Reverse phase
Mobile phase	Aqueous solution of buffer and solvent	Water, solvent, buffer and ion-pair agent
Increase *k*	Decrease salt	Decrease organic
Increase separation factor	Vary pH, change selectivity of counter-ion	Vary pH or concentration of ion-pair agent
Possible problems	Column variability and stability; less control over resolution; silica dissolution	Slow equilibration of column when mobile phase is changed

can influence retention by conventional reverse-phase mechanisms, therefore retention of such analytes can be modified by the organic modifier in the mobile phase. Care should be taken to avoid salt precipitation, especially with acetonitrile as modifier.

6 *Temperature*: as temperature is raised, so the kinetics of mass transfer are improved, the viscosity of the mobile phase decreases and buffer solubility increases. Care should be taken with exchangers at pH greater than pH ~8–9 as silica solubility also increases with temperature at high pH, especially in the presence of phosphate salts.

It has been shown that either ion-exchange HPLC (IEC) or ion-pair HPLC (IPC) may be used for the chromatography of ionic or ionisable compounds. There are, however, important differences and similarities between the two techniques, as shown in Table 4.3.

4.6 Normal-phase HPLC

Although reverse-phase and ion-pair HPLC are by far the most popular modes in the pharmaceutical industry, the selectivities available may be insufficient to provide adequate resolution. Normal-phase HPLC can be useful in these cases, since different retention processes provide different selectivity effects, as, unlike reverse-phase HPLC,

analyte–solvent interactions are relatively weak; analyte–stationary phase and solvent–stationary phase interactions predominate.

The solubility of many organic compounds, for example those with high log *P* values designed to cross the blood–brain barrier, is often greater in normal-phase HPLC solvents than in those used in other techniques. This mode may also be applicable to mass spectrometry.

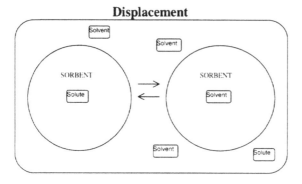

Figure 4.6 Schematic representations of normal-phase retention.

Normal-phase HPLC used to comprise unmodified adsorbents, such as silica or alumina, which can give impressive selectivities, but there are certain practical constraints upon their usage. Mainly, there is the need to exclude water from the system in order to obtain reproducible results. Fortunately the more recent cyano, diol and amino columns are more robust in this respect and are therefore recommended for HPLC, with the cyano column usually a good first choice.

In normal-phase HPLC, retention is governed by adsorption onto the stationary phase. For retention to occur, an analyte must displace one or more solvent molecules from the stationary-phase surface. This principle is shown in Figure 4.6. In addition to this 'displacement' effect, dipolar molecules can exhibit very strong interaction with polar sites on the stationary phase, an effect termed 'localisation'. For example, polar substituents, such as amines or carboxyls, on a benzene ring interact more strongly with the adsorption sites on the stationary phase than does the benzene ring itself.

Displacement: the competitive retention of solute and solvent molecules on the sorbent.

Localisation: the relative differences in retention due to different polar groups interacting with specific sites on the sorbent.

The two effects of displacement and localisation are the primary sources of mobile-phase selectivity and are the basis of the optimisation process. This is based on a

Table 4.4 Classification of solvents according to selectivity effects in normal-phase HPLC (adapted from Snyder *et al.* 1988, by permission of John Wiley and Sons)

Weakly polar (non-localising)	Polar (localising)	
	Basic	Non-basic
Dichloromethane*	Methyl tertiary-butyl ether*	Ethyl acetate*
Chloroform*	Ethyl ether	Acetone
Toluene	Isopropyl ether	Nitromethane
Chlorobutane*	Tetrahydrofuran	
Ethylene chloride*	Dioxane	
Acetonitrile*	Triethylamine	
	Dimethyl sulphoxide	
	Methanol*	
	Propanol*	

non-polar solvent, such as hexane, as the diluent (analogous to water in reverse-phase HPLC), plus three polar organic solvents to vary resolution and hence separation factor for the analytes. The three polar solvents are chosen, one from each of the groups in Table 4.4. While it is certainly unusual to refer to organic solvents as *basic* and *non-basic*, since they are not ionisable, the nomenclature is appropriate in this instance as it refers to their *relative* proton acceptor capabilities.

Solvents marked with * are preferred since they are relatively stable and do not significantly absorb UV light above 240nm. However, none of these solvents is stable indefinitely. In the presence of atmospheric oxygen many degrade to form peroxides and free radicals and it is important to take steps, such as fresh stocks and nitrogen purges, to prevent this occurring.

Hexane is useful for separations involving low UV detection, but is not miscible with all the above solvents, for example, acetonitrile. Therefore, when using hexane, co-solvents (e.g. dichloromethane) may have to be added to the mobile phase to ensure miscibility.

The first priority in normal-phase solvent optimisation is to obtain the correct solvent strength so that all analytes should elute within the range $3 < k < 20$. Once the solvent strengths of the various binary mobile-phase mixtures have been determined, the effect of different mobile phases on α can be examined in a manner similar to that for reverse-phase systems. It is recommended that the normal-phase HPLC optimisation in Figure 4.7 is followed, which is a triangulation optimisation using the non-localising, basic localising and non-basic localising solvents, dichloromethane, methyl tertiary-butyl ether and acetonitrile, respectively.

If this procedure fails to provide adequate selectivity and resolution there are other parameters that may be considered:

1   *Column type*: a cyano stationary phase is usually the first choice, but diol, amino and silica columns may be considered.
2   *Other solvents*: solvents from other selectivity groups will provide a different combination of the three selectivity effects.

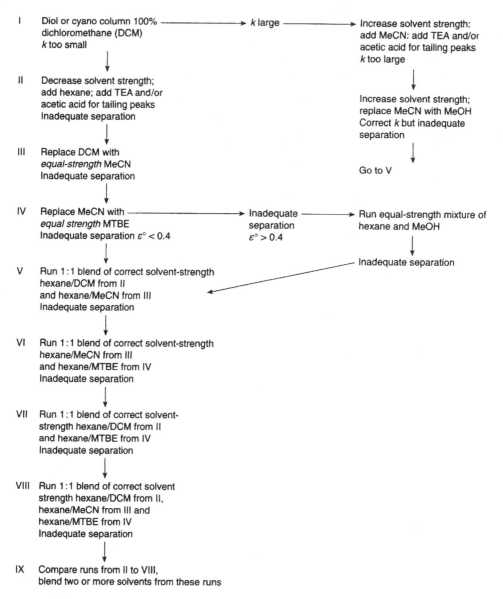

Figure 4.7 Flow sheet of normal-phase HPLC optimisation; see chapter glossary for explanation of abbreviations. (Adapted from Snyder *et al.* 1988, by permission of John Wiley and Sons.)

4.7 Chiral HPLC

The word 'chiral' is derived from the Greek word χειρ meaning hand, and refers to molecules of the same chemical structure but different orientation such that, just like hands, they are not superimposable with their mirror image. These non-superimposable forms of a chiral compound are called enantiomers. They exhibit the same chemical

Table 4.5 General principles of chiral separation (adapted from Snyder *et al.* 1988, by permission of John Wiley and Sons)

1 Use an HPLC column with a chiral stationary phase (chiral column)	This is the most popular approach due to convenience and the wide range of chiral columns available. There are few practical problems apart from achieving separation, usually by trial and error, and the relatively high price of the columns
2 Form diastereoisomers and separate by conventional HPLC	Most likely to be successful, but most inconvenient in routine applications. The assays are generally less precise and racemization is possible. The quantitative yield of the reaction may also vary. However, it does permit the full range of HPLC techniques to be used with the efficiencies afforded by such mechanisms
3 Add chiral complexing agent to the HPLC mobile phase	The least used approach; the chiral agent may be expensive and there is a limited variety of suitable reagents. The reagent may also interfere with detection, as in the case of crown ethers and UV detection

composition and constitution but different **sterical** configuration. Often chiral molecules possess an asymmetric carbon atom or another asymmetric centre such as phosphorus, nitrogen or sulphur, but this is not always necessary. The enantiomers of a chiral substance react identically *except* in a chiral environment, e.g. towards plane polarised light or chiral, optically active reagents and solvents or in biological systems.

The biological or pharmacological activity and efficacy of chiral molecules can be greatly influenced by their configuration. Often only one of the enantiomers is pharmacologically active, while the other may be less active, inactive or even toxic, so assays to determine enantiomeric purity are obviously required in the course of the development of such a chiral drug, and such assays are usually carried out chromatographically. However, the chromatographic separation of enantiomers cannot be achieved by conventional HPLC or GC because the enantiomers possess the *same* physical and chemical properties and are therefore unresolved on most systems; $\alpha = 1.00$. The chromatographic separation of enantiomers needs a different approach to non-chiral optimisations and can be achieved in three general ways, as shown in Table 4.5.

4.7.1 Chiral columns

Columns with chiral stationary phases are available in many different forms and this is important since the separation of a particular enantiomer pair depends to a very large extent on the chemistry of the analyte–chiral stationary phase interaction. However, changes in the mobile phase can also have a significant and sometimes unexpected effect on separation, since it not only affects the analyte/stationary phase interaction but may also alter the nature of the stationary phase itself, a phenomenon most relevant to protein columns.

Table 4.6 Examples of chiral columns and their analyte requirements (adapted from Snyder *et al.* 1988, by permission of John Wiley and Sons)

Column type	Examples	Sample requirement
Three-point interaction	Dinitrobenzyl (DNB)-phenylglycine; DNB-leucine	Two hydrogen-bonding groups in the sample molecule close to the chiral centre; an aromatic or other large group in close proximity
Protein	Bovine serum albumin (BSA); α_1-glycoprotein	One or more polar or ionic groups close to chiral centre
Cavity	Cyclodextrin; cellulose	Cyclic group close to chiral centre that fits into cavity
Substituted cavity	Cyclodextrin and cellulose carbamates and esters	One or more functional groups in close proximity to chiral centre
Macrocyclic antibiotics	Vancomycin; teicoplanin	One or more functional groups in close proximity to chiral centre

Due to the complexity of the sample/stationary phase interaction only rough guidelines on optimisation can be given. It is useful, perhaps, to divide chiral columns into a number of types, as shown in Table 4.6. Two attractive interactions are required for three-point interaction columns, usually hydrogen bonds, plus a third interaction that may be either attractive or repulsive. This type of column is usually based on an amide structure, such as 1,3-dinitrobenzyl phenylglycine where there are two carbonyl and one amine groups. This arrangement of functional groups gives many possibilities for hydrogen bonding with the analyte; all it needs is an aromatic ring plus two hydrogen-bonding groups, either donor or acceptor or one of each, close to the chiral centre. If the analyte has the appropriate requirements for separation by three-point interaction, but the enantiomers are not resolved on the first column tried, it is usually worthwhile trying a different chiral column, as chiral columns of similar structures can exhibit large differences in chiral selectivity for a given analyte. Therefore for chiral separations of this type, several different columns may be required for trial and error optimisation. Column choice is based on analyte functional groups, such as aromatics, amines, acidic, keto and hydroxyl groups. These three-point interaction columns were the first type to be produced, and are therefore more fully developed and, hopefully, more reliable. Peak shapes for this type of column are usually similar to those found in conventional reverse-phase HPLC.

Protein columns are another type of chiral column and use animal proteins as the chiral stationary phase. Carrier proteins, such as albumin or α_1-acid glycoprotein, circulate in the bloodstream and recognise a variety of different enantiomers by virtue of their configuration. These proteins may be immobilised onto a silica support to form the chiral column. Such columns have many different binding sites that are specific for a wide range of functional groups occurring on compounds that are transported in the bloodstream, giving these columns several distinctive features:

1   The separation of a particular enantiomer pair on a protein column is more likely than with other chiral columns as protein columns recognise hydrogen-bonding,

polar, ionic and hydrophobic groups in the analyte, as well as the three-dimensional structure of the sample molecule.

2   As there are many different binding sites, the concentration of each is low; therefore protein columns are easily overloaded. The maximum sample size is usually only a few nanomoles. This, of course, is total loading, not just the analyte. They are also less tolerant of a strong injection matrix as it is common, although not always necessary, for the mobile phase to comprise less than 5% of organic modifier.

3   In many cases, the strength of the binding of the analyte to the protein is quite variable due to the variety of interactions that can occur. It is not reasonable to expect the same column efficiency for this type of column as for others that run in the reverse-phase mode.

Although protein columns are more likely to produce enantiomeric resolution, the combination of broad peaks and limited sample sizes makes these columns less appropriate for routine use, but with the increase in prominence of LC–MS–MS, where detector sensitivity is very good, these limitations may be less significant than with optical detectors. Even the mass spectrophotometer, of course, needs enantiomeric resolution just as any other conventional detector, as the enantiomers have identical chemical properties.

Cavity columns exhibit chiral selectivity due to an asymmetric cavity in the stationary phase. To achieve enantioseparation the sample molecule must possess a cyclic substituent that 'fits' the size of the cavity and possesses one or more hydrogen-bonding substituents. Peak widths for cavity columns are typically between those of three-point interaction and protein columns, making them usable for routine use. However, care must be taken as they are highly sensitive to changes in flow rate and temperature.

An alternative form of chiral column has been produced with the three-point interaction moiety attached to a cellulose or β-cyclodextrin support. These columns are proving very popular, being made by a number of manufacturers, such as Daicel in Japan, Stagroma in Switzerland, Macherey-Nagel in Germany and Astec in the USA, among others. These cellulose columns are generally run in the normal-phase mode, although a few are available for use by reverse phase. The β-cyclodextrin hydroxypropyl-substituted columns may be run in reverse phase or in 95% acetonitrile, 5% methanol, with a trace of acetic acid *and* TEA as appropriate, for enantiomeric separation, termed the 'polar organic mode', or with 100% acetonitrile on *R*- and *S*-naphthylethylcarbamate-substituted β-cyclodextrins (Armstrong 1991).

The cellulose columns have the chiral phase absorbed at high temperature onto the support and as such are not particularly stable; a column lifetime of 1000 injections is the most that may realistically be expected and, at around £1000 each, this is an important consideration. The cyclodextrin-substituted columns are far more stable since the chiral phase is covalently bound onto the support; these may cost about £850. There is much literature to assist in the selection of these two column types, of which the following is only an example. Generally, non-aromatic compounds without polar groups are best suited to ester-type stationary phases, whereas compounds with amino or carboxylic groups are best suited to crown ethers or ligand exchangers, respectively. Aromatic analytes with a variety of functional groups are more readily resolved on these chiral conjugated cellulose or cyclodextrin stationary-phase columns with ester-, polymethacrylate- or carbamate-bonded phases.

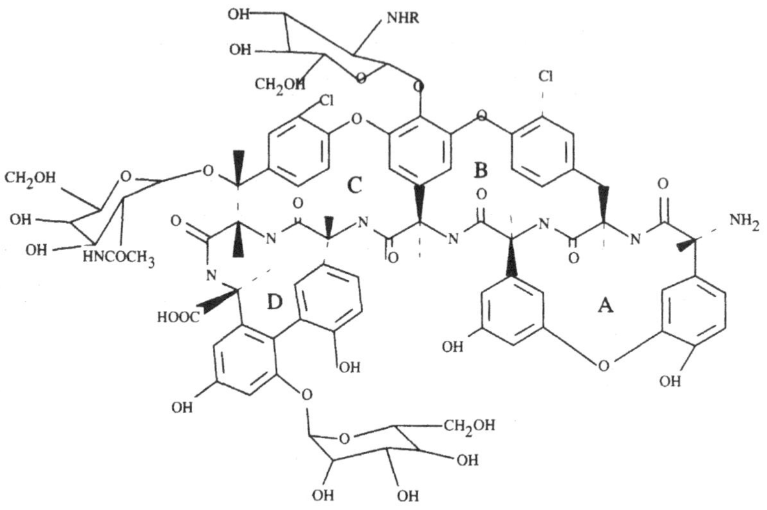

Figure 4.8 Chirobact T™, teicoplanin.

The macrocyclic antibiotic columns from Advanced Separation Technologies are relatively new to the market. They are presently of two types, vancomycin and teicoplanin, and offer complementary stereoselectivity in as much as enantiomers not resolved on one phase are usually resolved on the other. Both may be run in normal-, reverse- and polar organic-phase solvents, the first two of which are suitable for analysis by mass spectrometry. The structures of the bonded stationary phases are similar (the teicoplanin structure is shown in Figure 4.8):

- MW: ~1885
- chiral centres: 23
- sugar moieties: 3
- fused rings: 4(ABCD)
- R: methyldecanoic acid.

Teicoplanin has an unusual selectivity for a number of analytes, especially acidic compounds, small peptides, neutral aromatics and cyclic aromatic amines, as well as amino acids. As teicoplanin contains a variety of ionisable groups it is not surprising that the enantioselectivity can be modified in a number of ways and all the typical interactions defined for protein phases and other polymer-type stationary phases may be possible with this phase. This includes $\pi-\pi$ complexation, hydrogen bonding, dipole stacking, stearic interactions and inclusion complexation associated with cyclodextrin cavities.

4.7.2 Diastereoisomers

If chiral separations are required to be performed on conventional non-chiral columns they require the formation of diastereoisomers using an optically pure chiral

94

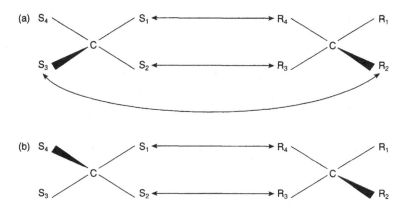

Figure 4.9 Complex formation with chiral reagent. (Adapted from Snyder *et al.* 1988, by permission of John Wiley and Sons.)

agent. Diastereoisomers differ in their *physico-chemical* properties and may therefore be separated by conventional chromatographic techniques.

It may not always be necessary to have a covalent bond between the chiral carbons in diastereoisomers to achieve enantioseparation. It may be possible to do this by forming diastereomeric complexes where the covalent bond may be replaced by a physical interaction between the enantiomer and the chiral reagent, as shown in Figure 4.9. If it is assumed that groups S_1 and R_4 are mutually attractive, as are groups S_2 and R_3, then in (a) there is the potential for attraction or repulsion between S_4 and R_1 and S_3 and R_2, respectively. Such attraction/repulsion is not possible in (b). This leads to the possibility of separation of these diastereomeric complexes by conventional HPLC. In order to form diastereomeric complexes, the two chiral centres must be reasonably close to each other, usually less than three bond lengths, and the nature of the third interaction must be different for the two enantiomers. This is referred to as the **Dalgleisch rule**. There is no guarantee that such complexes will always be resolved on conventional columns, or that all related compounds will behave in a similar manner. For example, a parent compound may be derivitised/complexed through its secondary amine with, perhaps, (–)-(1R)-menthyl chloroformate in the presence of triethylamine and acetonitrile, and the two enantiomers exhibit excellent selectivity on a C18 column, yet the de-ethylated metabolite, although it also reacts, exhibits no enantioselectivity.

There are a number of optically pure derivatising agents available from Fluka, among others, that may be used to form diastereoisomers. These are such reagents as isothiocyanates, orthophthaldehydes and chloroformates for the derivatisation of amines, and amines and alcohols for acids. With each of these it is, of course, important that the derivatising agent be available in an optically pure form, that the reaction kinetics are equal for both enantiomers and that the diastereoisomers so formed are equally and sufficiently stable. It may sometimes be useful for the derivative to add a detection aid to the analyte, such as a fluorenyl group with a chloroformate reaction for fluorescence detection; however, this is often of less importance than stable diastereoisomers possessing good separation factor ($\alpha > 1.2$) and peak shape in a short run time.

It is not unusual for chiral separations to be required where there is a large enantio-excess of one enantiomer against the other. In such cases it is preferable to have the smaller eluting first as it may be lost on the tail of the larger peak if the larger were to elute first. It may be possible to reverse the elution order by substituting the alternative enantiomer of the chiral derivitising agent.

4.7.3 Chiral complexing agents

Chiral complexing agents may be added to the mobile phase, but this is generally less popular than chemical derivatisation, and certain types of cyclodextrins may be added to the mobile phase to achieve separations similar to those using cyclodextrin stationary phases. It may, for example, be possible to saturate a C18 column with such a complexing agent and temporarily transform the column to one with chiral characteristics. This approach is quick and easy, if not particularly reproducible. The adsorbed chiral phase will, of course, gradually leach off with the mobile phase and require regular regeneration. It may be appropriate to use various metal–ligand complexing agents, such as 0.25mM aqueous $CuSO_4$ with an enantiomerically pure amino-acid column being used to separate amino-acid enantiomers.

4.7.4 Chiral summary

Overall, a defined strategy for chiral HPLC optimisation is far more difficult to produce than for conventional HPLC since enantiomeric separation is based upon all the factors present in conventional HPLC plus a consideration of the tertiary structure of the analyte. It is suggested that a thorough literature search be carried out first to ascertain whether any methods have been published for compounds with the same functional groups. If a method exists based upon a three-point interaction column, then it should perhaps be tried first, then one of the macrocyclic antibiotic columns. Failing that, then a substituted cellulose or β-cyclodextrin in either the reverse or normal phase. If none of these proves satisfactory, then a protein column or a non-substituted cavity column may be used, in each case adjusting the eluotropic strength of the mobile phase such that $3 < k < 20$. It is also worth pointing out that chiral columns are relatively expensive, at about £300–£1000 for a 150×4.6mm column, and that, as yet, they do not have the life span of conventional columns. Derivatisation to diastereoisomers should only be pursued if a satisfactory chiral column can not be found or if the separation is for a non-routine application.

The purpose of chiral analysis is to identify and quantify the analyte enantiomers. Ordinarily the quantification parameter of peak height is used since the achiral analyte will always have the same retention time and this parameter is less prone to errors in accuracy and precision due to an incorrectly assigned baseline. With chiral HPLC this is appropriate for direct quantification, but it is not applicable to use peak height to determine enantiomeric ratios because the enantiomers have different retention times. In a racemic mixture the earlier enantiomer will have a greater peak height than the latter since peak height is inversely related to retention time, as it is a function of the analyte concentration in the mobile phase. However, the total quantity or area of each enantiomer peak, from an injection of the racemate, will be equivalent since peak area is directly related to the mass of the analyte and is independent of retention time. However, if there are small errors in baseline construction using peak heights then the effect on the recorded values is minimal. This is not true for peak areas. Small errors in baseline setting, since these occur at the wide base of the peak triangle, have a disproportionately large effect on the integrated values, so for enantiomer ratio determinations the setting of the peak start/stop signals is quite critical.

The majority of our applications are concerned with the separation and quantification of potential drug substances, and/or their metabolites, in biological matrices. Such an exercise is typically carried out by HPLC with a column capacity of 20–30 peaks. It is often the case that the majority of development time is taken up with minimising the presence of interferences in the limited space on the chromatogram. One way that this is done is by maximising on the separation factor of the sample preparation process. Another is by LC–MS–MS where the multiple reaction monitoring is selective for the analytes. In chiral separations, since the parent and product ions are the same, time resolution between the enantiomers is required. If trace levels of enantiomers are to be resolved by optical detection, it may be necessary to utilise the full efficiency available on the analytical column. With chiral work this poses additional considerations: there are at least two enantiomers plus perhaps an internal standard to be resolved on a column with usually less efficiency, and therefore separation factor, than a typical C18 column, together with significant restraints on mobile-phase additives. Some chiral columns may not be tolerant of phosphate buffers, TEA or, in certain cases, even methanol. Thus major variables that would normally be used in controlling resolution and band broadening may be unavailable with these types. This reduces the peak capacity of the chiral column significantly. There are two solutions to this predicament:

1   Improve on the selectivity of the sample preparation such that the injection matrix is very highly enriched for the analyte and internal standard. This may be achieved with a double extraction, for example; a liquid/liquid extraction into ether followed by solid-phase extraction onto normal-phase cartridges.
2   Performing the analysis of the racemate in the conventional manner and collecting the eluate and applying this to a chiral separation technique. This may be done either by fraction collection or by column switching.

4.8   Column switching in HPLC

Multidimensional HPLC, or column switching, uses two or more orthogonal chromatographic selectivities to achieve separation of the analytes. Sample preparation by solid-phase extraction prior to HPLC is one form of multidimensional separation, although this usually involves low-resolution separation as the first step. Occasionally, multidimensional separation is used for the total analysis of a very complex sample, but more often it is used for the separation and analysis of one or more components in a complex sample matrix. In this case, it is necessary either to remove high concentration interferences in the first separation, or to achieve good resolution of the analyte(s) in the final separation. It has also been used successfully to concentrate the analytes from a dilute sample, as in the case of the Gilson ASTED^{TM} system, for example, where dialysates are concentrated onto a column which is then back-flushed to present the sample in a small volume to the analytical column.

Very many configurations are possible with this approach of using valves and multiple columns. It is quite possible to build a system comprising several pumps, columns, detectors and valves round one autosampler. When commencing such an assay development the end point should be kept in mind; is the assay for a very few samples with perhaps qualitative and semi-quantitative data required, in which case a complex system may be acceptable, or is it for a large number of samples where accurate quantification is required, in which case the maxim 'keep it simple' is highly appropriate. This is because with each extra step or process or column there is an increased source of error; a missed peak, a late eluter, a blocked pre-column, a temperature change, etc.

Of all possible configurations, Figure 4.10 is an example of a configuration for the determination of a specific analyte in mouse plasma on sequential CN and ODS phases, since a single phase can not afford sufficient selectivity in this matrix. Two HPLC columns and pumps are connected via a switching valve. On the first column the analyte and the internal standard are separated from the majority of interferences by dipole forces such that they elute incompletely resolved from each other. The total volume of the analytes is ~450μl as measured on exit from the detector. This material is switched to the second column, a narrow bore ODS running on a gradient. This provides band compression on the head of the column and complete separation of the analyte, the internal standard and all other switched material. A schematic representation of this is shown in Figure 4.11. The material from column 1 is diverted to column 2 while the remainder of the injection is cleared. Column 2 then resolves the switched material by gradient for quantification. Column 2 re-equilibrates during the injection cycle and first few minutes of the following injection to give a total run time of approximately 10 minutes.

The innovation of this application, using column switching with internal standardisation, has not appeared in the literature to date. The configuration shown here switches the analyte and internal standard together. If there was resolution of the internal standard from the analyte, with the internal standard eluting first, then the same format may be used, but it is desirable that the internal standard elute after the analyte since any chromatographed metabolites are likely to elute before the analyte in a reverse-phase system and so interfere with the internal standard. In this case the valve from the second pump would be switched to apply the analyte in the second loop to the second column while maintaining the flow path from the first column to the detector. Once the internal standard has passed the detector the other valve would be switched, bringing the second column in line with the detector to detect and quantify the doubly resolved analyte peak.

Column switching can be used with a variety of chromatographic modes, for example size-exclusion or ion-exchange HPLC followed by reverse-phase separations or vice versa. When all of the options – column, mobile phase and method – are considered, it is obvious that many different variables can be used to achieve the resolution of a given compound from the rest of the sample mixture. Any of the variables concerned with optimising reverse-phase, normal-phase, ion-pair, ion-exchange or gradient HPLC can be changed between the two separations, assuming mobile-phase compatibility, allowing the separate optimisation of band spacing in each separation.

A further point to consider with regard to trace level quantification is the dispersive effect of the system and solvents. With liquid chromatography the peak volume at

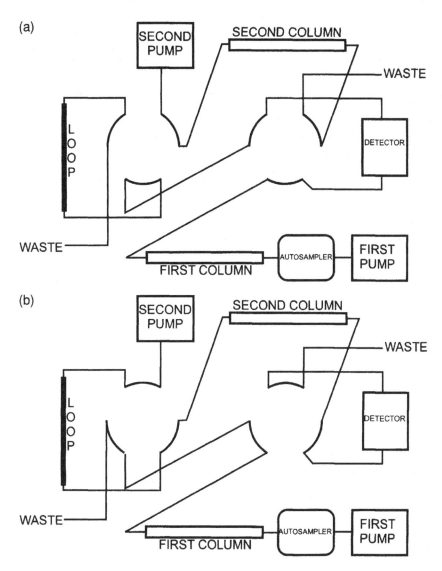

Figure 4.10 Column switching: (a) first column in line with detector; (b) second column in line with detector.

the detector is usually greater than the volume injected; an injection volume of, say, 10μl may easily result in a volume at the detector of 300–800μl. If this 300–800μl sample is going to be the injection for the second column, then the peak volume after the second injection is going to be even larger, with a concomitant decrease in sensitivity and resolution (Chow 1991). For this reason it is necessary for the mobile phase of the first column to be hypo-eluotropic to the second, so producing on-column band focusing at the head of the second column. If the sample can be loaded onto the first column using band focusing, and with a mobile phase hypo-eluotropic to the second mobile phase, absolute minimal band broadening should be produced

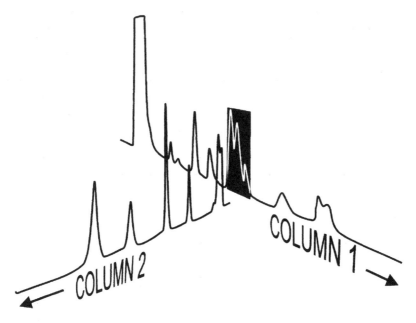

Figure 4.11 Schematic representation of the column-switching principle.

on the columns and throughout the system, giving rise to maximum sensitivity and resolution, assuming that appropriate tubing and connections are used.

4.9   Gradient reverse-phase HPLC

Since gradient elution separations are usually run in the reverse-phase mode they may be optimised in a similar way and the same conditions recommended for initial isocratic separations can be used. An initial concentration of organic-phase modifier is chosen to give k values in the range of 3–20, not neglecting the pH requirements of the mobile phase where ionised species are to be chromatographed. For isocratic separations this is done by trying different mobile-phase compositions. In gradient elution, the information from isocratic runs may be used to predict what concentration of organic modifier may be required to give a k of about 5 for all the analyte peaks. Once this range of modifier concentrations has been determined, they may be incorporated into a gradient programme. Once this is accomplished, resolution and run time can then be improved empirically. An alternative approach is summarised below:

- Step 1: select standard gradient conditions for first run. Use 15×0.46cm, 5µm C18 column of the highest silica purity currently available, as this generally provides the best peak shapes, 5% acetonitrile in 25mM ammonium acetate at pH 3.5 (with formic acid) to 100% acetonitrile in 25 minutes at 2ml/min flow rate, or 50 minutes at 1ml/min.
- Step 2: adjust the acetonitrile concentration range to minimise empty space at either end of the chromatogram and change the gradient time in proportion to this change.

- Step 3: if the chromatogram has many peaks with poor resolution at this stage, increase N by increasing the gradient time or the column length, and/or by decreasing particle size and flow rate while trying to keep the run time to an acceptable length.
- Step 4: once all the analyte peaks are resolved, adjust resolution as necessary by varying flow rate and/or temperature.

4.10 Column conditions

Optimisation in HPLC goes beyond the initial procedures outlined for the main mechanisms. Once the stationary and mobile phases and operating temperature have been selected, N can be optimised through column length, diameter and particle size – column conditions. It is a rule of chromatography that N will increase in proportion to the square root of the column length and with decreasing particle diameter. For each set of chromatographic conditions, there is an optimum flow rate. Higher flow rates lead to lower values of N, through impaired mass transfer leading to a loss in resolution and separation factor. Efficiency also decreases with an increase in mobile phase viscosity for the same reason. For reverse-phase systems, the higher viscosity of water (1.0cP at 20°C, cf. acetonitrile 0.36cP at 20°C) means that it is appropriate to operate at higher temperatures to maximise efficiency. A temperature of 45°C represents a good compromise in this respect. This is particularly important with methanol/water mixtures where there is an increase in viscosity to a maximum at 50 : 50 methanol/water producing high back-pressures.

When adjusting the number of theoretical plates to achieve the desired resolution, it is important to remember other variables that affect the viability of a method:

1 *Run time*: this should be as short as possible, and for an isocratic run in the range 10–20 minutes, although for a gradient run separating metabolites times up to 60 minutes may be acceptable. With quantitative LC-MS-MS run times are less than 10 minutes and, with acceptable ionisation, may be as low as 2 or 3 minutes.

2 *Pressure*: pressures less than 1000psi are desirable. Running columns at much higher pressures is likely to cause bed compression and void formation which produces channelling. The lifetime of the pump seals is also compromised, the more so if buffers are used in the mobile phase.

3 *Solvent usage*: for all analyses a reduction in the mobile phase consumed per run is desirable for both economic and environmental reasons. It may be possible to use narrower-bore columns. A 2mm i.d. (internal diameter) column running at the same linear velocity as a standard 4.6mm i.d. column at 1ml/min would have a flow rate of only 0.2ml/min, a saving on mobile phase of 80% with a corresponding increase in peak height of around fivefold.

4 *Injection volume and matrix*: to produce optimal peak shape and hence maximise efficiency, it is essential that the sample is adsorbed onto the stationary phase at the very head of the column in as small a band as possible. Sensitivity, efficiency and hence separation are all related to peak width, therefore injection volumes should be kept to a maximum of about 25μl for a 4.6mm i.d. column if injections are to be made in a matrix stronger than the mobile phase. Generally it

is better if injections can be made in, say, a 50 : 50 mixture of mobile phase and the appropriate, least eluotropic solvent component of the mobile phase (e.g. 50 : 50 mobile phase : water in reverse-phase and ion-pair systems, and 50 : 50 mobile phase : hexane in normal-phase HPLC) or, in the case of gradient HPLC, in the initial mobile phase. Under these conditions injection volumes of more than 100μl can easily be used without loss of column efficiency, providing the column is not overloaded.

5   *Column connections*: any dead space between the autosampler and the detector will lead to an overall reduction in system efficiency. Therefore ensure that all fittings are correct and that connecting tubing between the autosampler, column and detector is no more than 0.18mm i.d.; the red PEEK tubing is ideal in this application. The most critical connection in this respect is that between the column and the detector. With UV, also check that the detector flow cell volume is appropriate for analytical use.

Although the same plate number can be achieved by different combinations of column conditions, some conditions produce faster separations. Thus, when the required efficiency is small because the separation factor is large, shorter columns of 3μm particles are preferred. When the required efficiency is large, a 25cm 5μm column is preferable. Whereas columns of 3μm particles provide a higher efficiency than do columns with larger particles, they can become blocked more easily.

It is appropriate at this point to make a few observations regarding LC–MS–MS column conditions. The sensitivity, separation factor and specificity achievable on an LC–MS–MS system make it the method of choice for many pharmacokinetic analyses where the low analyte levels at $T_{24 \, hours}$ are needed for accurate AUC determination. Nevertheless, for optimum performance some modifications to UV assay conditions may be necessary. MS detection is dependent upon analyte concentration in the mobile phase as in any other system. It therefore follows that maximum sensitivity will be achieved with minimal peak elution volume. In electrospray, the optimum flow rate through the probe may be as high as 200μl/min. This is, or is close to, the optimum flow rate for a 2mm i.d. column. A 4.6mm i.d. column will therefore require 80% of the eluate, and therefore 80% of the analyte to be diverted to waste. Although typical chromatographic systems can run gradients quite satisfactorily at 1ml/min they are not really suited to doing so at 200μl/min. This means that the band compression utility present in gradient work allowing large sample volumes may not be applicable in LC–MS–MS. Clearly a different approach is called for. With the flow rate fixed at 200μl/min in the isocratic mode, column diameters of 1 and 2mm are assessed with methanol or acetonitrile mobile phases containing ammonium formate and/or formic acid to control peak shape. If available, 3μm particle size should be used with a column length of 5–10cm. Once acceptable peak shape and retention of $k = \sim 3$ have been achieved, then biological matrix extracts are applied to determine the level of quenching in the system. Ideally the percentage recovery by LC–MS–MS should be the same as that achieved through an optical detector. If quenching is observed, then the organic and salt modifiers are altered to change separation factor to restore the apparent recovery to its true value. Once this has been achieved then the concentration of salt ions in the mobile phased is reduced to the minimum that facilitates good peak shape. This is because the electrospray ion-source, on the Micromass

Quattro II for example, will not always completely ionise all the material presented in the mobile phase. If the salt : analyte ratio can be reduced, then a greater proportion of analyte molecules may be ionised. This dual approach of reducing salt ions *and* decreasing column diameter to decrease the peak volume to less than 100µl has been responsible for the establishment of validated assay with a 10pg/ml lower limit of quantification (LLOQ) from a 250µl plasma sample on a 150 × 1mm 5µm Prodigy ODS column at 200µl/min. This may be summarised for electrospray:

> *Optimum LC–MS–MS sensitivity in electrospray is obtained with the lowest salt concentration and column diameter compatible with sample type and volume.*

The specificity of mass spectrometry removes the need to have such a selective chromatographic method; however, although co-eluting material may not produce a direct response at the detector, it may well compromise the accuracy and precision as well as the LLOQ of a method. It therefore follows that, even with the popularity of LC–MS–MS, there is a place for chromatographic optimisation. An alternative to the approach outlined so far is to use computerised optimisation.

4.11   Computerised optimisation of HPLC

Several computer optimisation packages are available on the market, each with various strengths, weaknesses and prices. An example is the Hipac TQ™ system. This system may be used to optimise the chromatographic aspect of a method development, and may be used in either a binary, ternary, quaternary, pH or ion-pair mode.

Two initial runs are made on the user-selected column to ascertain the upper and lower limits of the eluotropic strength of the mobile phase, and these are used in the appropriate optimisation mode. The software then requires, for example, either 7 or 12 runs for the ternary and quaternary mode, respectively, to be made to adequately map out the 'chromatographic response surface'. Once this has been done and the data entered, then a simulated chromatogram may be produced and viewed on screen for any combination of mobile-phase parameters within the original range. The software optimises the conditions so that adequate separation is achieved in the shortest time, or with the most even spacing between the analytes, or with a minimum resolution value for the worst-resolved pair of peaks. With this, and indeed all such programs, it is necessary to track each of the peaks of interest, whether analyte or interference, so that their relative retention in each of the conditions may be entered into the calculations. This is a major drawback to the use of such packages in the pharmacokinetic laboratory. As the sensitivity needs increase, so does the number of endogenous interferences. This leads to a multitude of small peaks in the chromatogram in addition to the analytes. As chromatographic conditions are changed the analytes can be tracked but these very small peaks can not, and unless they are, computerised optimisation can not be employed effectively.

These packages are under continual development. Details of such can be found in many of the chromatographic journals and magazines.

4.12 Conclusions

This chapter is intended to provide a logical and systematic approach to the development and optimisation of HPLC separations for reverse-phase, normal-phase, ion-pair, ion-exchange and chiral HPLC, as well as some general considerations of chromatographic optimisation. In regard to the chromatography of pharmaceutical compounds from a biological matrix there exists a plethora of extraction/enrichment procedures, a variety of detection modes and advanced integration capabilities. None of these is any substitute for good chromatography. Getting the chemistry and the separation right are indispensable characteristics of modern analytical HPLC.

4.13   Glossary

α	separation factor, equal to k_b/k_a
AUC	area under the curve
BSA	bovine serum albumin
C1 to C18	carbon atoms in an alkyl chain
CN	cyano
DCM	dichloromethane
diastereoisomer	a molecule possessing two adjacent chiral atoms, it does not have a mirror-image and therefore has differing chemical properties
DNB	dinitrobenzene
ε°	solvent-strength parameter
enantiomer	one of a pair of compounds showing mirror-image isomerism
Hex	hexane
IEC	ion-exchange HPLC
IPC	ion-pair HPLC
isomerism	the possession by two or more distinct compounds of the same molecular formula
k	capacity factor for a given peak, a measure of relative retention
LLOQ	lower limit of quantification
MeCN	acetonitrile
MeOH	methanol
MTBE	methyl tertiary-butyl ether
N	column plate number, a measure of column efficiency
pK_a	negative logarithm of solute acidity constant
pK_b	negative logarithm of solute basicity constant
R	resolution, a measure of the separating power of the system
racemate	an equimolar mixture of two enantiomers
RT	retention time
TBA	tetrabutylammonium
TEA	triethylamine
THF	tetrahydrofuran
TLC	thin-layer chromatography
UV	ultraviolet

References

Armstrong, D.W., Chang, C.D. and Lee, S.H. (1991) (R)- and (S)-naphthylethylcarbamate-substituted β-cyclodextrin bonded stationary phases for the reversed-phase liquid chromatographic separation of enantiomers. *J. Chromatogr.*, **539**, (1), 83–90.

Chow, F.K. (1991) *Column Switching Techniques in Pharmaceutical Analysis*, New York: Par Pharmaceuticals Inc., pp. 41–61.

Snyder, L.R., Glajch, J.L. and Kirkland, J.J. (1988) *Practical HPLC Method Development*, Chichester: John Wiley and Sons. (Highly recommended reading. The main points on optimisation are included in these notes, now available as a second edition.)

5

HPLC DETECTORS

Richard F. Venn

5.1 Introduction

A chromatographic detector can be defined in a very broad way as any device that locates, in the dimensions of space or time, the positions of the components of a mixture that has been subjected to some form of chromatography, thus allowing an appreciation of the separation that has been attained. Such a broad definition allows for many types of detector and for our purposes the definition can be tightened to allow only such detectors as can give a measurable response to the concentration of analyte in the eluent from an HPLC column.

Chromatography literally means writing in colour, reflecting the first uses of the technique – writing out the 'signature' of a mixture in colour, the components separated on a substrate such that the individual components were visible to the eye: thus the eye was the first detection system used in chromatography. In fact our eyes have exquisite sensitivity as detection systems: able to detect a single photon and differences in colour represented by a wavelength changes of a few nanometres or less. When we use the modern techniques of chromatography in quantitative or qualitative analysis, we rely on detection systems to indicate the presence of the analytes of interest, and a detection system is central to any chromatography – without a detector there are no data. In qualitative chromatography this detection system is often still the eye, detecting coloured analytes undergoing TLC, column chromatography or electrophoresis. The eye, however, is limited to detecting those analytes which absorb light in the visible region of the electromagnetic spectrum; others go undetected or have to be chemically altered to make them visible. The other obvious limitation on the eye as detector is that concentration determinations are hard to obtain. An everyday example of the eye as a useful detector, though, is the pH strip, where reasonably high-resolution determination of pH can be achieved by comparison of test strips with fixed standard strips.

Detectors for gas chromatography (GC) were developed very rapidly; between 1956 and 1960 at least six detectors for GC were developed almost to their full potential; progress with HPLC detectors has been much slower and more difficult. This is mainly because a low concentration of analyte in a flowing liquid will not alter its physical characteristics so much as the same concentration will alter the properties of a gas. There is still no HPLC detector available that could be called truly universal; contrast this with the flame ionisation detector in GC. The most sensitive general HPLC detector is the ultraviolet (UV) detector; of the more specific detectors, the most sensitive are the fluorescence and electrochemical detectors.

The evolution of detectors and columns for HPLC have gone hand in hand; better LC detectors have allowed a better understanding of column separation theory, leading to increased column efficiency and smaller peak volumes, thus driving a need for yet better detectors with smaller flow-cells and better flow and optical geometry. Microbore columns have increased the demands on detectors, although there is evidence that the currently used flow-cells of approximately 8–12µl volume are small enough to provide uncompromised linearity and sensitivity with narrow-bore columns of 1–2mm diameter.

Without a detector there are no data and so HPLC systems rely on detectors: general detectors, responding to many analytes, or specific, responding to particular groups in the analytes of interest. The types of detector available rely on a number of physical principles, and can be categorised in a number of ways, for example by specificity, by sensitivity or by principle involved. Many of these categories are interrelated, and depend on the analyte involved.

5.2 Principles of detection

LC detectors can be classified in a number of ways; the most useful is a classification into two, the **solute-property** detectors, which function by measuring a physical or chemical property of the solute itself, such as fluorescence or UV absorbance, and the **bulk-property** detectors, which operate by measuring a bulk physical property of the column eluent, such as refractive index or dielectric constant. The principles involved in HPLC detection are as follows.

5.2.1 Solute-property detectors

Some property of the analyte is continuously monitored, in such a way that the presence in the column eluent of the analyte causes a concentration-dependent change in output signal.

- Absorption of energy, for example UV/visible absorbance, in which certain functional groups in the analyte absorb UV or visible light at defined absorbance wavelength. The difference in signal between light that has not passed through the flow-cell and light that has is the absorbance. This principle also holds for infrared (IR) detectors.
- Spontaneous release of energy, as in radioactive decay. The particles emitted by the analyte on radiochemical decay are collected by a scintillator, converted to photons and counted using a photomultiplier.
- Absorption and release, for example fluorescence or chemiluminescence, in which energy (UV or visible light, or chemical energy) is used to excite the analytes in the detector flow-cell. The analyte absorbs this energy and re-emits it as visible or UV light, which is then detected.
- Chemical reaction, followed by monitoring of some solute property conferred on the molecule by the reaction, for example fluorescence monitoring following pre- or post-column derivatisation. In this case, the final detector is still a

solute-property (for example UV) detector. Electrochemical detection also falls into this category, with the reaction (oxidation or reduction by application of a potential) taking place within the detector flow-cell.

The detector type chosen will depend on the functional groups in the analyte. For example, aromatic compounds will have a UV absorbance spectrum, allowing detection by a UV absorbance detector, and conjugated aromatic compounds may exhibit fluorescence as well as absorbance. The final choice of detector is a compromise between selectivity and sensitivity.

5.2.2 Bulk-property detectors

Bulk-property detectors rely on the continuous monitoring of some physical property of the column eluent via a suitable transducer; the output (voltage) being proportional to the property being measured and hence the concentration of the solutes being determined. Refractive index is probably the most commonly used bulk property; electrical conductivity and dielectric constant are less common. The presence in the solvent of a solute alters the bulk property being monitored, giving rise to an output signal related to the solute concentration.

The main limitation of bulk-property detectors is that of sensitivity. This is determined by the noise in the system, and is a much greater problem for such detectors than for solute-property detectors because the bulk properties being measured alter with temperature and pressure, whereas solute properties vary much less with these factors. Consequently, such detectors are susceptible to thermal fluctuation noise and pump pressure noise, requiring extremely well-controlled mobile-phase supply systems. They are thus only of use in isocratic HPLC systems, and cannot be used in gradient HPLC, or when temperature or flow programming is being used. When low sensitivity can be tolerated, however, or when there are no UV-visible chromophores, they are useful. A further limitation is that of linear dynamic range, which is often less than three orders of magnitude, whereas solute-property detectors often have a dynamic range of four or five orders of magnitude. They do, however, often operate as universal detectors, albeit at low sensitivity, and can be of use in preparative HPLC.

5.3 Selectivity in detectors

Bulk-property detectors are necessarily non-selective, since they rely on the solute to alter some property of the mobile phase. Solute-property detectors are more selective; UV detectors can be used at specific wavelengths, and diode-array UV detectors can obtain 'instantaneous' (within 1 second) spectra, thus increasing selectivity enormously. Many pharmaceuticals, however, absorb light at wavelengths very close to, or overlapping, those absorbed by endogenous compounds, and many have similar spectra to naturally occurring compounds, or metabolites of the analytes. In the analysis of such compounds in complex matrices (such as plasma or urine), then, selectivity with UV detectors will almost always need to be improved by selective extraction techniques and good chromatography. Fluorescence detectors are more selective, since far fewer compounds fluoresce naturally. Selectivity is also increased because there are two specific wavelengths, the excitation and the emission wavelengths, which confer further selectivity. Electrochemical detectors can be selective when used coulometrically since a 'scrubbing' electrode can be set at a potential just below the detection electrode, removing many interfering compounds. The voltage at which a particular group is oxidised or reduced also confers selectivity on electrochemical detectors. This voltage is related directly to the energy required to break or form the bonds involved and is thus specific to the compound; this, together with the limited number of 'active' compounds, confers specificity. However, both fluorescence and electrochemical detectors will usually also require extraction of the analyte(s) of interest from the biological matrix.

5.4 Detector response

5.4.1 Linearity

HPLC detectors for quantitative analytical studies need to have a linear response to the concentration of the analyte in the detection cell. The mathematical expression of a linear response is:

$$R = A + BC$$

where R is the response or output signal from the detector, A is a constant (offset), B is a constant (response factor) and C is the concentration of the analyte. For any detector, the range of R is limited by the internal electronics to some arbitrarily decided maximum, such as 1V or 10mV; A is usually adjustable within a defined range (again determined by the internal electronics of the device) and B is determined by a combination of the properties of the analyte and the detector. The concentration range through which the detector response is linear is limited by the principles by which the detector operates and the design of the instrument. It is a prerequisite of the detection system employed that it is linear over as wide a range as possible. The linear dynamic range should be 5–6 orders of magnitude, with a sensitivity of at least 10^{-12}–10^{-11}g/ml. Electro-opto-mechanical devices such as HPLC detectors are never absolutely linear, and the extent to which detectors approach the ideal of true linearity varies between instruments. In practice this non-linearity can be ignored since calibration curves and quality controls are routinely run in the analytical procedures normally used; the errors are usually small compared with the other experimental errors. However, if calibrationless quantitation is used (see below), any non-linearity in detector response could become significant, resulting in an increase in the inaccuracy of the method.

In the context of linearity and dynamic range of detectors, the same considerations need to be taken into account for integrators and data systems. A 16-bit analogue to digital converter, for instance, limits the resolution of the detector to 1 part in 65 535. Thus if the 1V output of a UV detector is used (as is normally the case) for the integrator, the smallest change in absorbance that can be detected is 1AU/65 535 = 15μAU. For a 12-bit converter, the figure is 244μAU. The former is close to the best of modern UV detector noise performance and the latter figure is greater than the noise quoted by most manufacturers of such detectors and will limit the resolution of the detector.

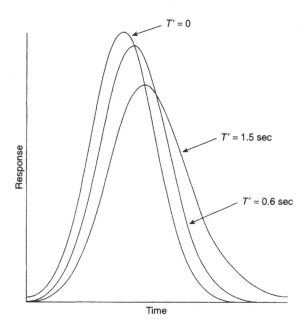

Figure 5.1 shows the following labels: T' = 0, T' = 1.5 sec, T' = 0.6 sec, with Response on the vertical axis and Time on the horizontal axis.

Figure 5.1 Effect of detector time constant, T', on response to a chromatographic peak.

5.4.2 Time constant

The time constant of a detector determines the time taken for the detector response to reach 67% or 90% of full scale, or to go from 10% to 90% of full scale, depending on which books you read! The time constant can be divided into different contributing factors. The detection device itself (photomultiplier tube, photodiode, electrode) has an inherent time constant, and the associated electronics responsible for amplification of the signal will also have a time constant. It is of importance to have a very rapidly responding detection device, the time constant of which can be manipulated electronically. A slow device cannot be speeded up; the converse is possible. The recorder and/ or integrator attached to the detector will also have time constants for response which, in some cases, can be manipulated. Figure 5.1 shows the theoretical effect of amplifier time constant on the peak shape of a peak eluting at the dead volume of a column.

A slow time constant will thus have the advantage of decreasing noise (electronic and/or chromatographic noise such as pump baseline noise), but this comes with the disadvantage of a reduction of sensitivity due to the slower response. The time constant for a detector is usually adjustable, and should be set such that the signal-to-noise ratio in the system being used is optimised. In order to quantify a peak, the time constant must be less than 10% of the peak width at half-height. Thus sharp, early eluting peaks require a smaller time constant than later, broader peaks. Modern, high-efficiency, short columns can severely stretch currently available detectors; some peaks may have widths at half-height of only 1 second – or even less when using 3cm columns – demanding time constants of less than 100ms. These are rarely available on current equipment. They also demand fast sampling rates from the attached integrators, and very fast chart recorders. It is in fact unlikely that a common chart recorder would follow such peaks, since their time constants are much slower; digital recorders would be more appropriate.

Some chromatography data systems allow for manipulation of an 'effective' time constant by 'data bunching'. In other words, broader peaks may be integrated by averaging consecutive data points in groups, increasing the effective time constant.

5.5 Detector types

5.5.1 UV–visible detectors

Principle of operation

UV–visible detectors operate on the principle that the analytes of interest absorb light in the UV or visible region of the electromagnetic spectrum. The eluent from the column is passed through a flow cell of specific design, and light of a particular wavelength and bandwidth passes through the eluent. Light emerging is compared in intensity to that entering the flow-cell; the difference is due to the light absorbed by the analyte plus any background absorbance due to the eluent or to imperfections in the flow-cell. A typical fixed-wavelength instrument is shown diagrammatically in Figure 5.2 and a variable-wavelength instrument in Figure 5.3. Common to each instrument is a light source, collimator system (to form a parallel light beam), flow-cell, reference system and detector. The source is usually a deuterium lamp, with a tungsten filament lamp for the visible wavelengths.

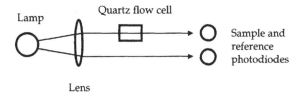

Figure 5.2 Diagram of a fixed-wavelength HPLC detector. The lamp is, for example, a mercury discharge lamp with a main emission band at 245nm.

New light sources have recently been described which rely on Cerenkov radiation with a strontium-90 source. Such sources are claimed to be more stable than deuterium lamps and to require less frequent renewal. Deuterium and other similar lamps deteriorate with time, becoming less stable in output and providing less useful light. The reference system is important for stability, allowing for corrections to variations in light output of the lamp to be made. The reference may be from either a half-silvered mirror providing some light from the optical path onto a reference photodiode, or the light beam may be split to pass through a reference flow-cell, which is normally filled with static mobile phase. Most detectors for HPLC are of the former kind.

Beer's law describes the relationship between the intensity of light transmitted through the cell and the solute concentration, and states that the absorption is proportional to the number of absorbing molecules in the light path. Thus:

$$I_T = I_0 \cdot e^{-klc}$$

where I_0 is the intensity of light entering the cell; I_T, the intensity of light transmitted through the cell; c, the concentration of the solute in the cell; k, the molar absorption coefficient of the solute at the wavelength used and l is the pathlength of the cell. If this equation is stated in the form:

$$I_T = I_0 . 10^{-k'lc}$$

then k' is the molar extinction coefficient, and taking logarithms and rearranging gives

$$\log(I_0/I_T) = k'lc = A$$

where A is the absorbance.

ΔA is used to define the detector sensitivity where ΔA is the change in absorbance that provides a signal-to-noise ratio of 2. Thus:

$$\Delta A = k'l\Delta c$$

and

$$\Delta c = \Delta A/k'l.$$

In other words, the smallest concentration of analyte that can be determined above the noise is inversely proportional to the molar extinction coefficient and the pathlength. Thus the greater the pathlength, or the greater the extinction coefficient, the smaller the minimum detectable concentration becomes. Pathlength is optimised by the manufacturers to suit the detector electronics, optical geometry, etc.

Design

FIXED-WAVELENGTH DETECTORS

Early UV detectors were of fixed wavelength, and these are still available and widely used. The wavelength is determined by the lamp used, which is usually a discharge lamp; low-pressure mercury discharge lamps have a major emission line at 254nm, zinc at 214 and 308nm and cadmium at 229, 326, 340 and 347nm. These lamps have other emission lines also, mercury being the cleanest. Usually filters are required in order to operate at a truly single fixed wavelength. These lamps are particularly useful in the low- to medium-pressure 'biotechnology' chromatography applications. Detectors are available that operate with a number of different plug-in sources and filters, offering a wide range of fixed wavelengths at a reasonable price. The advantages of a fixed-wavelength UV detector are that it is stable, can be used with gradient elution and has a fairly wide dynamic range of (usually) at least four orders of magnitude. Noise performance can be very good. The most common wavelength is 254nm, provided by a mercury discharge lamp. A disadvantage of UV detector cells is their susceptibility to temperature and to flow-rate changes. This can be overcome by thermostatting. Flow-rate changes are easily induced by imperfections in the HPLC pumps being used.

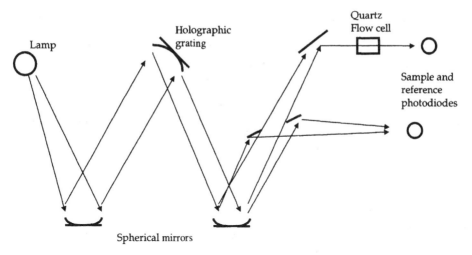

Figure 5.3 Diagram of a variable-wavelength UV detector.

VARIABLE-WAVELENGTH DETECTORS

As the name suggests, the wavelength at which the absorbance of the solute in the eluent is monitored can be varied in these detectors. Thus a light source is needed that can provide a continuous spectrum over a wide wavelength range; this is usually a deuterium discharge lamp; a xenon discharge lamp is less frequently encountered. The two types of variable-wavelength detectors available are the dispersion detector and the diode-array detector (DAD). The former is shown diagrammatically in Figure 5.3.

The major difference between them is that in the dispersion detector, essentially monochromatic light passes through the flow-cell, whereas in the diode-array detector, polychromatic light passes through the cell; this light is then dispersed by a holographic grating onto an array of photosensitive diodes. This difference is not trivial. Any fluorescent compound in the flow-cell will contribute to the light falling on the photodiodes in either detector. In the dispersion detector, monochromatic incident light is used and the fluorescent effect is generally small. In the diode-array detector, however, polychromatic incident light is used, and this increases both the probability of fluorescence and its intensity (see below). Thus measurement of transmitted light may, in fact, not be transmitted but include a significant proportion of fluorescent light, and thus a true absorbance spectrum may not be obtained. This effect may also impair the linear response of the detector.

A diagram of a DAD is shown (Figure 5.4), and it can be seen that the light passing through the flow-cell is polychromatic, i.e. light of all wavelengths is used to illuminate the eluent in the flow-cell. Light emerging from the cell is reflected via a holographic grating onto an array of photodiodes, each of which thus provides an output related to a specific wavelength range. A computer is normally needed to handle all the simultaneous data emerging from the detector. Typical uses of the detector would be to obtain instantaneous spectra of the analytes flowing through the cell without having to use stopped-flow, and to measure peak purity by measuring

113

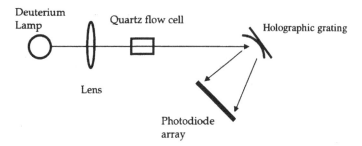

Figure 5.4 Schematic diagram of a diode-array detector. The lamp produces a broad-spectrum output to provide illumination at all wavelengths.

absorbance ratios at different wavelengths. As mentioned above, a possible disadvantage is the effect of fluorescence on the absorbance spectrum of a compound, which can be substantially altered as compared to the spectrum obtained by a spectrometer. The advantages of diode-array detectors are that peak purity can be confirmed, and co-eluting peaks which would otherwise be missed can be identified by their different spectra. This can help in the identification of metabolites. A disadvantage of diode-array detectors is the somewhat lower sensitivity usually obtained, although recently improvements have been made in DAD sensitivity; one recent introduction uses a 5cm pathlength to achieve this.

Flow-cell design

The design of the flow-cell is of great importance to the overall performance of the detector, and manufaturers each have their own, often patented, particular design. A longer pathlength will give a greater absorbance, but at the expense of greater susceptibility to refractive index perturbations and pump pressure noise. A longer pathlength demands a narrower cell cross-section if dispersion is not to be increased by an increase in flow-cell volume, and this makes severe demands on the optical system. One manufacturer has recently introduced a detector which has a 10mm pathlength 8µl flow-cell; a noise level of 2×10^{-5}A is claimed. As mentioned above, a 50mm 'light-pipe' flow-cell is used in one diode-array detector. Specially designed flow-cells are now also available for capillary microbore HPLC: the quartz capillary is bent in such a way that the UV light beam enters the capillary and travels along it for up to 13mm before exiting at another bend of the capillary to the photodiode detector. These are available as replacement cells for a number of commercial detectors and allow microbore LC to be used with a conventional detector.

Calibration

It is important that the response of the detector, its linearity and noise performance are checked regularly. This allows for quantification of analyte using a calibration line generated by spiking analyte into the matrix. Several recent papers have appeared (Torsi *et al.* 1989, 1990) suggesting the use of calibrationless quantitation. If the molar absorbtivity of the analyte is known, then by application of Beer's law the area of the peak can be related to concentration in the sample by allowing for flow

Table 5.1 Groups giving UV absorbance

Group	Examples	Wavelength	Absorbance
Acids and amides	$-CO_2H$, $-CONH_2$	$\approx$210nm	Medium
Aldehydes and ketones	(structure)O	270–300nm	Weak
Conjugated carbonyls	(structure)O	215–250nm	Very strong
Conjugated systems	(structure)		Strong
Aromatic systems		Varies	Strong

rate, pathlength and molar absorbtivity. This would also be capable of internal standardisation by the same means.

Advantages of UV–visible detectors

The advantages of UV detectors are that they are cheap, reliable and easy to set up at the correct wavelength which can be easily determined by UV–visible spectroscopy. Most compounds have some absorbance in the UV–visible region (Table 5.1 gives some typical absorbing groups), although it may not be strong enough to provide an analytical method, or it may be in an unsuitable part of the spectrum, for example in the 190–220nm region where many other interfering endogenous compounds also absorb. When this is the case, other detectors are used. UV detectors are available in a range of styles and sophistication, from fixed-wavelength to variable, multiwavelength, time-programmable devices that are able to scan peaks 'on the fly' to give a spectrum; diode-array detectors are able to do all this and more with the aid of a (usually) dedicated computer. Choose one to suit the need.

It should be noted that the UV-visible detector is designed to measure a change in light intensity at a specific wavelength. The detector is usually operating at low analyte concentrations, so the electronics are trying to distinguish between two very high light intensities. The noise in the system then limits the sensitivity of the detector. This is in contrast to, for example, a fluorescence detector, which is designed to detect a small increase in signal from (essentially) a dark background when the analyte is present in the eluent. It is theoretically possible to detect a single molecule by fluorescence.

5.5.2 Fluorescence detectors

Principles

When light is absorbed by a molecule (the phenomenon exploited in UV–visible absorbance spectrometry, and hence UV–visible detectors), one or more electrons in the molecule are raised to a higher energy state; the energy required to do this is specific to the absorbing molecule or group. Thus the wavelength required to excite the electron is specific, although the absorbance bands are quite broad; the energy conveyed by a photon is determined by its wavelength. Higher-energy photons have

115

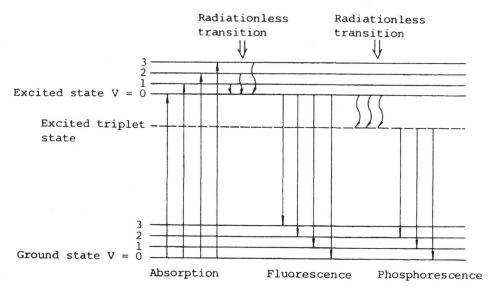

Figure 5.5 Jablonski diagram showing the atomic electronic events in absorption, fluorescence and phosphorescence.

shorter wavelengths (higher frequency) than lower-energy photons. The excited molecule then has a number of ways of returning to the ground state: by collisions with other molecules, for example. If this does not happen rapidly, another route is to emit energy in the form of electromagnetic radiation. Some energy is lost vibrationally in the internal energy states of the molecule, and thus the emitted radiation is always at a longer wavelength (lower frequency, lower energy) than the exciting wavelength. Figure 5.5 illustrates these energy-level transitions that are available to an organic molecule. It should be noted that it is not possible to predict with certainty that a particular structure will show fluorescence. A prerequisite for fluorescence is an absorbance in the electromagnetic spectrum, probably with appreciable conjugation. However, there are other ways for an excited molecule to dispose of its energy than fluorescence (chiefly by collisional loss), and these may quench the fluorescence of the molecule.

The fluorescence intensity is not quoted in specific units, but depends on the instrument. It can be expressed as

$$F = KI_0 Cl\varepsilon\phi$$

where F is the fluorescence intensity; K, an instrumental constant; I_0, the intensity of the excitation radiation; C, the concentration of the substance; l, the pathlength; ε, the absorbtivity of the substance; and ϕ is the quantum efficiency of the substance in the solution. The quantum efficiency is the proportion of excited molecules that release their energy as fluorescence, and this can vary in different solvents. This linear relationship between fluorescence and concentration only holds true at low concentrations.

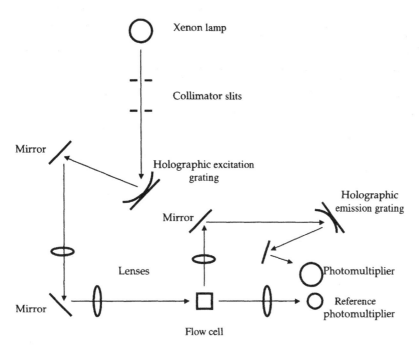

Xenon lamp

Collimator slits

Mirror

Holographic excitation
grating

Holographic
emission grating

Mirror

Lenses

Photomultiplier

Mirror

Reference
photomultiplier

Flow cell

Figure 5.6 Diagram of a dual-monochromator fluorescence detector for HPLC.

Design and use

A diagram of a dual-monochromator fluorescence detector is shown in Figure 5.6. The fluorescent light is usually observed at right angles to the incident light and this lowers the background noise. In some detectors an elegant design allows much more (up to one-half) of the emitted light to be collected by the photomultiplier. This is achieved by use of a parabolic mirror with the flow-cell at the focus. Incident light reaches the cell through a hole in the centre of the mirror and one-half of all emitted light can be collected. Light sources are usually xenon discharge lamps, sometimes augmented by mercury, although deuterium discharge lamps (which usually have a lower output power) are sometimes used. Lasers are increasingly being brought in as high-power sources – laser-induced fluorescence (LIF). These can increase sensitivity by several orders of magnitude. This can be seen from the equation shown for fluorescence emission; the greater the intensity of light at the appropriate wavelength used to excite the analyte, the greater the fluorescence signal, up to a maximum determined by the time taken to return to the ground-state, among other factors. The disadvantages of LIF are cost and relative inflexibility; only a relatively small number of wavelengths are available for excitation. Some detectors have variable-frequency pulsed light sources, allowing for complex electronics to 'gate' the collected light for long lifetime fluorescence (such as that from the lanthanides), thus decreasing background noise still further. The simplest kind of fluorescence detector has a fixed-wavelength light source with a detection device that responds to all emitted light. Filter fluorimeters use specific notch filters (allowing only a specific wavelength) or bandpass filters (low- or high-pass, which allow light below or above a specific

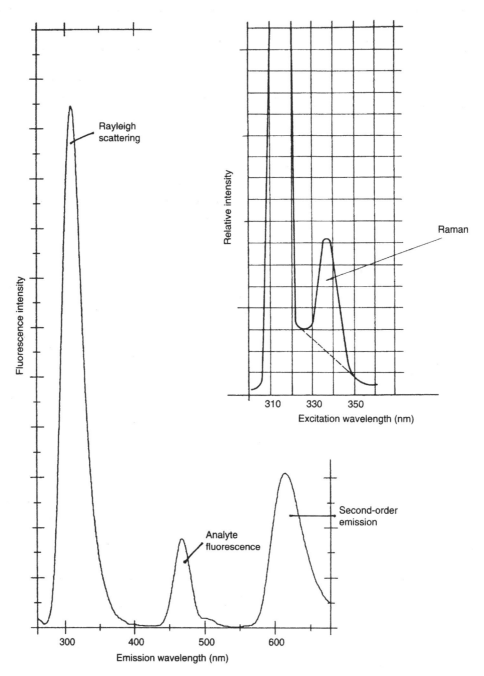

Figure 5.7 Fluorescence emission spectra, excitation at 310nm. Inset: the Raman band.

wavelength to pass) to increase specificity, and monochromators are used in the excitation and emission sides in the more sophisticated (and expensive) detectors. The most sophisticated allow emission and excitation wavelengths to be scanned under microprocessor control to obtain spectra, and also allow time-programmed wavelength changes during a chromatographic run.

A typical fluorescence emission spectrum is shown in Figure 5.7; note also the Raman line in the inset. This is a potentially interfering phenomenon which is a solvent-related light-scattering effect, as is the Rayleigh scattering. The latter occurs at the same wavelength as the excitation light; Raman scattering occurs at a longer wavelength, typically 20–50nm greater than that of the excitation. The wavelength of the Raman line varies with the excitation wavelength and with the solvent used. It is important when determining an analyte's potential for fluorescence detection to be aware of the Raman line, and also of the second-order scattering seen at double the excitation wavelength. Thus if the excitation wavelength being used is 310nm, the second-order band will be at 620nm, as seen in the figure. This is normally due to scattering at the holographic grating of the excitation monochromator. Third-order scattering is also sometimes seen at three times the wavelength. These signals can be misleading.

Fluorescence can be strongly pH dependent, and when determining the fluorescence of an analyte the ionic state of the molecule should be considered, and the pH dependency studied. It is often possible to enhance the fluorescence by orders of magnitude by pH manipulation. Using either the zero-order setting on the excitation monochromator, or the known UV absorbance maximum, scan up the emission wavelength to determine the fluorescence maximum. This can be done in the HPLC detector's flow-cell if a fluorescence spectrometer is not available. Having obtained a fluorescence emission maximum, scan the excitation wavelength to obtain optimum conditions. It may then be advantageous to re-scan the emission wavelength. Do this in basic, neutral and acidic solvents, and be aware that some solvents (for example those with high dielectric constants such as water and methanol) can cause severe fluorescence quenching.

Advantages

The advantage of fluorescence detection is its sensitivity, coupled with its insensitivity to instrumental instability such as pump pressure and flow rate. A further advantage is selectivity, gained by the use of two wavelengths coupled with the fact that comparatively few compounds fluoresce. In general, the fluorescence signal is superimposed on a very low background, whereas in absorbance the signal is superimposed on a high and sometimes noisy background. This is not always the case, however, when the emission wavelength is close to the excitation wavelength. In these cases, an optical filter can be added to cut down on the scattered light. The solvent-front injection artefact often seen in UV absorbance HPLC detectors is not normally seen with fluorescence detectors, although the injection solution may have a different background fluorescence to the mobile phase and this may cause an injection artefact to appear. Some groups that can give rise to native fluorescence are listed in Table 5.2; Table 5.3 lists some substituents and their effects. In general, rigid planar aromatic compounds are good fluorophores.

119

Table 5.2 Groups giving fluorescence

Group	Examples	λ_{ex}	λ_{em}	Comments
Indoles	Sumatriptan	$\approx$250nm	$\approx$340nm	
Aromatic rings	Tyrosine	200–230nm	250–350nm	Depending on number
Fused aromatics	Quinolines	250–350nm	280–450nm	and planarity of ring and system substituents

Table 5.3 Effects of substituents on fluorescence

Substituent	Effect on λ_{em}	Effect on intensity
Alkyl	None	Slight increase or decrease
OH, OCH_3, OC_2H_5	Increase	Increase
CO_2H	Increase	Large decrease
NH_2, NHR, NR_2	Increase	Increase
NO_2, NO	–	Total quenching
SH, halogen	Increase	Decrease
SO_3H	None	None

Other considerations

Fluorescence is sometimes temperature dependent, and lowering the temperature can increase the fluorescence yield, presumably by reducing collision losses. A 1°C rise in temperature can reduce the fluorescence by 1–10%, and the converse is also true. It is worthwhile investigating lowering the flow-cell and column temperature if sensitivity is a problem.

Some fluorescence detectors have been designed with diode-array detection, which give real-time emission spectra; these are not yet commercially available. They have the same advantages as UV-absorbance diode-array detectors, but suffer from the same drawback – a decreased sensitivity. Detectors with rapidly scanning holographic gratings are also available which give the possibility of diode-array-like emission spectra.

An alternative method of excitation of the analyte molecule is the use of chemiluminescence in which a chemical reaction passes energy to the acceptor (fluorescent analyte) molecule. Reactions that have been used are the peroxyoxalate, luminol and lucigenin systems. Reactions for the peroxyoxalate system are shown in Figure 5.8. The detector is usually a conventional fluorescence detector with the source turned off, and a post-column derivatisation system with reagent mixing and a reaction delay loop is necessary. Such systems require high-quality, low-pulsation reagent-delivery pumps for minimum noise. These chemiluminescence reactions are particularly suited for molecules with a low quantum yield.

5.5.3 Electrochemical detectors

Principles

The electrochemical detector is probably the most sensitive of the available HPLC detectors. These are specific in only responding to substances that undergo (elec-

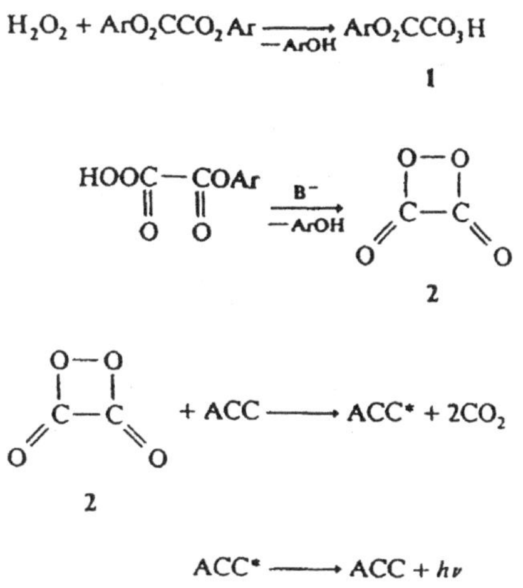

$$H_2O_2 + ArO_2CCO_2Ar \xrightarrow[-ArOH]{} ArO_2CCO_3H$$

1

$$\underset{\overset{\|}{O}\ \overset{\|}{O}}{HOOC-COAr} \xrightarrow[-ArOH]{B^-} \underset{O}{\overset{O-O}{\underset{\diagup}{C-C}}}\overset{}{\diagdown}_O$$

2

$$\underset{O}{\overset{O-O}{\underset{\diagup}{C-C}}}\overset{}{\diagdown}_O + ACC \longrightarrow ACC^* + 2CO_2$$

2

$$ACC^* \longrightarrow ACC + h\nu$$

Figure 5.8 Chemiluminescence reactions; ACC is the acceptor molecule (analyte).

trochemical) oxidation or reduction, at a particular voltage related to the activation energy for the oxidative or reductive process.

The equation describing an electrochemical reaction involving one electron is as follows:

$$MOL \rightarrow MOL^+ + e^-.$$

Each molecule oxidised gives up one electron to the working electrode; this passes through the external circuit and back into the solution at the counter-electrode. When molecules are reduced, an electron is added. Detectors can be grouped into two types, the dynamic detectors which involve electron transfer, such as the amperometric or coulometric detectors, and the equilibrium detectors which do not promote oxidation or reduction of the analytes. The field of electrochemical detection is still advancing fast, with active research in the areas of flow-cell design, new electrode materials and electronic instrumentation.

Dynamic detectors

Amperometric detection relies on the oxidation or reduction of an analyte at a potential corresponding to the energy required to free or add an electron from the molecule. Coulometry is merely an extension of amperometry, in which all, rather than a few percent, of the analyte is oxidised or reduced in a stoichiometric manner. The charge transferred (measured in coulombs) is directly proportional to the number of molecules oxidised or reduced. The noise in such systems is similar to that of normal amperometric detectors, but they have the advantage that 'scrubber' electrodes can be inserted upstream of the detector electrode in order to oxidise and completely remove potentially interfering analytes at lower potentials. This increases the selectivity of the detector considerably. There are at least two kinds of detector cell, the flow-through and the wall-jet cells.

Electrode arrays allow the detection of several analytes at once by operating each electrode in the array at a different potential. There are two types, serial and parallel.

Eight or 16 electrodes in a ring in a wall-jet electrode assembly have been operated successfully, and ESA offer an 8 or 16 serial array with four electrodes in each of two or four blocks. Such an array, which operates under software control from a PC, is able to produce simultaneous data on peaks eluting sequentially or co-eluting, much like a diode-array UV detector. In the multi-electrode ECD, each electrode operates at an individual potential.

Equilibrium detectors

Equilibrium detectors are generally designed as bulk-property detectors, and measure the conductance of the flowing solvent stream. Changes in the conductivity induced by solutes are recorded. The detector and its associated electronics are designed such that there is infinitesimal current flow in the cell. Combination detectors have been described recently which measure simultaneously conductivity and pH, or conductivity and absorbance, in very low-dead-volume cells.

Other considerations

Derivatisation for ECD, pre- and post-column, is a topic that has been published on but has not become popular. This is probably because fluorimetry is preferred if derivatisation is necessary.

Electrochemical detection has a poor reputation for robustness and this is largely due to the need for long periods of stabilisation, and for polishing of glassy carbon electrodes. Porous carbon coulometric electrodes solve some of the problems, but short electrode life can still be a problem. However, some pharmaceuticals are assayed using HPLC-EC, and such systems are quite capable of supporting phase I and phase II studies. Table 5.4 lists some of the groups that are electrochemically active. Quality and robustness of electrochemical detectors (in particular, the flow-cell electrodes, which are susceptible to fouling) has limited the usefulness of these detectors in the past to non-routine applications. However, the newer generation of cells are much more robust and several applications have been published for the routine analysis of pharmaceuticals in plasma using electrochemical detection.

5.5.4 Multifunctional detectors

It is possible, and often done, to put two different detectors in series to determine two or more analytes of interest having different characteristics; or it may be that in method development UV absorbance, for example, may be used when the final method will require fluorescence, and the two detectors are put in series. The disadvantages are physical size, coupled with the extra band-broadening introduced by the first detector and associated tubing. Several companies have sought to overcome these problems with multifunctional detectors such as a UV–fluorescence detector in which the incident light passes through the flow-cell; transmitted light at the relevant wavelength is detected for the absorbance measurement and emitted fluorescence at a longer wavelength is collected at right-angles to the incident light. Other such detectors include a UV–conductivity detector, and a UV–RI detector, combining bulk- and solute-property detection, and a detector combining UV absorbance, fluorescence

Table 5.4 Groups conferring electrochemical activity

Group	Examples	Voltages
Aromatic alcohols	Tyrosine	+800–1000mV
	Tyramine	
	Morphine	
	Adrenaline	
	Dopamine	
Aromatic amines	Chloroaniline	+900–1000mV
	p-Phenylenediamine	
	Sulphonamides	
Indoles	Melatonin	+600–900mV
	Tryptophan	
	Tryptamine	
Phenothiazines	Promethazine	+900mV
	Perphenazine	
	Chlorpromazine	
Purines	Xanthine	+800–1000mV
	Guanine	
	Theophylline	
Thiols	Cysteine	+800mV
	Penicillamine	
	Glutathione	
Others	Ascorbic acid	+800mV
	Nitrophenols	−100–500mV
	Quinone	−400mV
	Amides	
	Secondary amines	1000–1600mV
	Tertiary amines	

and conductivity. None of these detectors has proved popular with users, however, and they are generally not commercially available.

5.5.5 Radiochemical detectors

Principles

Radio-detectors for HPLC have to detect the disintegrations occurring in a flowing eluent stream due to the solute (analyte) decomposition. It takes any given molecule 0.6 seconds to flow through a 10μl flow-cell at a flow rate of 1ml/min. The number of decomposition events that occur in a cell depends on the volume of the cell, the flow rate, the peak width at half-height and the activity present in the analyte peak. In order to increase the probability that a decay event will occur in the detector cell, the cells are usually much larger in volume than those in UV or fluorescence detectors, and it is necessary to make a compromise between resolution of the detector and efficiency. The decomposition events are detected by scintillation counting, which is of less than 100% efficiency in the best liquid-scintillation counters – normally 30–55% for tritium, and 80% for carbon-14.

Design

Radio-detectors for HPLC fall into two categories, the solid scintillant detectors and the homogeneous detectors.

Solid scintillant (heterogeneous) detectors

In these detectors, the flow-cell is constructed such that the eluent passes by a maximum surface of a solid scintillant. Particles emitted by the decay process interact with the scintillant (usually a polymer with scintillators incorporated in the polymer) which converts the decay event into a number of UV photons. An alternative arrangement uses powdered scintillant packed into a transparent tube. The scintillant in these detectors is typically cerium-activated lithium glass, yttrium silicate or calcium fluoride. The final detection uses two photomultiplier tubes and sophisticated electronic circuitry to eliminate chemiluminescence. The flow path is usually a spirally arranged tube with a volume of 0.3ml. The disadvantages of such detectors are: low efficiency, especially for tritium, whose emitted particles are very weak – many particles do not even reach the scintillants; large cell volume, leading to loss of resolution; and potential contamination of the flow-cell and scintillants by adsorbed compounds which may give a high background signal.

Liquid scintillant (homogeneous) detectors

The eluent from the HPLC column is mixed, via a low-volume mixing tee, with liquid scintillant, and the homogeneous mixture is passed into what is effectively a flow cell in a liquid scintillation counter. In this way tritium counting efficiencies can be improved, at the cost of being unable to recover the sample. It can be very difficult to achieve good counting efficiencies for tritium, and can take much time optimising flow rates and eluent : scintillant solution ratios.

5.5.6 Other detectors

Optical activity

Polarimetric, optical rotation detectors and circular dichroism detectors have been described, but currently most chiral detection relies on the use of chiral columns to obtain a resolution of enantiomers (for further information see Fielden 1992).

Atomic spectroscopy

This has been used for HPLC detection but will not be covered here (for further information see Scott 1986; Fielden 1992).

Mass spectrometry

The use of mass spectrometers as detectors for HPLC is thoroughly covered in Chapters 11–13.

HPLC-NMR

This is a new field, and commercial instruments are just becoming available. Such instruments lack sensitivity, but offer an enormous potential in metabolite identification. It is possible to use stopped-flow or real-time HPLC with these machines. See Chapter 14 and the emerging literature for further information.

Infrared detectors

IR detectors have been available and can be used for CH_2 groups at 3.4µm or for C=O monitoring at 5.75µm. In general the constraints on the mobile phase necessitate non-aqueous (normal-phase) chromatography. There are some off-line fourier-transform IR (FT–IR) detectors for HPLC available, in which the eluent is sprayed and dried in a spiral onto a slowly rotating IR-transparent disc. The disc is then taken to an off-line FT–IR spectrometer and the relevant spectra read. Such instruments can be extremely useful for identification of closely related compounds, but the inherent sensitivity is generally low. A real-time FT–IR detector for HPLC has been under development for some years but is not yet generally commercially available. This operates in a similar way, but the LC eluent is passed through a heated nebuliser similar to those used for HPLC–MS–MS and then deposited on a moving plate which is then subjected to FT–IR.

Light-scattering detectors

These are detectors in which the HPLC eluent is evaporated in a drift tube or nebuliser, leaving particles of analyte passing through the detection cell in a stream of inert gas. A laser light source is used to illuminate this flow and scattered light from any particles is measured. There are now several such detectors available and they act as universal detectors. They are especially useful when the molecule of interest has no chromophore or other solute property suitable for detection. Their drawbacks are that only volatile components can be used in the HPLC mobile phase and that their sensitivity is limited (to about 1ng of material).

Nitrogen detectors

Detectors that combust the HPLC eluent and then allow reaction of oxidised nitrogen with ozone are available. The detection relies on the chemiluminescence of the reaction with ozone. Their advantages are that any compound containing nitrogen can be detected and that the response is stoichiometric and thus the need for calibration curves can be reduced. However, the mobile phase must not contain any nitrogen (thus acetonitrile cannot be used) and contamination of methanol, for example, with acetonitrile is very hard to eliminate.

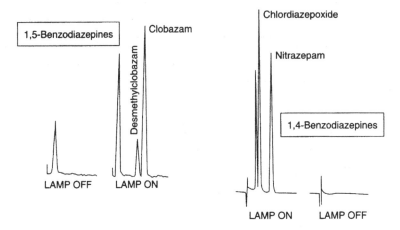

Figure 5.9 Effect of post-column UV irradiation on the detection of benzodiazepines.

5.6 Sensitivity considerations

Several techniques are available for improving the sensitivity of an assay. Chromatographic methods include using mid- or narrow-bore columns and improving peak shape. Others are as follow.

5.6.1 Irradiation

Sensitivity can often be increased in UV detection by post-column irradiation using a UV lamp. The column eluent is led into a serpentine-coiled or knitted Teflon tube which surrounds a UV source. The eluent is thus irradiated for a time determined by the flow rate and the length and diameter of the tube. The irradiated eluent is then passed to a UV, fluorescence or electrochemical detector. For some compounds the increase in sensitivity is very large, due to photochemical reaction of the analyte giving rise to a highly absorbing product. With others the effect is smaller or even productive of a lowering of sensitivity. It is always worth checking if an increase in sensitivity or selectivity is required. Figure 5.9 shows the advantages for detection of benzodiazepines.

5.6.2 Pre-column derivatisation

The advantages of using pre-column derivatisation for increasing the sensitivity of an assay are that the chromatography, particularly of basic compounds, can be improved; less complex equipment is required than for post-column derivatisation, and the wavelength of absorbance can be shifted into the visible band where very few impurities or interfering peaks absorb. This latter is also true for post-column derivatisation.

The disadvantages are that pre-column derivatisation is not always amenable to automation (although several methods exist, for example for derivatisation using fluorenylmethylchloroformate (FMOC-Cl), which are automated, using sophisticated autosamplers that are capable of mixing, diluting and specific needle positioning). A further disadvantage is the danger of there being insufficient reagent due to sample matrix reaction. Derivatisation reactions are also difficult to implement quantitatively at very low concentrations of analyte (sub-nanogram per millilitre). Pre-column derivatisation can be followed by a variety of detectors, in particular UV–visible and fluorescence detectors.

The functional groups that are usually derivatised are amines (primary and secondary), alcohols, phenols, thiols, carboxylic acids and carbonyl compounds. There are too many reagents available to list here (reviewed by Imai and Toyo'Oka 1988). One particular reagent that is worth singling out is FLEC (9-fluorenylethylchloroformate) which is available as one or other of an optically active pair. When reacted with mixtures of optically active amine- or alcohol-containing analytes, diastereomers are formed which can be resolved from each other on achiral reversed-phase HPLC columns. The 'Beam-Boost' UV-irradiation device (Section 5.6.1) has also been used as a precolumn derivatisation tool.

5.6.3 Post-column derivatisation

Post-column derivatisation followed by UV, fluorescence or ECD requires another pump capable of producing a very pulse-free flow, a mixing tee and a reaction coil. The post-column pump is often a syringe pump. It also requires a reactive group on the analyte which can be conveniently derivatised. The earliest use of post-column derivatisation was in the analysis of amino acids from protein hydrolysates, in which the eluates from ion-exchange chromatography were derivatised with ninhydrin (absorbencies at 440 and 570nm) and quantified using visible light absorbance detectors. Many reagents are now available (reviewed by Imai and Toyo'Oka 1988). The sensitivity can be increased by several orders of magnitude using these methods. The 'Beam-Boost' device (Section 5.6.1) is a special case of post-column derivatisation.

5.7 Selectivity

The selectivity of an analytical method can often be improved through the choice of HPLC detector; for example UV–visible detection is probably the least selective; if the analyte fluoresces, then use of the fluorescence detector will offer an immediate improvement in selectivity (and probably sensitivity). Similarly, if the analyte is electroactive, electrochemical detection will also offer a selectivity improvement.

The techniques described in Section 5.6 will also often offer better selectivity. If selectivity is still a problem, with co-eluting peaks, then improvements in the sample preparation stage or the chromatography will be necessary. Solid-phase extraction can be used in place of liquid–liquid extraction, for example, or column-switching techniques can be employed. A column-switching assay with a limit of detection of 50pg/ml using UV detection has recently been described (Wang et al. 1994). This was achieved with a highly absorbent compound with strong carbonyl conjugation, and column switching from cation-exchange to narrow-bore reversed-phase chromatography.

5.8 Detector problems

Most problems with HPLC detectors manifest themselves as baseline problems. These fall into the following categories:

5.8.1 Noise due to bubbles

Bubbles in the flow-cell of a detector form as a result of the outgassing of dissolved air from the compressed solvent when the pressure returns to atmospheric pressure. The flow-cell is a common place for this to happen. Bubbles are seen as very sharp, irregular spikes on the baseline; occasionally a bubble will lodge in the cell and cause a sudden change in the baseline which remains then high (or low). There are several solutions; a back-pressure regulator is the simplest and most effective. Thorough vacuum or helium degassing of the solvents is also essential for high-sensitivity work; dissolved oxygen may present chemical problems by quenching fluorescence or by interfering with electrochemical processes.

5.8.2 Spurious peaks

Spurious peaks are those not associated with the analyte or matrix but appear in a blank or mobile-phase injection. They may be due to late-eluting peaks from a previous run (in which case they will be much broader than normal), or to impurities in the solvents – especially when a gradient is being run. They may also arise from impurities in the autosampler vials, or from other sources – such as contaminated nitrogen in sample evaporation apparatus. Meticulous attention to detail is usually the answer, or a change of solvent supplier or grade.

5.8.3 Baseline instability

Baseline instability may take several forms. Short-term noise may arise from detector electronic instability, requiring attention from the service engineers. The lamp in a UV–visible or fluorescence detector may be failing in its output and this will increase the noise, decreasing the sensitivity. There may also be a regular noise superimposed on the baseline, and if this can be correlated with the pump pistons' movement, then the pump requires attention; seals and/or pistons may require replacement, or there may be an air bubble trapped in the pump cylinder. Regular checks on baseline noise of an HPLC system should form part of the operating routine of the laboratory. Drift is more difficult to deal with because its origins are less easy to trace. Detector electronics or lamp output changes may cause drifting baselines, but these will usually stabilise after the equipment has warmed up for 30 minutes or so. Alterations in the composition of mobile phase will also cause drift – the cause here could be helium sparging which removes volatile solvents and alters composition, or the gradual dissolution of atmospheric oxygen into a mobile phase. This can be a major problem when using electrochemical detection. A gradient will almost always cause baseline changes.

5.9 Appendix

5.9.1 Buying a detector

Test several detectors using the same system and sample analytes. Obtain chromatograms from your samples at the correct conditions (wavelengths, voltages, etc.) for your analysis. Determine the noise (peak-to-peak height over a time equal to at least the width at half-height of your analyte peak) just before or just after the analyte retention time. Determine the limit of detection for your analyte (amount of analyte multiplied by three times the noise divided by analyte peak height) for a signal-to-noise level of 3.

Be aware that response time, sensitivity settings, photomultiplier voltages, filters, etc. can all impact the performance of a detector and different detectors may have non-equivalent settings. Some fluorescence detectors perform better than others in certain regions of the spectrum. Some have limited lower excitation wavelengths. Some will give you an on-screen emission or excitation spectrum. Choose the detector that gives you most sensitivity compatible with your application. It may be that you can sacrifice sensitivity for flexibility. Wavelength scanning, time programmes, wavelength changes and autozero may be more or less important.

5.9.2 Which detector to use?

This is not just a sensitivity/selectivity decision. With the availability of LC–MS–MS methods with short run-times, it may be advantageous when high throughput is required to choose MS directly.

Determine sensitivity required:

- Obtain UV spectrum and calculate absorbtivity (molar or A/mg/ml) at λ_{max}.
- Run LC at λ_{max}. Determine whether sensitivity/selectivity sufficient.
- If not: check fluorescence (emission scan at UV λ_{max} followed by excitation scan at fluorescence emission maximum). Run LC at fluorescence $\lambda_{max(em\ and\ ex)}$. Determine whether sensitivity/selectivity sufficient.
- If not fluorescent: check electrochemical activity. Obtain cyclic voltammogram. Run LC at optimum voltage. Determine whether sensitivity/selectivity sufficient.

5.10 Bibliography

Dorschel, C.A., Ekmanis, J.L., Oberholtzer, J.E., Warren, F.V. and Bidlingmeyer, B.A. (1989) LC detectors: evaluation and practical implications of linearity. *Analytical Chemistry*, **61**, 951–968.

Fielden, P.R. (1992) Recent developments in LC detector technology. *Journal of Chromatographic Science*, 30, 45–52.

Frei, R.W. and Zech, K. (eds) (1988) *Selective Sample Handling and Detection in High-performance Liquid Chromatography, part A, Journal of Chromatography Library*, Vol. 39A, Amsterdam: Elsevier.

Gertz, C. (1990) *HPLC Tips and Tricks*, LDC Analytical.

Huber, L. and George, S.A. (eds) (1993) *Diode Array Detection in HPLC, Chromatography Science*, Vol. 62, New York: Dekker.

Imai, K. and Toyo' Oka, T. (1988) Design and choice of suitable labelling reagents for liquid chromatography, in Frei, R.W. and Zech, K. (eds) *Selective Sample Handling and Detection in High-performance Liquid Chromatography, Part A, Journal of Chromatography Library*, Vol. **39A**, 209–288.

La Course, W.R. (1997) *Pulsed Electrochemical Detection in High-Performance Liquid Chromatography*, New York: Wiley.

Patonay, G. (1992) *HPLC Detection, Newer Methods*, New York: VCH.

Radzik, D.M. and Lunte, S. (1989) Application of liquid chromatography/electrochemistry in pharmaceutical and biochemical analysis: a critical review. *CRC Critical Reviews in Analytical Chemistry*, **20**, 317–357.

Scott, R.P.W. (1986) *Liquid Chromatography Detectors, Journal of Chromatography Library*, Vol. 33, Amsterdam: Elsevier.

Steiner, F. (1991) Applications of narrow-bore columns in HPLC, Hewlett-Packard.

Torsi, G., Chiavari, G., Laghi, C., Asmudsdottir, A.M., Fagioli, F. and Vecchietti, R. (1989) Determination of the absolute number of moles of an analyte in a flow-through system from peak-area measurements. *Journal of Chromatography*, **482**, 207–214.

Torsi, G., Chiavari, G., Laghi, C. and Asmudsdottir, A.M. (1990) Responses of different UV-visible detectors in HPLC measurements when the absolute number of moles of an analyte is measured. *Journal of Chromatography*, **518**, 135–140.

Wang, K., Blain, R.W. and Szuna, A. (1994) Multidimensional narrow bore liquid chromatography analysis of Ro 24-0238 in human plasma. *Journal of Pharmaceutical and Biomedical Analysis*, **12**, 105–110.

6

GAS CHROMATOGRAPHY:
WHAT IT IS AND HOW WE USE IT

Peter Andrew

From earlier chapters about HPLC you will be familiar with the general concept of chromatography and with the theory that underpins it. When a gas is used as the mobile phase, the process is described as gas chromatography (GC). The other (stationary) phase is usually a liquid although not apparent as such as it is invariably very viscous and spread very thinly over the walls of a capillary tube. The word 'liquid' is nowadays generally omitted from the earlier title of gas–liquid chromatography. Only a simple explanation of why GC works will be included here because there are many textbooks devoted to the subject, most of which carry a detailed theory of the chromatographic process as it applies in GC. The book *Analytical Gas Chromatography* (Jennings 1987) is particularly recommended. Capillary GC will be described here. Packed columns are now largely defunct due to the considerable improvement in efficiency that the open-tube systems provide. Although capillary GC continues to be used as a term of reference, purists argue that the newer wide-bore tubes can no longer be described as capillaries. Open-tubular (OT) GC is the more correct term and will be used here.

6.1 Why gas chromatography works

First a simple analogy . . .

Imagine a beaker of water to which is added some acetone and propanol. After a few seconds, for each of the organic liquids, an equilibrium will have been established between the concentration in solution and the concentration in the vapour phase above the water in the beaker. It is very likely that at any given instant there is more acetone in the vapour than propanol because acetone is a lot more volatile, i.e. has a lower boiling point. Put another way, acetone partitions itself to a far greater extent in the gas phase above the liquid in the beaker than does the propanol. If one were then to quickly 'blow away' the mixture above the surface, a new equilibrium would be established as before, but with a tiny difference. There would be slightly different amounts of each compound present because of the relative amounts of each lost during the blowing away. If this process was continued many times there would occur a disproportionate loss of acetone over that of propanol. A practical example of this phenomenon is the change in fragrance of a blended perfume that occurs as the top of the bottle is repeatedly removed during use. It is this difference in the extent of partitioning between a gas and a liquid phase that causes separations to occur in GC.

Using this analogy in a GC environment, imagine a mixture of acetone and propanol introduced into the injector end of the GC column which is continuously swept with carrier gas and maintained at a temperature where each solute will be immediately vaporised to the gas phase. Both solutes will immediately partition between the moving gas and the immobile liquid phase. In other words some molecules will dissolve in the stationary phase (cf. the water in the beaker analogy above) and some will remain in the gas phase.

All other things being equal, the lower boiling acetone molecules that are dissolved in the liquid phase will vaporise before (and more frequently than) the molecules of the higher boiling propanol. As they enter the mobile phase they will be carried down the column and over the virgin stationary phase, where they will redissolve. A fraction of a second before the propanol molecules revaporise to be carried downstream by the carrier gas, the acetone molecules move again; hence the more volatile acetone molecules continuously increase their lead over the less volatile propanol molecules and separation is achieved.

Although this concept may prove helpful in visualising how a multiplicity of vaporisations and re-solutions on the part of individual solute molecules achieves the desired degree of separation, it must be stressed that this oversimplification presents an inaccurate picture. The chromatographic process is dynamic and continuous rather than occurring in discrete steps and at any one time some of the molecules of each solute are in the mobile phase and some are in the stationary phase.

To summarise, solutes can only progress through the column when they are in the mobile phase, they undergo no separation in the mobile phase nor do they do so in the stationary phase; rather, solute separation is dependent on the difference in solute volatility which, in turn, influences the rates (or frequencies) of solute vaporisations and re-solutions. Both column temperature and the nature of the stationary phase markedly influence solute volatility and hence the separation. These factors will be considered in more detail later.

6.2   Factors that affect the chromatography

We have seen that separation is achieved by exploiting the difference in extent of partition of solutes between the two phases. In turn, the extent of partition is influenced by the solubility of the solutes in the stationary phase and their relative volatility at the operating temperature. Thus by varying one or both of these circumstances one can start to develop ways of changing, improving or hastening the separation by changing various operating parameters. Let us consider several aspects in turn, while relating them to experimental variables that we can control.

The tendency for solutes to remain in the stationary phase is a consequence of vapour pressure and solubility. Increasing the temperature causes solute vapour pressure to increase, with a consequent increasing tendency to partition into the mobile phase. This, in turn, increases the rate of forward movement and solutes reach the detector sooner.

The solubility of solutes will change with stationary phases of different polarities, leading to changes in retention or resolution. By considering the interactions that may occur between the functional groups in the solute molecules and those of the stationary phase (if indeed there are any), a stationary phase may be chosen which changes the partition characteristics of one or more of the solutes, leading to improved separation. More of this later when we consider the nature of the stationary phase.

The 'amount' or 'extent' of stationary phase – the liquid spread over the surface of the column – to which the solute is exposed will have a profound effect on the elution time. Where several solutes are introduced together, the extent of exposure provides the opportunity for differences in their respective partition characteristics to become apparent and thus to aid separation. This aspect may be related to column dimensions, largely column length and film thickness but also includes column diameter.

The exposure of solutes to the liquid surface, considered in terms of time, will also profoundly affect retention and separation. Exposure may be considered as largely a function of carrier flow rate – the faster the carrier movement through the column, the greater the forward movement of solutes during the period they spend in the mobile phase, and hence the sooner they emerge into the detector. For solute mixtures, the greater the rate of forward movement, the less time there will be for differences in partition to aid separation, so resolution of solute mixtures will be reduced.

From these observations, certain experimental parameters can be identified as having specific effects on retention and resolution and so provide the chromatographer with a choice of operating conditions with which to achieve a separation. Several of them have strong similarities to HPLC.

6.3 Choices in GC

6.3.1 Stationary phase

Stability and non-volatility are the key features of a good stationary phase. Loss of stationary phase through the lack of either invariably ends with contamination of the detector, which causes baseline instability, increased noise and reduced sensitivity. The stationary phase should also be chemically inert – there is no place for chemical reactions within the chromatographic column. Stability and non-volatility are markedly increased by chemically bonding the phase to the silica surface of the column and by introducing a degree of cross-linking within the bulk of the liquid to stiffen and immobilise it. Most popular phases are now available in this form and, as evidence of their stability, columns can be rinsed with a suitable organic solvent to remove accumulated debris, particularly associated with the injection region.

Before considering what phases are used, let us consider why a choice of different phases is desirable. We have established that partition is the key to separation and, in the absence of other factors, it is differences in volatility that cause separations to occur. On this basis, two compounds with the same boiling point will not be separated. If there were an additional feature that influenced the vapour pressure of one over the other, then separation would be achieved. The introduction of functional groups into the stationary phase will provide that 'little extra' that will cause changes in retention to occur for some molecules relative to others. The interactive forces that

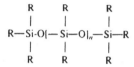

Figure 6.1 Structure of dimethyl polysiloxanes.

occur between solute and these functional groups are additional to the dispersion interactions but remain weak (induced dipole–induced dipole, hydrogen bonding) by chemical bond standards.

By far the most common type of material used as GC liquid phases are the polysiloxanes, with the dimethyl polysiloxanes predominating (Figure 6.1). They are chemically inert, have no functionality, very low vapour pressure and their viscosity is almost unaffected over a wide range of temperatures. They have many names, ranging from the old elastomer notation, E301, SE30, to the modern 0V1, 0V101, DB1, CP-Sil 5. Suppliers' catalogues describe the liquid-phase options, usually with the general name and also by their own notation. Functionality can be introduced into the siloxane chain by replacing a proportion of the methyl groups with, for example, phenyl or trifluoropropyl groups, in different combinations and proportions. The change in polarity provides different selectivities and a number of special phases have been designed for well-defined tasks, including several methods of analysis specified by the American regulatory body, the Environmental Protection Agency (for example the GC determination of polychlorinated biphenyls in soil).

Much more polar phases are provided by polyethylene glycols (PEGs), some of which can now be chemically bonded to the silica column surface. This has enhanced their utility, as previously they did not have a particularly useful upper operating temperature limit. However, all the polar phases, including the more polar siloxanes, are less thermally stable than the non-polar liquids, so have a reduced temperature operating range. This limits their use for pharmaceutical compounds, which tend to require fairly elevated (250°C+) temperatures.

Fortunately, in capillary GC there has not been the great proliferation of phases that was deemed necessary with packed-column GC. This is because OT columns are so much more efficient and capable of such tremendous resolution that a wide choice of phases of differing selectivity is not necessary. This is particularly true in the bioanalytical field, where usually one is looking for only one or two components of interest and these are usually significantly different from one another (parent drug plus a metabolite or two and maybe an internal standard at the very most). Thus a DB1 (100% methyl) or DB5 (5% phenyl, 95% methyl) column is usually more than sufficient to provide the required selectivity. Table 6.1 is taken from the Chrompak catalogue. It describes most of the phases currently available and illustrates just how helpful the suppliers can be as they try to win your custom. All the columns described are referred to as wall-coated OT (WCOT) columns. Two alternatives are support-coated (SCOT) and porous-layer (PLOT) columns. The former provides enhanced sample capacity, albeit with reduced efficiency and the latter, special applications usually associated with very volatile compounds.

Table 6.1 Stationary phase equivalency and composition: liquid phases (WCOT)

Stationary phase	Chemical composition	Replacement for	T_{min}	T_{max} iso/progr.
CP-Sil 5 CB	100% dimethyl siloxane (Bonded CP-Sil 5)	DB-1, Ultra #1, BP-1, SE-30, OV-1, OV-101, SP-2100, 007-1	−25°C	300/325°C
CP-Sil 8 CB	95% dimethyl, 5% phenyl (Bonded CP-Sil 8)	DB-5, Ultra #2, BP-5, SPB-5, SE-52, OV-73, SE-54, 007-2	−25°C	300/325°C
CP-Sil 19 CB	85% dimethyl, 7% cyanopropyl, 7% phenyl, 1% vinyl (Bonded OV-1701)	DB-1701, BP-10, SPB-7, OV-1701, SP-10, SP-2250, 007-1701	−25°C	275/300°C
CP-Sil 43 CB	50% dimethyl, 25% cyanopropyl, 25% phenyl (Bonded OV-225)	OV-225, DB-225, BP-15, SP-2300, 007-OV-225-B	45°C	200/225°C
CP-Sil 88	100% cyanopropyl	SP-2330, SP-2340, Silar-10 C	50°C	225/240°C
CP-Wax 52 CB	polyethylene glycol (Bonded Carbowax 20-M)	DB-Wax, Supelcowax-10, BP-20, Carbowax 20M, Superox 20M	20°C	250/275°C
CP-Wax 58 CB	polyethylene glycol, nitroterephthalic acid ester (Bonded FFAP)	FFAP, DB-23	20°C	250/275°C

6.3.2 Mobile phase

The nature and choice of the mobile phase in GC is simple in the extreme. Unlike HPLC where it is carefully tailored to, and influential in the separation process, this is not the case in GC. The choice of mobile phase is restricted to only one of three gases. The mobile phase, or carrier gas, as it is more often termed, must be inert, dry and, for capillary columns, oxygen-free. Current practice employs either nitrogen, helium or hydrogen, with filters to remove water and oxygen. With hydrogen, it is essential to employ a concentration sensor to shut down the supply in the event of a leak into the oven.

Although any one of the three gases will suffice, there are advantages in choosing hydrogen over the other two. This is because hydrogen is much 'thinner' so the diffusivity of solute molecules is much greater than with the other two. Efficiency is increased threefold compared to the use of nitrogen. This manifests itself practically by enabling the analysis time to be reduced by a factor of three. A further advantage of hydrogen is that almost optimum efficiency prevails over a much wider range of flow rates than for helium or nitrogen. Nevertheless, helium is a good practical alternative to hydrogen especially at low flow rates where the advantages of hydrogen are less. Nitrogen is seldom used in OT GC.

6.3.3 Column length

The longer the column, the greater the separating power. Although a longer column offers increased performance it is offset by band broadening (i.e. the peaks are wider). Improvement in separating power is a function of the square root of the increase in length, i.e. doubling the length only gives 1.4 times the separating power. Pressure drop across the column also increases as the length increases and can become disadvantageous. For most pharmaceutical applications, 15m is sufficient.

Performance of the column is judged in terms of plate number as in HPLC. The term was originally coined in the early days of GC when it was used to monitor, and could be related to, distillation of oil and petroleum products. A typical 25m WCOT GC column gives about 50 000 plates. This compares very favourably with a typical 10cm HPLC column which only exhibits about 6000 plates.

6.3.4 Column diameter

Columns vary from 0.1mm to just over 0.5mm in diameter. There is a little confusion in the terminology, as with HPLC. They are all loosely termed capillaries but the more correct term is 'open tubular' and this is used in the catalogues in such abbreviations as WCOT and PLOT. The typical diameter for 'normal' capillary GC is about 0.3mm but the practical constraints of sample introduction (not least because the traditional Hamilton syringe will not fit down such a column!) has led to increasing popularity of the so-called megabore 0.5mm (530µm) columns originally pioneered by Hewlett Packard. Megabore columns bridge the gap between the old packed column systems and modern true capillary GC and can generally be fitted into old hardware without much trouble.

Column diameter primarily affects sample capacity, both in simple terms of column volume, which governs the useable gas flow it can accommodate, and the film thickness, which tends to a minimum as the diameter reduces.

6.3.5 Film thickness

Film thickness complements column diameter for varying speed and sample capacity. The thicker the film, the greater the retentivity and the longer solutes take to move through the column. Separating power is increased. However, efficiency is reduced as thicker films are employed. Thus the fastest columns have narrow diameters (about 0.15mm) and thin films (0.1µm) but limited sample capacity. The typical capillary column will have a diameter of about 0.3mm and will usually have a film thickness of about 0.1–0.2µm. Although a similar film thickness is available for megabore columns, these usually have (very) thick films of between 1 and 5µm.

There remain two vital operating parameters which control the chromatography. They are flow rate and temperature. As with the other variables that govern separation, although we consider them individually, they all impact on one another to some extent and all need to be kept in mind when considering one specific parameter.

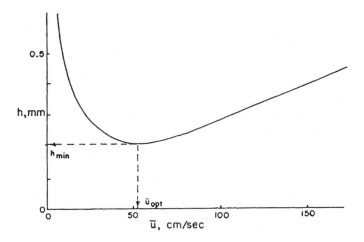

Figure 6.2 Theoretical van Deemter curve exhibiting an h_{min} of 0.2mm at an optimum average linear carrier gas velocity ($\bar{u}_{opt}$) of 54cm/s.

6.3.6 Flow rate

The rate of carrier flow through the column profoundly affects the time a solute takes to reach the detector and, as a consequence, the extent to which partition can occur during that journey. The theoretical relationship between flow rate (more correctly, carrier velocity) and efficiency is described by the van Deemter equation. Solutions of the van Deemter equation (see Chapter 3, p. 62) are often presented graphically via the construction of van Deemter curves, which are very helpful in understanding the relationship between the two (Jennings 1987). In practice, the optimum conditions prescribed by the van Deemter equation are often ignored in the modern analytical laboratory. The typical carrier flow rates used (0.5–2.5ml/min, equating to an average linear flow of say 20–90cm/s) are generally higher than the theoretical optimum but still provide very good efficiency. The choice of carrier gas can minimise the consequence of working at these higher flow rates. A typical van Deemter plot is illustrated in Figure 6.2; h is the height of one theoretical plate and u is the linear gas velocity. The *actual* flow rate is not critical. In practice one tends to note the typical operating head pressure and set the flow by that. The expected flow is confirmed and, provided it approximates to that nominally specified in the method, will suffice. The chromatogram is the ultimate yardstick of acceptable operating conditions.

6.3.7 Temperature

The first requirement of any GC separation is that solutes are introduced to the column in the gas phase. The higher the temperature, the greater the tendency for the solute to occupy the mobile phase and the more rapid its conveyance to the detector. More importantly, minor changes in temperature have a marked effect on the partition ratio, so choice of temperature is a key parameter in obtaining suitable retention and separation. With a fairly complex mixture or with solutes of widely differing vapour pressures, a temperature gradient can be employed to provide optimum conditions for all solutes (similar to a solvent gradient in HPLC).

However, temperature is also used to aid the transfer of the injected sample from the injection region into the column (strictly it is the amount of heat available that aids transfer but this is regulated via the temperature settings). There are a number of methods of sample introduction, all of which affect the temperature that is experienced by the column and which, in turn, affect the chromatography.

6.3.8   Some rules of thumb

GC in bioanalysis will probably employ a 10–15m column coated with a 0.1μm thick film of predominantly dimethyl polysiloxane, e.g. DB5. If a 0.32mm diameter column is used, the flow rate will be about 50cm/s, equating to about 1.5ml/min. If a 0.5mm megabore column is used, flow rates will be higher, up to 15ml/min, with sample introduction by direct septum injection into the column. Temperature settings will depend on the injection method but chromatography will probably require temperatures typically of at least 250°C.

6.4   GC hardware

Like HPLC, GC requires a system that enables a mobile phase to pass through a 'column' of stationary phase such that solute mixtures can be separated and detected as they emerge. Thus there is much in common with HPLC equipment although at first glance this may not be apparent. This is because in GC the mobile phase is a gas, through which solute molecules can very readily, and hence quickly, diffuse. Connections between the injector and column and column and detector must be kept as short as possible to minimise band broadening and re-mixing of separated components. Hence one large 'box' accommodates, at least in part, column, injection and detector systems. In HPLC this could and often does spread over the whole bench! The solutes must be introduced to the column in the gas phase and maintained thereafter at whatever elevated temperature is required. Again, the one large, insulated box is the solution, even though local areas occupied by the injection system and detectors may be at different, but constant, temperatures. Figure 6.3 illustrates the component parts of a typical HPLC system, alongside a GC assembly, which will enable you to appreciate the functions of the various components of the GC system. Several key points can be noted:

- The equivalent to the HPLC pump is situated in the factory where the carrier gas was put into the cylinder. Simply opening the gas bottle is all that is needed to set the mobile phase in motion. Furthermore, there is less emphasis on the absolute carrier flow rate in GC.
- Unlike LC, temperature is a key variable in GC. Hence the emphasis on temperature control, both for the column and other associated parts. The column oven is very versatile, being capable of multiple gradients in order to accommodate sample transfer into the column and then to ensure adequate progress of solute mixtures through the column.

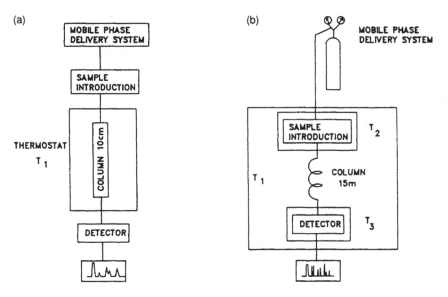

Figure 6.3 Diagrammatic comparison of (a) HPLC and (b) GC assemblies.

- The column is very different in both appearance and geometry than its LC counterpart. It comprises a long fused silica tube with an external diameter of between 0.5 and 1mm. Because the silica is susceptible to atmospheric corrosion which weakens the crystal lattice and makes it more liable to breakage, the outer surface is protected with a layer of polyimide. Unlike the former glass capillaries, a fused silica capillary is intrinsically straight and flexible so columns made from it are much easier to handle. Connections to injector and detector are made directly, there is no intermediate piping, using graphite or vespel ferrules which can deform sufficiently to effect a seal. Vespel ferrules may be used repeatedly. Both are stable to 350°C.

6.4.1 Pneumatics

The flow of carrier gas through the column is usually arranged by setting a head pressure that will give the required linear flow. The flow may be measured by injecting a compound that is non-retained, usually a few microlitres of methane. The time taken for the peak to appear is used to calculate the speed of travel along the column length. This linear flow through the column is an average flow. The actual flow reduces as it progresses down the column and further slows as a temperature gradient is applied, so for comparative purposes it should always be measured under the same temperature conditions. The bulk flow (in ml/min) can be measured so that it may be re-set conveniently in the future. Total carrier flow into the GC may be greater than the flow through the column by a considerable amount if split or splitless injection is used, see below.

Gases supplied to the detectors are also controlled by pressure regulation. For example, with the flame ionisation detector, hydrogen as fuel will require some unspecified but regulated pressure to give a flow of typically 20ml/min. Similarly, compressed air is pressure regulated to about 10 times the hydrogen flow. This ratio is not critical and more detail may be found in the manufacturer's manual. Flows are set with a simple bubble flowmeter and stopwatch. Once one has a feel for the required pressure on any one GC, the various gas flows may be set very easily and quickly.

6.4.2   Sample introduction

All GC equipment can be purchased with an autosampler. All mimic the human act of taking a microlitre syringe, filling it, reducing the syringe content to a pre-set volume and then, by piercing a rubber septum, placing the sample solution into an injection zone. A typical injection volume of 1–2μl, when flash vaporised, occupies a new volume of about 2ml, sufficient to more than fill the entire capillary. Thus it is necessary either to provide an 'expansion chamber' from which the vaporised sample can enter the column in a controlled way or avoid flash vaporisation altogether. It is at this stage that capillary GC may appear complicated to the newcomer.

We will consider four ways of sample introduction:

- split injection;
- splitless injection;
- programmed temperature vaporisation (PTV);
- direct on-column injection.

The first three constitute the expansion chamber approach, with the first two representing the traditional way of injection, fitted as standard to most gas chromatographs. The PTV injector is an add-on of more recent development which can provide a number of different options for sample transfer.

Split injection

The injection device is usually placed on the top of the GC. It has separate temperature controls which are adjusted to ensure that as the sample enters it is vaporised very rapidly. A temperature between 200°C and 300°C is typical. The injector has two outlets, one to the column and one to atmosphere. By setting a high, fixed column head-pressure this will ensure a constant carrier flow through the column with the remainder exiting directly to atmosphere. In the split injection system, the supply of carrier gas entering the injector (our 'expansion chamber') and mixing with the volatilised sample may then flow down one of two paths – into the column or out to atmosphere. By measuring the flow through the column and the flow out to atmosphere, the proportion of sample entering the column (the split ratio) can be determined. By varying the volume of carrier entering the injector, the split ratio may be adjusted to suit the assay requirements.

Increasing the amount of carrier entering the injector will not cause more carrier to flow down the column, as this is governed by the fixed head-pressure. Instead, the extra carrier will flow out to waste, taking a proportionally greater amount of sample with it. Thus, by first fixing a suitable head-pressure to achieve good chromatography, adjustment of the amount of carrier entering the injector (via a mass-flow controller) will enable the user to split the sample, some entering the column, the remainder flowing to atmosphere. A typical split ratio is about 1 : 100. For most bioanalytical applications isothermal conditions will suffice, i.e. the column oven is maintained at a constant temperature. Figure 6.4b illustrates the flow conditions utilised in the split mode.

Splitless injection

The above injection system is the simplest and most popular but it is less useful when maximum sensitivity is required – the usual case with bioanalysis! Splitless injection attempts to minimise the loss of sample by arranging that most of the sample enters the column. This will take a little time, typically 30–45 seconds, so the chromatographic conditions must be such that solutes dissolve in the stationary phase at the head of the column and stay there, i.e. the oven temperature must be low in order to minimise partition. Once the sample introduction process is complete, the temperature of the column is increased to a temperature that gives a chromatographic run time that is acceptable.

With the possible exception of liner dimensions, the injection system hardware and operating temperature are exactly the same as for the split mode, with the simple addition of a three-way valve which can close the exit vent. Just prior to injection, the exit valve is closed so that after injection the vaporised sample can only enter the column. After about 30–45 seconds, when most of the sample will have left the injection zone and entered the column, the exit valve is re-opened and the remaining sample is flushed away. Figure 6.4a illustrates the gas flow path during the injection period. At any time other than the immediate post-injection period, the flow path and control conditions are as for the split mode. The operation of the valve and initiation of the thermal ramp are all fully automatic on most modern GC equipment.

The programmed temperature vaporisation injector (PTV)

This system, usually fitted as an additional external module, receives the injected sample as a liquid. Unlike the injectors described above which are maintained at one constant temperature, a variety of temperature options can then be used, dependent on sample or assay requirement. For example, once the sample solution is injected into the PTV, the solvent may be back-flushed away by gentle heating under a stream of gas. After some predetermined interval of time, the analyte is vaporised and carried into the column under controlled thermal conditions. As with the splitless mode, the sample is concentrated at the top of the column under the influence of a low oven temperature. A thermal ramp is then initiated to effect chromatography. All the necessary PTV valve closures and thermal conditions are pre-set and executed automatically.

All three of the systems described above require that the sample is vaporised. This may cause problems with the analyte if it is thermally labile, especially with the split and splitless modes where a substantial thermal shock is applied to the analyte

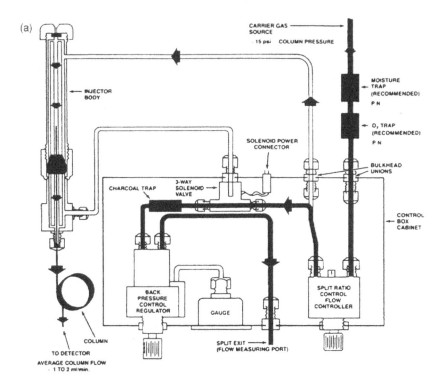

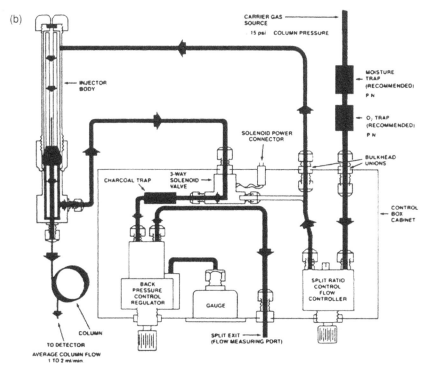

Figure 6.4 Pneumatics – split and splitless modes. (a) Splitless mode during injection cycle. (b) Split mode (splitless post-injection cycle).

molecules in what may be considered a reactive environment (a very hot glass liner). A further potential problem with all three systems, particularly in bioanalysis, is an accumulation of involatile material in the glass liner arising from the injection of 'dirty' samples. This deposit can then act as an absorptive site for analytes, producing tailing peaks. The glass liner in most systems is easy to remove and clean but remains an inconvenience.

An alternative to flash vaporisation is to introduce the sample solution directly into the column. This is described, not unexpectedly, as direct on-column injection. Much milder thermal conditions prevail with this procedure, but the problem of deposits in the injection region can still remain a problem.

Direct on-column injection

This injection procedure is becoming increasingly common, particularly as 0.5mm columns are now more widely used. The procedure was late in arriving for most people because capillary columns require extremely thin needles to fit into the column and, as such, they are not strong enough to penetrate the traditional rubber septum. Hair-like silica needles were developed, accompanied by septumless injection heads that utilised electrically operated seals. Automated sample injection was not practical. The advent of 0.5mm OT columns meant that the normal Hamilton microlitre syringe could be used in the conventional way once suitable needle guides were developed. The subsequent introduction of simple column-joining systems has enabled short lengths of 0.5mm column to be used ahead of narrower-diameter capillaries. Standard autoinjectors can be used with these systems. The capillary must be at a temperature that does not cause rapid vaporisation of the sample solution upon injection (hence the often-seen term 'cool on-column injection'), as there is just not the volume within the capillary to accommodate the vapour. A thermal ramp is initiated post-injection as per the splitless mode. Should deposits in the injection region become a problem, chemically bonded phase columns can be rinsed with a suitable solvent or, alternatively, the first few centimetres of column can be discarded.

A final note about sample introduction in GC: in bioanalysis, one of the best features of HPLC is that whatever volume is left from the sample extraction, typically up to 200µl but occasionally more, can all be introduced into the LC column, maximising assay sensitivity. This is *not* the case in GC. Injection volume for typical plasma extracts cannot usually exceed about 3µl, representing at most 10–20% of the available final volume of sample extract. Retention gap technology, pioneered by Konrad Krob (1987) may increase the usable volume for some analytes 10–20-fold, but at present this approach has not been widely applied. Retention gaps can serve in the same way that guard cartridges are used in HPLC. The short length of fused silica tubing may be easily changed as it becomes fouled with deposits, etc. Only Carlo-Erba have an autosampler capable of making large-volume injections under the necessarily slow delivery conditions. This inability to load most of the sample extract into the GC may be perceived as a drawback of the technique. However, it has led manufacturers to designing autosamplers and vials that enable sample extracts to be redissolved in volumes of less than 25µl.

FLAME IONIZATION DETECTOR

ELECTRON CAPTURE DETECTOR

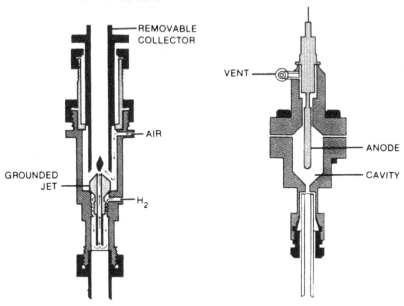

NITROGEN-PHOSPHORUS DETECTOR

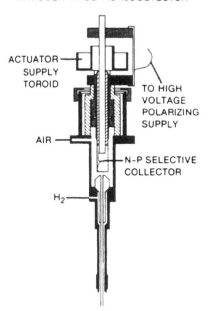

Figure 6.5   Cross-sections of GC detectors used in bioanalysis.

6.4.3 Detectors

Although there are about 10 different detectors that can be used in GC, we shall only consider those used in bioanalysis; they are:

- flame ionisation detector (FID) (Figure 6.5);
- thermal ionisation detector (TID or NPD) (otherwise referred to as the nitrogen detector) (Figure 6.5);
- electron capture detector (ECD) (Figure 6.5);
- mass selective detector (MSD).

For each detector, operating requirements, utility, advantages and disadvantages will be reviewed briefly.

Flame Ionisation Detector (FID)

This detector was the first successful universal detector to be developed and remains the most widely used. It works via combustion of the eluting band in a hydrogen flame. The mechanism is believed to involve the formation of CH radicals which combine with oxygen to yield an acetaldonium ion and an electron. One reaction is believed to occur for every 100 000 C atoms present. Hydrogen as fuel is required over the flow range 15–30ml/min depending on manufacturers' recommendations. Air flow rate is typically 10 times this flow. The FID requires a constant temperature so has its own temperature control, usually set somewhat above the column oven temperature (plus 20°C is typical).

The advantages of this detector are:

- It is robust and very easy to operate, with high sensitivity.
- The response is very rapid and the effective detector volume is very small so the detector can tolerate widely differing carrier flow rates (i.e. make-up gas not required).
- It has a very wide linear range, 10^7-fold.
- Universal for organic compounds, with a similar response for most analytes.

Its disadvantages are:

- it is non-selective;
- destructive; and
- (• universal).

Thermionic detector (TID)

Also more commonly called the nitrogen or the nitrogen phosphorus detector (NPD), the construction is similar to that of the FID. The same controller, called an electrometer, is used for both. The difference is the inclusion of a small refractory bead of an alkali metal salt (usually rubidium chloride), just above the jet tip. The bead can be either heated by the hydrogen flame or heated electrically in a hydrogen atmosphere. The mechanism is not fully understood. Different responses and different selectivity can be achieved with different polarity applied to the collector. The detector can be further tuned by varying the distance of the bead from the jet, choice of bead temperature and hydrogen flow rate. A detailed description of the detector is given by Dressler (1986). Operation is generally like that of the FID, with modifications particular to each manufacturer's recommendations.

The advantages of this detector are:

- It is selective for compounds containing nitrogen or phosphorus.
- It is 10–100 times more sensitive than FID.
- It has a wide linear range, similar to that of FID.

Its disadvantages are:

- it has a variable response, dependent on operating conditions and nitrogen environment and content;
- the bead life can be short and variable; and
- it is destructive.

Electron capture detector

Unlike the two ionisation detectors described above, this detector senses a reduction rather than an increase in standing current. In the absence of the analyte, β-particles from radioactive nickel-63 interact with atoms of the carrier gas to produce bursts of electrons. These are collected and constitute the background current. If an electron-capturing analyte enters the detector, the resultant reduction in electrons causes a reduction in the background signal. Due to various alternative processes which can limit the sensitivity and range of the detector, several modes of operation are generally available via the controller. The one generally preferred and most widely used is the 'pulse mode with constant current' (Dressler 1986).

Although no additional fuel gases are required, the relatively large detector volume may necessitate make-up gas (nitrogen or helium) in order to achieve optimal sensitivity. In its response, the detector is especially affected by temperature, so requires its own accurate temperature control. Sensitivity is enhanced at higher temperatures so the detector is usually operated at about 300°C. The detector is sealed and requires no maintenance. Contamination may be removed by maintaining it for a short period at a high temperature, otherwise it has to be returned to the manufacturer for cleaning.

The advantages of this detector are:

- It is selective for halogens, nitro groups, peroxides, quinones.
- It is extremely sensitive (1pg).
- It is non-destructive.

Its disadvantages are:

- a limited dynamic range, 10^4-fold;
- it is prone to contamination; and
- it is radioactive; leakage must be checked regularly.

Mass selective detector

This represents the ultimate GC detector, both in terms of sensitivity and selectivity. Although the GC can be interfaced to very sophisticated mass spectrometers, a range of simpler instruments is available from Hewlett Packard, Finnigan-MAT, Varian, VG and Perkin-Elmer. They are relatively simple and, in MS terms, relatively inexpensive instruments. They are based on quadrupole or ion-trap technology and use electron-impact ionisation or chemical ionisation. Emphasis is on GC detection rather than MS interpretation and they are used to monitor specific ions for quantitative purposes. For some simpler applications, they may be used to assign molecular weights and provide tentative structures from the mass spectra obtained. Narrow-bore, thin-film capillaries operated with low carrier flow rates are generally preferred, to minimise the load on the vacuum pumping system.

6.5   Derivatisation for GC

There are two reasons for contemplating derivatisation of the analyte in GC. The first is to improve the chromatography. Even with chemically bonded phases and a so-called inert silica, some functional groups can interact with so-called 'active sites' on the silica surface, to cause tailing of the band. Such groups include primary and secondary amines, carboxylic acids and some phenolic hydroxy groups. The second reason for derivatisation is to enhance detectability and/or selectivity. Often derivatisation improves thermal stability and increases volatility.

There is a multitude of ways of making derivatives, but not that many that give reliable, quantitative yields (Blau and King 1977; Knapp 1979). Suppliers of derivatising reagents (particularly for silylation) provide much free literature about experimental conditions. Very briefly, the more common reagents are described below.

1   *Diazomethane*: a very useful gaseous reagent which, when dissolved in ether, quantitatively converts carboxylic acids to their methyl esters and phenolic OH groups to aryl ethers. It is usually prepared as required, although it can be stored for up to 4 weeks at −20°C. Although it has a terrible reputation for explosive instability, is relatively safe if handled in specialised glassware in a fume hood in small amounts. The derivatisation reaction proceeds quickly when added in modest excess to the substrate (typically dissolved in methanol or acetone) at 0°C with elimination of gaseous nitrogen. As the sample warms to room temperature, excess diazomethane decomposes slowly, with liberation of nitrogen.

2   *Silylation, using a variety of reagents*: there are a number of commercially available reagents and reagent cocktails for the preparation of trimethylsilyl derivatives (Pierce 1979; the Pierce general catalogue). These reagents react with the very groups that are the cause of peak tailing, to give derivatives that are more volatile than the underivatised compounds. Two reagents will be mentioned. The first, N,O-bis(trimethylsilyl)acetamide (BSA), reacts quantitatively under mild conditions with alcohols, amines, phenols and carboxylic acids to give trimethylsilyl ethers and esters. Reaction is usually carried in a solution of pyridine or dimethyl formamide (DMF), and is usually complete after 5–10 minutes with gentle warming (about 40°C). Another reagent, trimethylsilylimidazole (TMSI), reacts similarly

except with amines groups, which are unaffected. Water must be excluded from the reaction media for all silylation reactions.

3   *Acylation*: this offers both improved chromatography, as for silylation, and by the incorporation of halogen (usually fluorine or chlorine) atoms, improvements in sensitivity via the use of the electron-capture detector or negative-ion chemical ionisation mass spectrometry (NICI). Trifluoroacetic acid anydride, pentafluoropropionic acid anhydride and heptafluorobutyric acid anhydride may all be used to give quantitative conversions of amines, alcohols and phenols. Simply adding the reagent to the dry residue is often sufficient, although triethylamine may be used as a catalyst. Excess reagent may be evaporated off and the resultant product redissolved in a suitable solvent. When ECD or NICI is used, an additional clean-up procedure is normally required.

6.6   A GC strategy for bioanalysis

By far the greater part of bioanalysis involves extraction of the biological matrix to isolate the analyte in a form suitable for the final chromatographic determination. The process of isolation is largely independent of the chromatographic end point so a GC finish need not impact to any great extent on the procedure chosen. The only GC requirement is that the final extract is obtained in a relatively volatile solvent and, because of injection volume constraints, in as small a volume as is quantitatively practical (about 50µl). Thus a single-stage liquid/liquid extraction interfaces readily with GC. Similarly, solid-phase extraction can be accommodated by using methanol as the final eluent. This can be reduced in volume or evaporated entirely to allow the residue to be taken up in a small volume of a 'GC friendly' solvent, e.g. acetone, methanol or, if derivatisation is desired, the reagent solution.

6.7   Bibliography

Blau, K. and King, G. (1977) *Handbook of Derivatives for Chromatography*, London: Heydon.

Dressler, M. (1986) *Selective Gas Chromatographic Detectors*, Amsterdam: Elsevier Science Publishers.

Jennings, W. (1987) *Analytical Gas Chromatography*, London: Academic Press.

Knapp, D. (1979) *Handbook of Analytical Derivatisation Reactions*, New York: Wiley.

Krob, K. (1987) *On-column Injection in Capillary Gas Chromatography*, Heidelberg: Huethig.

Pierce, A. (1979) *Silylation of Organic Compounds*, Rockford, Illinois: Pierce.

Sandra, P. (1985) *Sample Introduction in Capillary Gas Chromatography*, Vol. 1, Heidelberg: Huethig.

Schomberg, G. (1990) *Gas Chromatography – A Practical Course*, New York: VCH.

Suppliers' catalogues: for chromatographic information, J&W Scientific and the Chrompack guide to chromatography; for derivatisation, the Pierce handbook and general catalogue.

7

THIN-LAYER CHROMATOGRAPHY

Hugh Wiltshire

7.1 Introduction

Thin-layer chromatography involves the equilibration of a compound between a mobile phase and a thin (e.g. 0.25mm) stationary phase bonded to a flat plate (usually glass). The mixture is applied in a volatile solvent and the plate is developed with a suitable mixture of solvents. The mobility of each component is measured with respect to the solvent front and the ratio of the distance moved by the compound to that of the solvent front, the 'Rf', should be constant for a given combination of stationary and mobile phases.

The stationary phase used for the majority of TLC applications is 'normal' rather than 'reverse' and the technique uses quite different physico-chemical properties for the separation of compounds than does reverse-phase HPLC. The stationary phase is usually silica, which is characterised by high polarity and the presence of many silanol (Si-O-H) groups which form hydrogen bonds and polar interactions with the mobile phase and analyte. Silica is acidic (forming -Si-O$^-$ ions) and can form ionic interactions with basic drugs. The chromatography of these compounds is more difficult and generally involves the use of strongly acidic solvent systems (e.g. 4 : 1 : 1 n-butanol : acetic acid : water) or the addition of aqueous ammonia to reduce the ionic interactions between the analyte and the solid phase, and hence to reduce 'streaking'.

TLC can also be carried out on alumina or cellulose plates, which can have advantages with particular compounds. Alumina, unlike silica, is basic and is therefore less likely to cause 'streaking' with basic compounds as there will be no ionic interactions. Cellulose, on the other hand, is neutral but possesses a high proportion of alcoholic (C-O-H) groups. It is especially effective for the separation of polar compounds such as amino acids with highly polar solvent systems.

Reverse-phase chromatography, in which the silanol groups of the silica are replaced by organic chains such as octadecyl (-Si-O-C$_{18}$H$_{37}$), makes use of dispersive forces for its selectivity or, when used with ion-pairing reagents, ion-exchange phenomena. Polar interactions are minimal and the selectivity is quite distinct from that of normal-phase TLC.

Greater resolution can be obtained by using silica with smaller particle sizes, but this 'high-performance' TLC (HPTLC) has not, to my knowledge, proved to be significantly more useful than ordinary TLC.

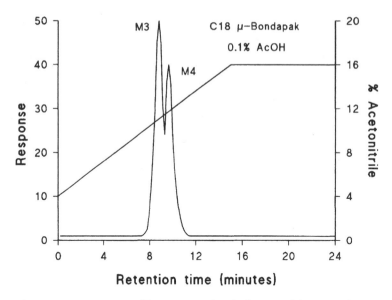

Figure 7.1 Separation of hippuric acid and phenacetylglutamine by HPLC.

7.2 Uses of TLC

It is possible to use HPTLC with densitometry measurements for quantitative analysis, but this is unlikely to offer any advantages over HPLC.

7.2.1 Preparative TLC

One of the more important uses of TLC in drug metabolism is for the isolation of less-polar metabolites. A few milligrams of material can be applied in a volatile, non-polar solvent to a thin 20cm × 20cm plate, or up to 50mg to a thick one (preferably with a concentration zone of coarser silica). The plate is then developed with a suitable solvent and the compounds of interest located using UV light (which produces a greenish colour with a fluorescent material in the silica, quenched by UV-absorbing molecules) or by radiochemical detection. The bands of interest are scraped off and the metabolites extracted with a polar solvent such as methanol. Two metabolites of the acetylcholinesterase (ACE) inhibitor, cilazapril, were not cleanly separated by reverse-phase HPLC (Figure 7.1) but had significantly different Rf values on TLC (Figure 7.2). Preparative TLC proved to be the method of choice for separating these two conjugates which were isolated from baboon urine. (Note that M3 is more mobile than M4 on both HPLC and TLC!)

The stationary phases of modern TLC plates contain a binder that makes them hard and resistant to abrasion. It also means that it is relatively difficult to elute the metabolites once the zones of interest have been scraped off the plates. If methanol is used for this purpose, some of the silica will be dissolved and so it will be necessary to evaporate off the solvent and re-extract the compounds of interest with a less polar solvent, such as ether or 10% methanol in ethyl acetate. Preparative TLC is

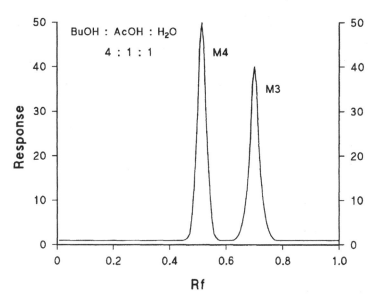

Figure 7.2 Separation of hippuric acid and phenacetylglutamine by TLC (for structures see Figure 7.5).

therefore more suitable for relatively non-polar compounds which can be extracted more easily from the silica. Elution of metabolites from the silica tends to add impurities from the plates and reverse-phase HPLC is usually more appropriate for the final purification step.

7.2.2 Metabolic profiling

The other important use of TLC is as an analytical method for the profiling of metabolites. Samples of biological material containing a radiolabelled drug and its metabolites are applied to a TLC plate which is developed with an appropriate solvent. The spots are visualised by autoradiography (by placing the plates in contact with a suitable X-ray film and leaving it for between a few days and a few weeks before development and fixing) and their relative quantities determined using a suitable radiochemical detector, such as a linear analyser.

It is important to use autoradiography as well as a linear analyser because the resolution and sensitivity are so much better and because the two-dimensional picture can often distinguish between spots that appear to be single with a one-dimensional view. Two-dimensional radiochemical scanners can be used instead of autoradiography, but the lower resolution obtainable is a major disadvantage.

TLC with radiochemical detection has two distinct analytical uses. When carrying out a complex metabolic investigation, it is generally necessary to analyse each separation/purification step to determine the efficiency of the stage. As reverse-phase HPLC is most often used for the actual separation, it is usually much better to apply a technique like TLC, which uses quite different physico-chemical properties to separate compounds, for the analysis. Figure 7.3 shows the initial gradient HPLC separation of the metabolites of Ro 31-6930 into eight fractions. Although M4 appeared to

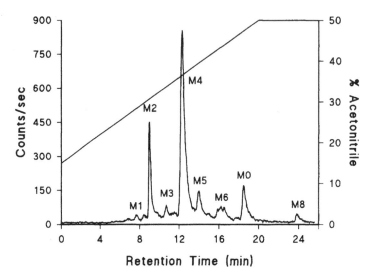

Figure 7.3 Separation of the urinary metabolites of Ro 31-6930 by gradient HPLC on a C18 column at pH 3 (0.1% AcOH).

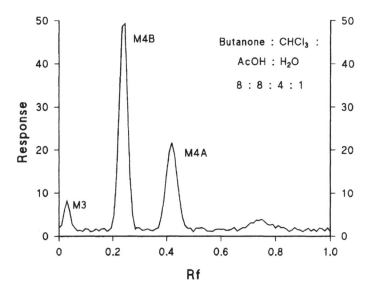

Figure 7.4 Separation of metabolites 4A and 4B of Ro 31-6930 by TLC.

consist of a single component, TLC analysis clearly showed that it was composed of two major metabolites (Figure 7.4).

The second analytical application of TLC is for the comparison of metabolic profiles in biological material (e.g. urinary profiles of different species). About 10 samples can be applied to a standard TLC plate and the analysis of all 10 carried out in parallel. As all the chromatograms can be visualised on the same plate, comparison between samples is simplified. HPLC analysis, on the other hand, has to be carried

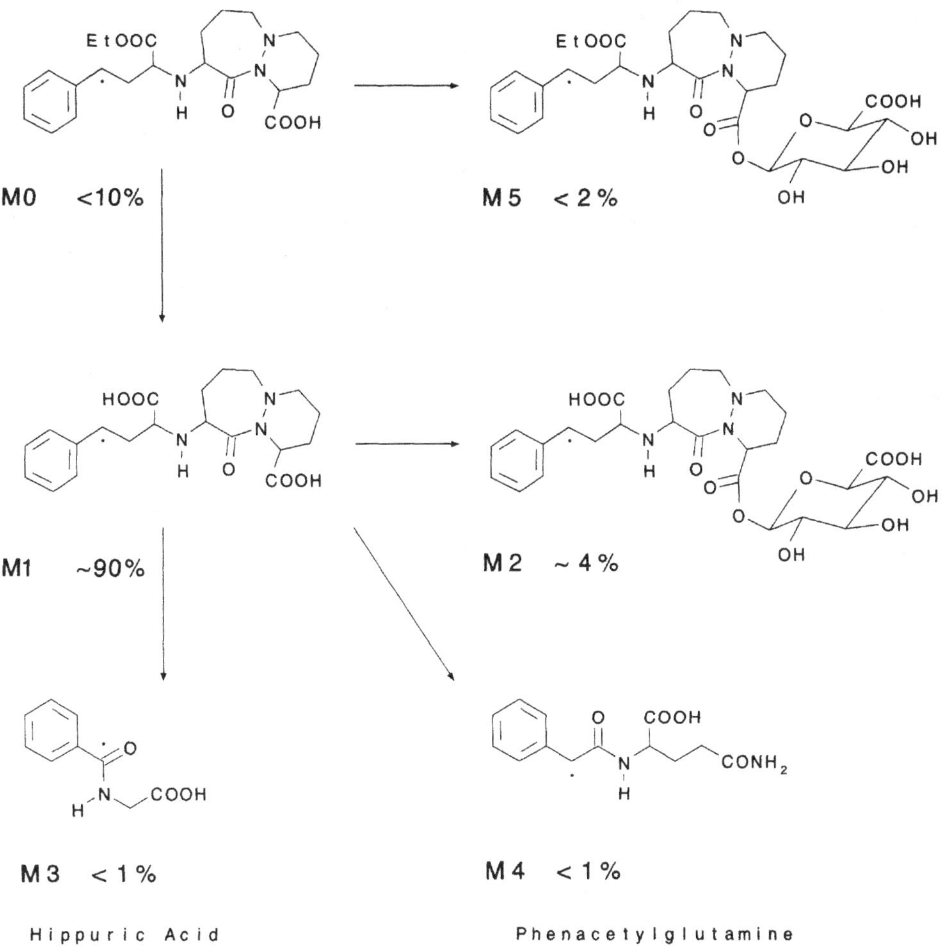

Figure 7.5 The metabolism of cilazapril.

out sequentially. Radiochemical detection of ^{14}C-labelled spots on TLC plates is also considerably more sensitive than radio-HPLC because it is possible to count the disintegrations for hours (or weeks with autoradiography) rather than seconds. The detection of tritium is more difficult because of the low energy of the electrons emitted. Enhancers can be used which will increase the sensitivity of detection of this isotope on TLC plates, but HPLC with homogeneous scintillants may be preferable.

Profiling of complex metabolic patterns (e.g. that of cilazapril, Figure 7.5) can be simplified by 'two-stage' chromatography. A preliminary separation of the metabolites can be performed using, for example, solid-phase extraction with batchwise elution (e.g. 2.5%, 5%, 10%, 20%, 50% acetonitrile in water) of the components of interest. Each fraction is then analysed by TLC with radiochemical detection. The relative recoveries of radioactivity in the fractions are determined by liquid scintillation counting of aliquots and the relative abundance of each metabolite in the fractions by use of a linear analyser (Figures 7.6, 7.7).

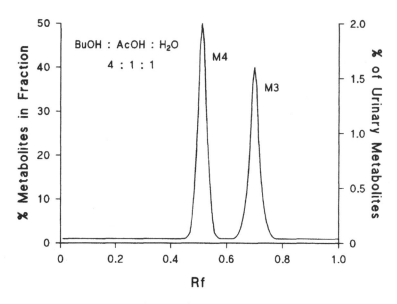

Figure 7.6 Separation of metabolites of cilazapril in the 2.5% MeCN fraction.

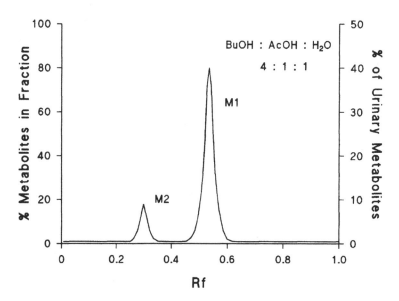

Figure 7.7 Separation of metabolites of cilazapril in the 20% MeCN fraction.

This approach is particularly effective when the low-abundance metabolites can be separated from the major radioactive components by solid-phase extraction. In the above example, metabolites M1 and M4 have similar Rf values (0.55 and 0.50 respectively) and the large spot of M1 (80% of total metabolite abundance) would have swamped the small one of M4 (2%) if a preliminary separation step had not been used (Figure 7.8).

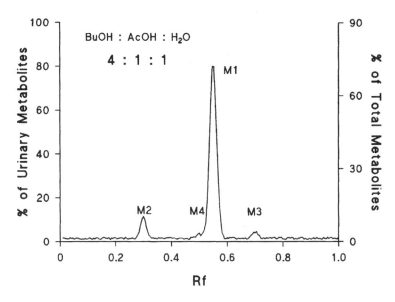

Figure 7.8   TLC separation of metabolites of cilazapril.

In an alternative 'two-stage' analytical technique, the sample is applied to the corner of a TLC plate and developed first in one direction and then, using a different solvent, at right angles (Figure 7.9). The classical two-dimensional analysis which has been used for peptides on a paper support uses chromatography in one direction and electrophoresis in the other. The quantitative determination of the amount of radio-activity in each spot will require either location by autoradiography and the scraping off of each metabolite followed by liquid scintillation counting, or the use of a sophisticated two-dimensional radio-scanner.

Whichever method is used for 'two-dimensional' profiling, it is important to use different techniques that make use of different physico-chemical properties of the analytes for the different stages of the analysis.

7.2.3   'Rules of thumb'

One might expect the order of elution of a number of metabolites on a reverse-phase HPLC column to be the opposite of their Rf values on normal-phase TLC. Although this is often the case, there are exceptions.

Small molecules tend to be relatively less polar on TLC.

The smaller the molecule, the less likely is it to exhibit large dispersive forces (which require a correspondingly large non-polar region in the molecule) with reverse-phase material and the more important the polar interactions with solvent and stationary phase. Small molecules therefore appear to be relatively more polar on reverse-phase material. Thus the two 'small' metabolites (M3 and M4, Figure 7.5) of cilazapril were removed from a C18 Bond-Elut with 2.5% acetonitrile in water,

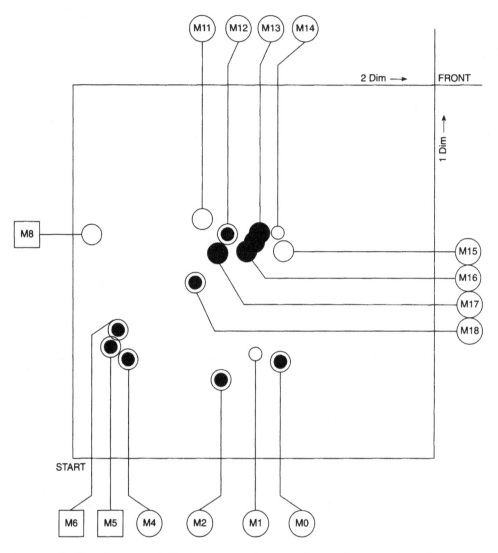

Figure 7.9 Two-dimensional thin-layer autoradiogram of the basic metabolites of Ro 11-1163. Solvent for dimension 1: *n*-butanol : acetic acid : water, 4 : 1 : 1; solvent for dimension 2: dioxan : water : ammonia, 90 : 7.5 : 1.5.

whereas the 'large' metabolites required 20% of the organic solvent – and yet the Rf values of M1 and M4 were similar on TLC (Figures 7.6–7.8).

Conjugates are relatively polar on TLC.

The presence of large numbers of silanol groups in normal-phase silica means that interactions with polar compounds are particularly strong. As a result, conjugates such as glucuronides are difficult to move from the baseline. Under acidic conditions, the acyl glucuronide of a non-polar organic acid will often behave similarly on reverse-phase

156

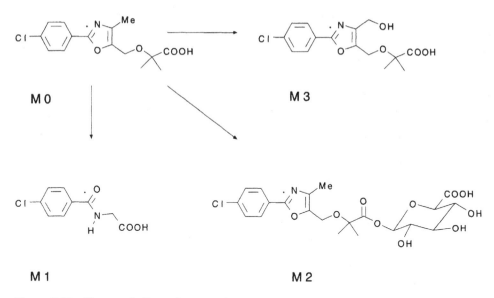

M 0

M 3

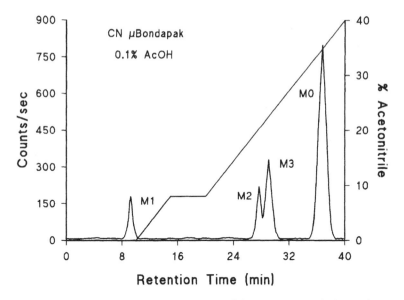

M 1

M 2

Figure 7.10   The metabolism of romazarit.

CN μBondapak

0.1% AcOH

M1

M2

M3

M0

Retention Time (min)

Figure 7.11   Gradient HPLC separation of the urinary metabolites of romazarit.

HPLC to the parent drug. This is because the main interactions between the analytes and stationary phase are dispersive forces resulting from the 'dissolution' of the identical non-polar parts of the molecules into the organic chains attached to the silica. On normal-phase silica, the strong hydrogen bonding between the glucuronic acid moiety and the silanol groups will ensure a large separation between drug and metabolite.

Figures 7.10–7.12 illustrate these two points with reference to some of the metabolites of romazarit (Figure 7.10). The 'small' chlorohippuric acid (M1) is relatively

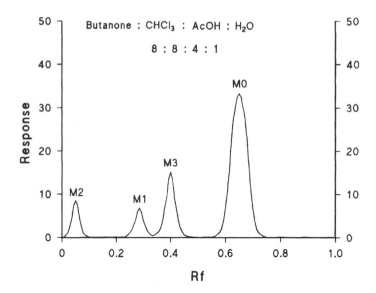

Figure 7.12 Separation of the urinary metabolites of romazarit by TLC.

polar on reverse-phase HPLC (Figure 7.11), whereas the acyl glucuronide conjugate (M2) is the most polar on TLC (Figure 7.12) although barely separable from the alcohol (M3) on HPLC.

7.3 Some recommended solvent systems

Acidic solvent systems are generally needed to separate acidic metabolites, and the converse is often true for basic compounds. Indeed, the behaviour of different metabolites on TLC in the presence of different solvent systems can give information about the nature of the compound. The following systems have proved useful in a number of metabolic investigations:

- 4 : 1 : 1 *n*-butanol : acetic acid : water
- 8 : 8 : 4 : 1 methyl ethyl ketone : chloroform : acetic acid : water
- 4 : 1 chloroform : methanol (+ drops of acetic acid)
- 80 : 20 : 3 dichloromethane : methanol : aqueous ammonia
- 4 : 1 *n*-butanol : ammonia
- 80 : 15 : 5 ethyl acetate : methanol : aqueous ammonia.

7.4 Detection of compounds on TLC plates

Although radiochemical detection and the quenching of the fluorescent indicator are likely to be the most widely used methods for the visualisation of metabolites on TLC plates, spray reagents may sometimes be helpful.

The classical method for the detection of amino acids is to spray the plate with a 1% solution of ninhydrin in acetone, followed by heating, which produces purplish-blue spots. Allowing the plate to stand for some time in the presence of iodine vapour has more general application; compounds with readily oxidisable groups, such as amines, form brown spots. Similar compounds can be detected, following oxidation by chlorine gas, with the starch/potassium iodide reagent which gives blue spots.

Detailed lists of spray reagents are contained in standard texts on TLC (see below).

7.5 Bibliography

Kirchner, J.G. (1967) Thin-layer chromatography, in Weissberger, A. (ed.) *Techniques of Organic Chemistry*, Vol. XII, New York: Interscience.

Stahl, E. (1965) *Thin-Layer Chromatography*, Berlin: Springer-Verlag.

8

CAPILLARY ELECTROPHORESIS: AN INTRODUCTION

Peter Andrew

8.1 Introduction

Capillary electrophoresis (CE) is a new analytical technique. Because much of the instrumentation is based on chromatographic equipment, a very comprehensive kit has become available in a very short time, providing chromatogram-like output of needle-shaped peaks with an apparently amazing sensitivity. This chapter aims to provide an insight into how and why it works and to indicate its usefulness in bioanalysis.

Electrophoresis and several allied techniques have been in use as qualitative tools for many years, serving rather specialised interests associated with peptides, proteins and other charged macromolecules. The format of the technique, dictated by the need for stabilisers such as paper or polyacrylamide gel, is not readily adaptable to on-line sample injection and detection. Then, in 1976, an alternative procedure, involving the use of a small capillary tube as a compartment in which to carry out zone electrophoresis was first reported by Everaerts *et al*. Developments were rapid and free-solution capillary electrophoresis (FSCE) as practised today was described by Jorgenson and Lukacs in 1981. A later modification, by Terabe *et al*. (1985), described as a chromatographic process within the electrophoretic environment, involves the use of micelles as an additional mechanism to aid separation. This procedure, a modification of FSCE, is known as micellar electrokinetic capillary chromatography (MECC). The first commercial equipment appeared in 1988–89.

Capillary electrophoresis can be described as the separation of a mixture in a capillary tube, by virtue of differing ionic mobilities induced by the application of a high potential along the capillary. Sources of general information are Jorgenson and Lukacs (1983), Ewing (1989), Altria *et al*. (1990) and Perrett and Ross (1992). The two reviews by Altria *et al*. (1990) and Smith and Evans (1994a) are particularly comprehensive.

8.2 How capillary electrophoresis works

The hardware for CE is basically very simple, as may be seen from Figure 8.1. Each end of a capillary (usually made of fused silica) is immersed in its own reservoir, containing a buffer solution (electrolyte) and a high-voltage electrode. The capillary is filled with the same electrolyte as in the two reservoirs. An on-line detector is situated towards one end of the capillary.

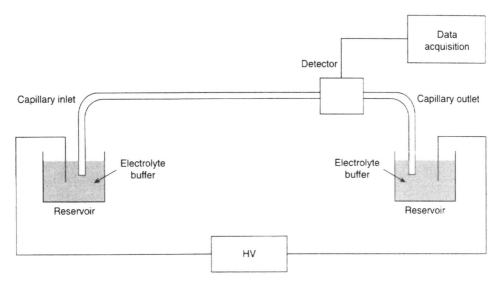

Figure 8.1 Schematic illustrating the components of capillary electrophoresis instrumentation.

A sample mixture is introduced at one end (usually the anode) and under the influence of an applied field (typically 20kV), charged species in the sample move along the capillary. The analyte migration time, i.e. the time to get from the start of the capillary to the detector, is made up of two contributions. The first arises from the migration due to the charge on the molecule – the electrophoretic mobility – and occurs because negatively charged analytes migrate towards the anode while those having positive charge move towards the cathode. The second contribution is due to the movement of the buffer solution along the capillary, induced by the high potential applied across the silica capillary. This is known as electro-osmotic flow (EOF). The flow carries with it the analytes in the mixture and augments or diminishes their rate of progress along the capillary towards the detector.

The rate of EOF varies with the pH of the buffer solution, so changing pH will obviously have an impact on migration times and hence resolution. The pH will also affect the extent of ionisation of analytes which will, in turn, affect migration time. A number of other variables can also be invoked, such as the addition of small amounts of an organic solvent, which will change the migration velocities of the analytes in the mixture and hence may improve the separation. As the separate bands move along the capillary, they will eventually reach the detector, producing 'chromatogram-like' peaks which can be quantified in the usual way. In CE parlance the output from the detector is referred to as an electropherogram.

The addition of a suitable surfactant, causing micelle formation, can be used to modify both the rate of movement of uncharged species (which otherwise depend solely on EOF to reach the detector and as a consequence undergo no separation) and to provide a further extremely versatile separation mechanism (MECC). Why these phenomena work in this way will be explored below.

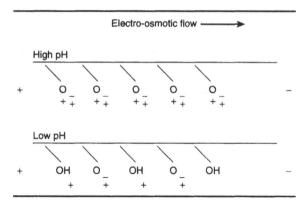

Figure 8.2 Ionisation at the capillary wall at high and low pH.

8.3 Why capillary electrophoresis works

8.3.1 Electro-osmotic flow

EOF is a fundamental process which drives CE. It is a consequence of the charge on the surface of the silica capillary arising through ionisation of silanol groups in contact with the buffers in the electrolyte. The degree of ionisation is a function of pH, being minimal at about pH 2 and maximal at about pH 9. The negatively charged capillary wall will attract positively charged species from solution, producing an electrical double layer (Figure 8.2). When a potential is applied, it causes this annulus of positive charge to move towards the cathode, with the effect that the whole of the liquid bulk eventually moves with it. For a typical 50μm i.d. capillary at pH 9 filled with 20mM borate, linear velocity is around 2mm/s, equivalent to a volume flow of 4nl/s. At pH 3, the EOF is much lower, about 0.5nl/s. EOF is usually greater than the electrophoretic mobility of the analyte so that, regardless of charge, sooner or later the analyte will always arrive at the detector.

The outstanding feature of EOF is the nature of the flow profile, illustrated in Figure 8.3. In HPLC, flow down the column is characterised by a parabolic profile, so there is great variation in linear velocity of analyte molecules within each band. This is a major cause of band broadening. In contrast, the electrically driven EOF has a flat flow profile, with virtually zero contribution to band broadening. As a consequence, bandwidths in CE are extremely narrow, providing efficiencies which, judged in chromatography terms, are far superior to those of HPLC and GC.

8.3.2 Free-solution capillary electrophoresis

This is CE in its simplest form. Analytes migrate at rates proportional to their charge to mass ratio, so if they are all singly charged, separation will be based on size. For a mixture of a drug and its metabolites, it may not be easy to predict relative sizes, so one falls back to the usual situation of trial and error. Figure 8.4 illustrates the way the separation develops.

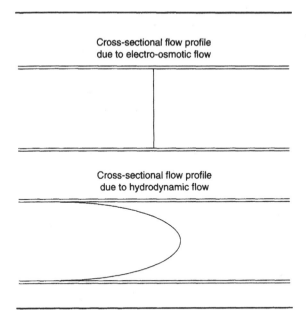

Figure 8.3 Flow profiles in the capillary.

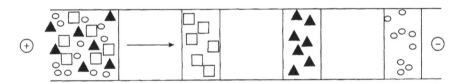

Figure 8.4 Illustration of separation development.

The experimental variables which may be used to achieve a separation are described below. Considerably more detail may be found in the general papers listed in the bibliography.

Applied voltage

The greater the applied field strength, the faster the rate of migration. However, there is a practical limit to the voltage that can be applied, due to heat generation in the capillary (Joule heating) which must be dissipated quickly to avoid a changing environment in the capillary. Up to 30kV is typical, with an upper practical limit of about 50kV.

Choice of buffer and ionic strength

A wide variety of buffers and buffer mixtures has been used in CE. Phosphate and borate buffers are commonly used in 20–100mM concentrations. The concentration affects the rate of EOF because increasing ionic concentration in the electrolyte reduces

163

the extent of the electrical double layer and hence its capacity to drive the flow. Also high buffer concentrations increase Joule heating.

Choice of pH

This is a very important parameter because pH affects the extent of ionisation of the analyte in solution and hence the rate of migration. Total ionic suppression of an analyte will cause migration to depend solely on EOF as its means of transport. In this situation, there is no separative mechanism at work, so a mixture of neutral analytes would arrive at the detector together.

Addition of organic modifiers

In broad terms, organic modifiers (e.g. methanol, isopropanol) reduce EOF by diluting the ionic strength of the electrolyte. Increasing the amount of organic modifier will increase migration time to the detector, so the peak will take longer to appear. Note, however, that acetonitrile is odd in that its effect on retention time is the reverse of the above, i.e. migration time is reduced so the compound appears sooner. The reason for the anomalous behaviour of acetonitrile is not fully understood. Inclusion of small amounts of organic modifiers in MECC (see below) produce much greater effects on the retention times than those observed in FSCE.

A different type of organic modifier is exemplified by the cyclodextrins – carbohydrates that have open, cage-like structures. They may be added to the electrolyte to effect separation of enantiomers. The cyclodextrin interacts with each enantiomer differently, by virtue of the different spatial orientation of the chiral centres, with the consequence that one enantiomer is complexed by the cyclodextrin to a greater extent than the other, leading to differences in their electrophoretic mobility, and hence separation. A range of cyclodextrin additives has been used with subtly different molecular dimensions. By careful selection of the cyclodextrin, separations of a wide variety of enantiomeric types can be achieved. The success of many chiral separations in CE (including those that have been unsuccessful using other chromatographic techniques) has led to considerable coverage, particularly with modified cyclodextrins (reviewed by Guttman 1995; Guttman *et al.* 1996).

8.3.3 Micellar electrokinetic capillary chromatography

In simple terms, this is FSCE with a small amount of a surfactant included in the electrolyte. However, careful consideration of the role that surfactants have indicates that an additional and far-reaching mechanism occurs with these additives.

An anionic surfactant (sodium dodecyl sulphate (SDS) is a commonly used example), when added to the electrolyte in amounts above a certain concentration (the critical micelle concentration), aggregates to form micelles – roughly spherical structures with the hydrophobic tail groups towards the centre and the charged groups along the outer surface (Figure 8.5). The electrolyte can then be considered as being composed of two phases: aqueous and micellar. The surfaces of the SDS micelles have a high negative charge, giving them a large electrophoretic mobility towards the anode. However, in most circumstances EOF is greater, so the net migration of the

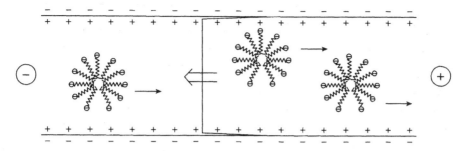

Figure 8.5 Schematic illustrating the relative movement of micelles relative to EOF.

micelle is towards the cathode, i.e. towards the detector. The result is a faster-moving aqueous phase within which is the slower-moving micellar phase (Figure 8.5).

Analytes can partition between the two phases, resulting in migration times dependent on differential solubilisation by the micelles. The MECC system was originally developed for non-ionic analyte mixtures which are not separated in FSCE. Retention of such compounds is based on hydrophobic interactions; the more hydrophobic the analyte, the greater will be the interaction with the micelle and the slower will be its progress towards the detector. Addition of organic modifiers to the mobile phase decreases the extent of partition of non-polar or hydrophobic materials into the micelle phase, increasing analyte mobility (i.e. they reach the detector sooner) thus providing a further aid to selectivity. The technique can also provide enhanced selectivity for ionic species and many of the CE separations now appearing in the literature are based on this mechanism.

8.3.4 Electrochromatography (electrically driven HPLC)

The application of capillary electrophoresis has been extended recently to include capillaries filled with conventional HPLC packings (Tsuda 1992; Smith and Evans 1994b), typically 3μm silica-based material such as ODS. The term electrochromatography has been used to describe this new technique. Two variants may be used, depending upon whether the mobile-phase flow is created by external pressure (as in conventional HPLC) or by endo-osmotic flow alone.

Pressure-driven electrochromatography uses a conventional HPLC pump to drive the flow. Electro-osmotic flow will still occur under the influence of the applied field and, furthermore, the surface of the silica packing will also be modified by an induced charge. Thus, pressure-driven electrochromatography achieves a separation through the usual HPLC mechanisms with the additional 'dimension' of differing electrophoretic mobilities of the analytes. However, the benefit of plug flow (which is largely responsible for the very high efficiencies achieved in CE) will be lost because the movement of the supporting buffer through the packed bed produces the hyperbolic flow profile characteristic of conventional HPLC. Selectivity is enhanced but efficiency remains typical of HPLC.

On the other hand, electrochromatography with endo-osmotic flow retains the very high efficiencies associated with CE (the square flow profile becomes slightly convex in a packed capillary but the effect is inconsequential). In general, retention tends to be somewhat slower than in CE and to adjust for this, higher field strengths are used. In turn, this may cause bubble formation at the electrodes with consequent disturbance of the detector. Application of pressure, sufficient to prevent bubble formation, is the usual remedy, so often this 'mixed' approach provides the best result.

Although this new form of chromatography is attracting widespread interest, its general utility is only just starting to emerge.

8.4   CE hardware

The equipment required to carry out capillary electrophoresis is basically quite simple and the earliest instruments were easily constructed in the laboratory. Figure 8.1 illustrates the components required to perform capillary electrophoresis.

Commercial developments appeared quite quickly in the late eighties, against a background of sophisticated, highly automated LC and GC equipment, with the result that even the earliest kit provided fully controlled autoinjection, capillary rinsing, voltage gradients, a choice of detectors and the usual outputs for integrator starts, etc. Specific aspects will be considered in turn.

8.4.1   The capillary

The fused silica capillary is analogous to the column in chromatography. It is manufactured in the same way as GC fused silica capillary tubing and comes with the usual external polyimide coating. With the exception of occasional specific applications, see below, the inner surface is untreated. The length is about 50cm, depending where the detector is situated. The capillary is located in a thermostated zone, as a stable temperature is essential for reproducible migration times. As may be expected, the capillaries in use are circular in diameter, but rectangular and oval profiles have been advocated to improve sensitivity (Section 8.4.3).

The capillary is filled with the electrolyte, a cocktail of ions and other compounds in solution, which impact selectively on the analytes as they travel towards the cathode. Internally coated silica is now available. C8 and C18 hydrophobic coatings, bonded to the surface silanols, reduce and stabilise EOF (no longer influenced by pH) and minimise adsorption of such materials as proteins.

8.4.2   Sample introduction

Sample introduction requires that a reproducible volume of sample solution is somehow introduced into the capillary at the opposite end from that of the detector. Thus the filled capillary must be removed from the buffer reservoir and dipped into the sample solution. Thereafter there are a number of ways that an aliquot of the solution can be 'persuaded' to enter the capillary. Three ways have been used on commercial instruments, each will be briefly described. Typical injection volumes are 10–40nl.

Gravimetric (syphonic)

The earliest instruments used this procedure. The sample solution, with the capillary dipped in it, is raised above the level of the other reservoir (at the detector end).

Gravity causes sample solution to flow in as electrolyte solution siphons out at the other end of the capillary. Accurate elevation and timing provide reasonable reproducibility. This method is now largely superseded.

Electrokinetic

After the capillary is dipped into the sample solution, a potential is applied across the capillary (typically 5kV) for about 5–10s, which causes charged species to migrate into the capillary. A significant contribution will also be made by EOF which will be representative of the bulk sample constituents. By operating at low pH, which suppresses EOF, this method may selectively sample only charged species, with amounts of each depending on their relative electrophoretic mobility. The bulk distribution remaining in the sample is insufficiently disturbed to see differences occurring upon repeat injection. Once again good reproducibility is possible but samples and calibrants must be matrix matched for satisfactory quantification. Having set the applied electroinjection potential, the operator specifies the time it is applied in order to set the injection volume.

Differential pressure

By applying either pressure or vacuum to the sample surface or receiving reservoir, flow may be induced into the capillary from the sample solution. This is the principal method used at present in such instruments as Beckman P/ACE and ABI 270A (all three of the methods are available on the Dionex CES1). Once again the operator sets the injection volume as a time, which is the time that the differential pressure is applied to the sample solution.

In all these processes, the amount of sample introduced into the capillary is very small. To increase the amount of analyte by simply lengthening the time of injection would not be practical as the band containing the analyte ions would be too wide and would severely compromise efficiency. However, using a technique described as **field amplification** (Chien and Burgi 1992) greater amounts of the analyte ions can be introduced into the capillary.

If sample is introduced hydrostatically into the column in a buffer solution that is of the same constitution as the support buffer but is more dilute, the induced electric field strength in the more dilute solution will be greater. The electric field strength, induced by the applied potential, is what causes the ions to move through the column, their electrophoretic mobility being proportional to the field strength. Thus the analyte ions will move faster in the introduced aqueous region than in the support buffer. Once in the support buffer, they move very much more slowly and in effect 'stack up' to give a narrow band at the buffer boundary. This process of ion concentration is limited because of the inevitable difference in endo-osmotic flow induced in the two regions by the electric field. At the concentration boundary, these differences generate a laminar flow which starts to disrupt the ion band, leading to band broadening. Thus, stacking and band broadening work against each other. Hence there is an optimal plug length of the dilute buffer that can be used to maximise sample loading.

This principle can be applied to electrokinetic injection of the analyte if it is presented to the capillary in a more dilute buffer solution than the support buffer filling the capillary. The resultant enhanced electric field at the point of entry into the capillary causes rapid migration of sample ions into the capillary. Application of field amplification can result in 10- to 100-fold enhancement of the detector signal. The technique is comprehensively described by Chien and Burgi (1992).

Once the sample aliquot is introduced, the capillary is removed from the sample solution and re-immersed into the electrolyte reservoir. The potential is applied and analytes begin their journey at a rate consistent with their charge and mass. Before considering how we detect each analyte band, there remains one feature of CE sample introduction that sets it apart from analogous chromatographic techniques. This is the cleaning of the capillary between 'injections'.

Like HPLC or GC, retention time or more correctly in CE, migration time, is very important because it is the parameter that one almost always uses to identify the peak of interest. EOF plays a key role in migration time and this, in turn, is dependent on the nature of the capillary wall. If the capillary surface becomes fouled (as it often can become when proteinaceous material is injected), the charge on the wall is reduced and hence EOF slows down. Also, adsorbed material can act as a secondary adsorption medium and compromise electropherographic quality.

This situation is easily remedied in CE by flushing the capillary between injections with dilute sodium hydroxide solution, which strips off the top surface (the amount of silica removed is very small and the operation can be repeated many times without detriment to the capillary). The capillary is then rinsed and refilled with buffer, ready for the next injection. All this is done automatically, taking a couple of minutes only.

8.4.3 Detectors in CE

The first detector in CE was a conductivity detector, which is still used for special applications involving quantification of simple inorganic ions such as chloride or sulphate. The predominant detector, based on HPLC technology, is the UV detector. Fluorescence detection is also available from certain manufacturers, notable particularly in that laser-induced fluorescence is an option.

There is not a separate detector cell in CE, the capillary dimensions and band volume make this impractical. Instead, the bands are detected within the capillary itself. With UV and fluorescence detectors, a small 'window' is created in the translucent silica wall by removing the polyimide coating. Bands are detected as they pass this window.

UV detector

Essentially this is the same detector as used in HPLC. Microfocusing or fibre optics ensure good sensitivity despite the apparent difficulty of a 50μm pathlength. All the usual features are available, including full-spectrum scanning with diode-array based systems. The electropherogram looks just like an HPLC or GC trace, so routine integration may be employed. The ultra-short pathlength across the capillary means that low (e.g. 200nm) UV wavelengths may be used, boosting sensitivity, despite being close to the optical cutoff of the electrolyte.

Fluorescence detector

Not as widely available, this detector offers much improved sensitivity when appropriate analytes are examined. It has been used extensively for amino acids for which there are established fluorescent-labelling procedures. Recently a laser-induced fluorescence monitor has been introduced which offers considerably more sensitivity. However, the spectral output may not favour all applications and indirect fluorescence techniques may need to be employed.

Electrochemical detection

This has been reported in the literature (Wallingford and Ewing 1988) and is notable in two respects. The first is that the detector is 'off-column', being situated in a second length of capillary coupled to the first by a piece of porous glass capillary. The joint is immersed in the second reservoir along with the ground electrode and is thus electrically conducting. The EOF generated in the first capillary continues to flow through the second capillary, bringing the analytes to the detector. The second notable feature is the micromanipulation involved in inserting the carbon-fibre electrode, only 10μm diameter, into the 50μm capillary! Electrochemical detectors are not yet commercially available.

Mass spectrometry

Finally, the ubiquitous mass-spectrometer is featured again, not simply as a detector for CE but also because CE provides a convenient way of sample introduction into the MS.

8.4.4 Sensitivity in CE

In absolute terms, the amount that is detectected as a typical UV-detected peak will comprise only a few picograms of material, which when compared to that comprising the usual HPLC peak may seem fantastic. However, the actual volume of the band, usually 10–50nl, means that in concentration terms, CE still has some way to go to routinely match useable HPLC detection limits. At present, it should be possible to routinely measure, by CE, concentrations of the order of 100ng/ml and above, provided the compound has a reasonable chromophore. Much development work is being carried out by manufacturers to improve sensitivity; the development of the UV Z-cell to increase path length is one example.

8.5 Use in bioanalysis

The aqueous environment of CE and the ability to 'start afresh' after each sample injection, by virtue of the rinsing and refilling of the capillary, would appear to make CE an ideal tool for bioanalysis However, despite these apparent advantages, only a few of the many CE papers have dealt with bioanalytical applications. More are now starting to appear. There is little to be gained by simply using CE as a final stage after a plasma extraction procedure, rather than an HPLC finish, as the technique seems less robust at present, less sensitive in concentration terms and subject to variable interferences despite an apparently satisfactory (in LC terms) extraction clean-up.

Successful applications include electrophoresis of plasma, urine and bile, where concentrations are relatively high (10μg/ml plus). Plasma gives the 'worst' electropherogram with large, broad protein peaks obscuring the latter part of the electropherogram. Urine gives the 'busiest' electropherogram, usually containing many peaks but of narrow width, leaving sufficient clear baseline for peaks of interest. Judicious use of MECC may be used to retain endogenous peaks (i.e. reduce their electrophoretic mobility) until the peak of interest has been eluted.

8.6 Bibliography

Altria, K., Rogan, M. and Finlay, G. (1990) Electrokinetic separation techniques – A review. *Chromatography and Analysis*, August.

Chien, R. and Burgi, S. (1992) On-column sample concentration using field amplification in CZE. *Anal. Chem.*, 64, 489–496.

Ewing, A.G. (1989) Capillary electrophoresis. *Anal. Chem.*, 61, 292–303.

Guttman, A. (1995) Novel separation schemes for capillary electrophoresis of enantiomers. *Electrophoresis*, 16, 1900–1905.

Guttman, A., Brunet, S. and Cooke, N. (1996) Capillary electrophoresis separation of enanttiomers using cyclodextrin array chiral analysis. *LC–GC International*, February, 88–100.

Jorgenson, J. and Lukacs, K.D. (1981) Zone electrophoresis in open-tubular glass capillaries. *Anal. Chem.*, 53, 1298–1302.

Jorgenson, J. and Lukacs, K.D. (1983) Capillary zone electrophoresis. *Science*, 222, 266–272.

Perrett, D. and Ross, G. (1992) Capillary electrophoresis: powerful tool for biomedical analysis? *TRAC*, 11, 156–163.

Smith, N. and Evans, M. (1994a) Capillary zone electrophoresis in pharmaceutical and biomedical analysis. *J. Pharmaceut. Biomedical Anal.*, 12, 579–611.

Smith, N. and Evans, M. (1994b) Analysis of pharmaceutical compounds using electrochromatography. *Chromatographia*, 38, 649–657.

Terabe, S., Otsuka K. and Ando, T. (1985) Electrokinetic chromatography with micellar solutions and open-tubular capillaries. *Anal. Chem.*, 57, 834–841.

Tsuda, T. (1992) High performance electrochromatography. *LC–GC International*, 5, 26–36.

Wallingford, R.A. and Ewing, A.G. (1988) *Anal. Chem.*, 60, 1072.

9

IMMUNOASSAY TECHNIQUES

Richard F. Venn

9.1 Introduction

Immunoassay is a powerful technique for quantitative and qualitative analysis of a wide range of analytes. These include peptides, proteins, simple organic molecules, environmental pollutants and pharmaceuticals. Immunoassay has been used since the mid-fifties as an analytical tool and has developed during that time into a mature analytical methodology which is, however, still developing. Its advantages are sensitivity, throughput and, often, selectivity. Immunoassay is capable of very low limits of quantification and allows a very high sample throughput. Selectivity is usually good, but is dependent, however, on the analyte in question; this can limit its usefulness where there are very similar, structurally related molecules present in the matrix.

This chapter will deal briefly with the pros and cons of immunoassay, what the practical aspects are, when to consider using immunoassay and will give some practical examples of immunoassays used in the pharmaceutical industry.

9.2 Definitions

- **Antigen**: a substance capable of reacting immunospecifically with an antibody. A substance capable of triggering an immune response.
- **Hapten**: designates any substance, large or small, which does not elicit an immune response by itself but can be shown to interact with an antibody.
- **Immunogen**: a substance able to provoke an **immune response**, whatever its specificity.
- *A substance may be an antigen BUT not necessarily an immunogen.*
- **Antibody**: a large protein consisting (usually) of four peptides arranged in two pairs which has two combining sites capable of interacting specifically with an antigen.
- **Antiserum**: serum collected from an animal that has mounted an immune response to an immunogen.
- **Titre**: the dilution of antibody in an immunoassay that provides a 50% (or 33%) response.
- **Epitope**: the region of the antigen that binds to an antibody. The smallest part of a molecule that the antibody can combine with.

9.3 Theory

Immunoassays are based on the ability of antibodies to recognise and bind specific epitopes on a molecule with a high affinity. The immune system (part of an organism's defence mechanism) is capable of generating high-affinity antibodies against almost any molecular conformation. Not all molecules are large enough to stimulate the mechanism, since it requires at least two recognition sites to mount a response. It is possible to fool the system by artificially increasing the size of the molecule, by covalently coupling multiples of the compound to a large protein for example. This conjugate is then used to stimulate a response in the chosen species. The recognition/binding events are determined by physico-chemical laws, in particular the law of mass action.

9.3.1 Mass action

In an equilibrium such as that in antibody–hapten binding, the system can be described by the following equations:

$$Ab + L \Leftrightarrow AbL$$

$$k = \frac{[AbL]}{[Ab].[L]}$$

$$\text{initial association rate} = k_1[L].[Ab]$$

$$\text{initial dissociation rate} = k_2[AbL]$$

$$\text{at equilibrium, } k_1[L].[Ab] = k_2[AbL]$$

$$\frac{[AbL]}{[L].[Ab]} = \frac{k_1}{k_2} = K_A$$

$$K_D = \frac{[L].[Ab]}{[AbL]} = \frac{k_2}{k_1}$$

where K_D (molar) is the equilibrium dissociation constant.

The ligand, L, is the small molecule which is recognised by the biological receptor, in this case, and antibody. If we include a labelled ligand, L*, which has an *identical affinity* for the antibody, the law of mass action demands that *at equilibrium*, the proportion of label bound to the antibody with respect to the unlabelled ligand will be exactly equal to the ratio of labelled ligand to unlabelled ligand. Thus, if there is no unlabelled ligand, all the bound ligand will be labelled: if the concentration of unlabelled ligand equals that of the labelled ligand, then 50% of bound ligand will be labelled, and so on. This is illustrated in Table 9.1 and Figure 9.1. Note the logarithmic nature of the x-axis and that such competitive assays produce sigmoidal calibration lines, not linear ones. The useable area of the curve is the centre (almost linear) part over about two orders of magnitude.

Thus, if we set up an equilibrium between antibody and analyte, including a constant, small concentration of labelled analyte in every assay tube, incubate, then

Table 9.1 Theoretical binding data for a competitive immunoassay in which the labelled analyte concentration is at 0.05nM and the analyte concentration added (to construct the calibration curve) increases from 0 to 200nM

Label (nM)	Analyte (nM)	Total (nM)	Label/total	Bound	Label bound	Counts
0.05	0	0.05	1.000	0.980	0.9804	10204
0.05	0.01	0.06	0.833	0.984	0.8197	8597
0.05	0.02	0.07	0.714	0.986	0.7042	7442
0.05	0.05	0.1	0.500	0.990	0.4950	5350
0.05	0.1	0.15	0.333	0.993	0.3311	3711
0.05	0.2	0.25	0.200	0.996	0.1992	2392
0.05	0.5	0.55	0.091	0.998	0.0907	1307
0.05	1	1.05	0.048	0.999	0.0476	876
0.05	2	2.05	0.024	1.000	0.0244	644
0.05	5	5.05	0.010	1.000	0.0099	499
0.05	10	10.05	0.005	1.000	0.0050	450
0.05	20	20.05	0.002	1.000	0.0025	425
0.05	50	50.05	0.001	1.000	0.0010	410
0.05	100	100.05	0.0005	1.000	0.0005	405
0.05	200	200.05	0.0002	1.000	0.0002	402

'Total' is the sum of the label and added analyte concentrations, 'bound' is the fraction of antibody binding sites occupied by the analyte and 'label bound' is the fraction of label added to the assay bound to the antibody sites. 'Counts' is arbitrarily arrived at by assuming a non-specific binding of 400 and a total count of 10 000 from a label concentration of 0.05nM.

separate bound analyte from free and determine bound (or free; or both) a calibration curve can be generated: the more unlabelled (unknown) analyte present, the less label will be bound. For this determination we require a sensitive means of quantifying bound ligand – usually high specific activity radiolabel. We also require a means of separating bound analyte from free, unbound material. This is usually done on the basis that antibodies are large molecules (140 000Da) and the analytes we are interested in are small (< 1000Da, usually).

In order to develop an immunoassay, an equilibrium must be set up between antibody binding sites and the analyte. It is this equilibrium that is at the heart of all immunoassays. In competitive assays, a labelled analyte competes with unlabelled (standard, quality control or unknown) analyte for a limited number of antibody binding sites. In non-competitive assays, the antibody binding sites are non-limiting.

9.3.2 Competitive assays

A low, fixed concentration of labelled analyte or analogue is incubated with the antibody and known (for the calibration curve) or unknown concentrations of analyte. Unlabelled analyte competes with the labelled analyte for the binding site of the antibody. Thus, at high concentrations of unlabelled analyte, little labelled material remains bound to the antibody, and vice versa. The incubation is allowed to equilibrate and then bound and free separated. Either the bound or free labelled analyte is determined in the appropriate instrument and a calibration graph constructed of bound (B) versus analyte concentration.

(a)

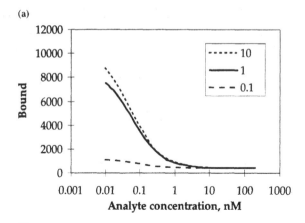

(b)

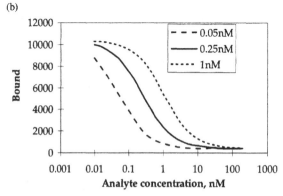

Figure 9.1 Theoretical calibration curves for a competitive immunoassay derived from data similar to that in Table 9.1: (a) showing the effect of a change in affinity of the antibody over two orders of magnitude; and (b) showing the effect of altering the concentration of labelled analyte on the sensitivity of the assay.

9.3.3 Non-competitive assays

If the molecule of interest is large enough to contain two epitopes that do not overlap and are physically removed by a minimum distance, then a two-site immunoassay can be set up. This requires two antibodies, one for each epitope. One antibody is bound to a physical support (e.g. a 96-well plate, latex beads, etc.). The analyte-containing matrix is introduced with buffer, allowed to incubate for a period and this is then followed by addition of the second (labelled) antibody. A sandwich is thus built up and, if equilibrium is reached and there are sufficient reagent concentrations, a straight-line calibration curve should be achieved over many orders of magnitude. This non-competitive immunoassay method is preferable for many reasons over competitive immunoassays, but is only possible currently with large (generally protein) molecules. The advantages are the large, linear calibration range as well as an increase in sensitivity. In the non-competitive immunoassay, for every analyte molecule present, one labelled antibody will bind (ideally), allowing extremely low limits of detection if, for example, fluorescent labels are used.

9.4 Requirements for immunoassay

9.4.1 Antibody

For small molecules it is necessary to couple the molecule or an analogue to a larger molecule, such as keyhole limpet haemocyanin (red blood cells and beads have also been used), in order to make them immunogenic. The antisera produced will also therefore contain antibodies against the carrier moiety. This conjugate is then introduced to the animal with an adjuvant (normally oil based, and whose function is to stimulate the immunogenic response). At 2- to 4-weekly intervals the animals are given further immunisations and a small sample of blood taken for testing. These techniques give rise to polyclonal antibodies (which arise from many different immuno-competent cells and thus have many different binding sites). Monoclonal antibodies are produced by specialised techniques which use only a single cell line to produce antibodies and thus express a single binding site.

Species used to produce antisera

- *Sheep or goats*: these are large animals so large volumes of antisera can be obtained. More immunogen is required, however, than with other species, which could cause a problem. The titre finally obtained is not always as high as with other species. Some species are reportedly more 'wild' than others, increasing the chances of raising good antibodies. Sheep may not be more expensive to buy and maintain than rabbits.
- *Rabbits*: these are commonly used, small, easily handled and not too expensive to buy and maintain.
- *Guinea-pigs*: these are rather uncommon for immunoassay due to limited blood volumes available, but are used when rabbits do not respond.
- *Mice*: these are generally too small to use for polyclonal antibodies, but are the commonest source of spleen cells for monoclonal antibody production.
- *Chickens*: these are beginning to be used as a convenient source of egg IgG (sometimes referred to as IgY). Being removed phylogenetically from the mammalian species above, they often mount a response to otherwise poorly or non-immunogenic conjugates.

9.4.2 Label

The label has to be of high-enough quality, both in purity and specific activity, to enable minute quantities of label to be quantified: K_D may be 0.01–10nM so the need is to be able to determine binding from less than 0.01nM in (say) 1ml. This is equivalent to 0.1pmol.

Radiolabels

For radiolabels (e.g. ^{3}H) a good specific activity would be 50Ci/mmol which is equivalent to 50μCi/nmol or 50nCi/pmol; thus 0.1pmol would be 5nCi, giving $5 \times 10^{-3} \times 2.22 \times 10^6$ (= 11 100) disintegrations per minute (dpm). Binding depends on antibody and label concentration, and the maximum will normally be 50% of total, leaving 5550dpm. This is equivalent to 2500 counts per minute (cpm) if 50% tritium efficiency is achieved (and it rarely is). This will decrease as the analyte concentration increases. Thus, long counting times (in the order of several minutes per sample) are necessary to achieve good precision. These long counting times can severely limit the throughput of an immunoassay.

Radiolabels are usually made by custom synthesis, for example at Amersham International or NEN. For greatest sensitivity, 2–3 tritium atoms should be incorporated. This can be very expensive if complete synthesis is required. If tritium exchange is possible, the cost is much reduced. ^{14}C is not usually an appropriate label due to its low specific activity. ^{125}I can be used but requires frequent relabelling due to its very short half-life (60 days). It also requires an easy site for labelling (usually tyrosine on peptides). However, due to its very high specific activity, shorter counting times such as 1min can often be used.

DISADVANTAGES OF RADIOLABELS

- Health and safety issues. There is a requirement to maintain accurate records of all radiochemical acquisitions, disposals, etc., and this can add to the costs of an immunoassay. There is also the need for laboratory and personnel monitoring.
- Disposal is becoming almost impossible in the US, and is closely monitored in the UK. Records must be kept.
- Shelf-life; radioactive decay. This is not an issue for tritiated compounds, but for ^{125}I it can be a problem, requiring frequent re-labelling. Radiochemical molecular disintegration is a much greater problem for tritiated compounds.
- Long count times (2–10 minutes/tube) are often necessary. When large numbers of samples are being assayed, this can cause a long delay and become a rate-limiting factor in sample throughput.

Enzymes

Enzymes can be coupled to the hapten to allow amplification of the signal by production of a coloured substrate. These are popular systems, particularly for non-competitive assays. However, dynamic range is limited by Beer's law (Chapter 5) and a range of 0–2.5 absorbance units (A) is usually the maximum attainable. Enzymes are also very bulky and can interfere with the conformation and kinetics of binding of a small analyte molecule. A further drawback of enzyme systems (ELISA: enzyme-linked immunosorbent assay) is that enzymes are very sensitive to changes in temperature, pH, etc., and this can lead to imprecision and/or inaccuracy in the assay unless conditions are very carefully controlled. These assays are usually carried out in the 96-well format. Plates are read very quickly (a few seconds per plate).

Time-resolved fluorescence

Certain chelates of the lanthanide metal ions exhibit very long-lived fluorescence under certain circumstances, and this property is used in the DELFIA system for time-resolved fluorescence assays. DELFIA (dissociation-enhanced lanthanide fluorescence immunoassay) uses an EDTA-like chelate to label a hapten. This label then competes in the normal way with the unknowns or standards. Bound lanthanide ions are then released into another fluorescent chelate (enhancement). Plates are read quickly (1 second/well).

The usual lanthanide employed is europium (Eu) as it gives the greatest sensitivity.

ADVANTAGES OF Eu-LABEL

- Disposal is not a problem. Lanthanides are non-toxic and easily disposed of. Care needs to be taken not to contaminate the workplace with high levels of lanthanides, however, if labelling is being carried out. This can lead to severe problems with high fluorescence backgrounds.
- Safe to handle. Non-toxic.
- Long shelf-life; no radioisotope decay and no self-decomposition. Labelled compounds within Pfizer have shown no sign of decomposition or degradation.
- Short count times (1 second/well). No bottlenecks. The plate fluorimeter is idle for all but 30 minutes each day!

Others

There are a number of other labelling/detection systems, in particular fluorescence polarisation assays, but these will not be dealt with here.

9.4.3 Separation

Because the immunoassay technique is driven by the law of mass action, it is very important to be aware of the equilibrium process. Anything that can disturb the equilibrium can affect the precision and accuracy of the assay. Thus removing the free label/analyte mixture from the incubation medium will inevitably disturb the equilibrium. The complex formed will immediately start dissociating, and if there is a significant time lag between the first and the last samples to be processed, there is the risk that this dissociation will affect the results. What constitutes a 'significant time lag' will depend on the strength of the binding between the antibody and the analyte. A rule of thumb (borrowed from receptor analysis) is that if separation takes place in less than $0.15 \times$ half-life of dissociation of the complex, less than 10% of the complex dissociates. Table 9.2 gives some typical data. Antibody–antigen dissociation constants are typically 10^{-9}–10^{-11}.

The most commonly used separation techniques are dextran-coated charcoal, second-antibody precipitation and immobilisation on a 96-well plate.

Dextran-coated charcoal

Finely ground charcoal is coated with dextran by stirring for 20 minutes. A small aliquot of the slurry is added to each assay vial. After mixing and centrifugation, the supernatant (containing bound analyte/label) is removed and counted. If the assay uses a ^{125}I label, the charcoal pellet can be counted since the gamma-counters used do not rely on homogeneous liquid samples as do the liquid scintillation counters used

Table 9.2 Times $(0.15t_{1/2})$ in which less than 10% of an antibody–antigen complex will dissociate

K_D (M)	$0.15t_{1/2}$
10^{-12}	1.2 days
10^{-11}	2.9 hours
10^{-10}	17.0 min
10^{-9}	1.7 min
10^{-8}	10.0 s
10^{-7}	0.10 s
10^{-6}	0.01 s

A weak antibody exhibiting a K_D of only 10^{-9}M will require very careful optimisation of separation methods in order not to introduce errors from beginning to end of the assay.

for ^{3}H or ^{14}C. Free analytes are able to pass through the dextran coating and be adsorbed by the charcoal, whereas antibody-bound analyte cannot.

Second antibody

A second antibody (e.g. goat anti-rabbit antibody) which has been raised against the species used to obtain the primary antisera is added (often with polyethylene glycol to enhance precipitation of the antibody–antibody complex) to the incubation. Antibody-bound analyte precipitates with the primary antibodies which have been complexed with the second antibody and, after incubation, the vials are centrifuged and the supernatant (containing free analyte) aspirated and counted.

Immobilisation

A second antibody (e.g. goat anti-rabbit antibody) which has been raised against the species used to obtain the primary antisera is bound onto, for example, each individual well of a 96-well plate. The incubation with primary antibody, analyte and label is carried out in the wells of the plate. Primary antibody binds to the plate-bound antibody, and analyte binds to the primary antibody. Thus at the end of the incubation, separation is effected by removal of the incubation medium and washing each well. Any antibody-bound label remains on the plate when the reagents are washed off and the label is then detected by addition of substrate (for an enzyme label), enhancement of fluorescence (DELFIA) or scintillation or gamma-counting in the case of a radiolabel.

9.5 Practical aspects

9.5.1 Preparation of hapten–carrier protein conjugates

Typical pharmaceutical compounds have a low molecular weight of less than 1000Da. Such molecules can be usually made antigenic (by conjugation to a larger protein) but are not immunogenic unless they covalently bind *in vivo* to a plasma protein, for example (this is often the mechanism for chloracne). Therefore there is a need to conjugate the hapten to a large molecular weight carrier protein to elicit an immune response (e.g. KLH, keyhole limpet haemocyanin). It is also coupled to a second carrier (e.g. BSA, bovine serum albumin) for use in detecting antibody. Haptens are covalently coupled to carrier proteins via amino or carboxyl groups using reagents such as 1-ethyl-3-(3-dimethylaminopropyl)carbodiimide hydrochloride (EDC). If suitable groups for conjugation are not present in the compound, the hapten can be added to to the required molecule via a spacer arm (e.g. 2–6 carbon chain). You will need the help of a trained and docile chemist to do this (usually). After coupling, the conjugate is then purified by either dialysis or gel filtration.

Carrier protein molecular weight:

- KLH: 4.5×10^5 to 1.3×10^7 Da.
- BSA: 68 000Da.
- Ovalbumin: 45 000Da.

9.5.2 Immunisation

Following conjugation, the concentration of the conjugate is determined, e.g. by the Coomassie Blue protein assay or by measurement of the absorbance at two UV wavelengths. This can also give information about the number of haptens coupled to each protein molecule. The optimum is 10–20 haptens/protein molecule. The conjugate is normally diluted with phosphate-buffered saline (PBS) to give a concentration of approximately 600µg/ml.

An emulsion is prepared for injection (two rabbits) consisting of 0.5ml conjugate with 1ml PBS and 1.5ml Freund's adjuvant. The adjuvant is used to stimulate the animal's immune response; initial immunisation is usually with 'complete' adjuvant and further boosts with 'incomplete' adjuvant. Each rabbit is given 1.5ml emulsion, equivalent to 150µg conjugate per rabbit.

Typical immunisation protocol

1 Pre-immunisation bleed, week 0;
2 primary injection, week 0;
3 booster injections, weeks 2 and 4;
4 first test bleed, week 5;
5 booster injections, weeks 6, 8, 10, etc.;
6 test bleeds, 1 week after each booster.

Excessive administration of the antigen may cause immunological tolerance:

- induction of the inability to form antibodies; or
- a cellular reaction towards a specific antigen.

Once a suitable titre has been achieved, a larger bleed (20ml for rabbits) may be taken, or the animal may be bled out (50ml).

9.5.3 Antibody detection

Antibody can be detected using europium-labelled protein G (a bacterial surface membrane protein which binds to the Fc region of IgG of most animal species by a non-immune mechanism). Microtitre plates are coated with hapten–BSA conjugate (0.5 or 0.05μg/well). This method does not require a specifically labelled analyte or hapten.

Procedure

1 Serial doubling dilutions of antisera are prepared;
2 sample volume = 200μl diluted sample added to wells;
3 incubated at room temperature for 90–120 minutes;
4 wells washed and 200μl Eu^{3+}-labelled protein G added to each well;
5 incubated at room temperature for 30 minutes;
6 wells washed and 200μl enhancement solution added to each well;
7 read in fluorimeter.

High fluorescence counts indicate a high concentration of antibody.

To check for specificity, perform antibody titre determinations with and without the addition of free hapten (100ng/well). If anti-hapten antibodies have been produced, reduced counts will be observed.

If ^3H-labelled analyte is available, a charcoal assay procedure can be carried out instead.

Figure 9.2 shows the work-flow for competitive immunoassays using radio-immunoassay or the DELFIA technique using europium-labelled tracer.

9.5.4 Antibody titres

Radioimmunoassays

A working dilution of label added to appropriate tubes in duplicate or triplicate to give approximately 10 000dpm/100μl added. This is then incubated with serial dilutions of antiserum (100–100 000) to determine the titre (50% binding). This is then used for subsequent assay development. If sensitivity is not an issue, then the radiolabel concentration may be increased to shorten counting times. This will be at the expense of sensitivity (Section 9.3.1).

DELFIA

Similar experiments are carried out to determine the titre. However, in this case, the optimum label concentration should also be determined for the sensitivity required. Titres obtained by DELFIA are usually greater than those obtained using radiolabelled material.

9.5.5 Calibration curves

Calibration curves are prepared in buffer and the intended matrix (e.g. human plasma) for comparison. These should initially be over a wide range of concentrations to

(a) RIA Techniques

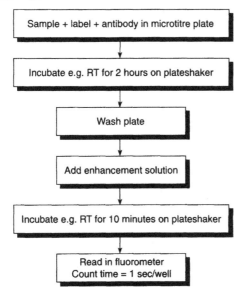

(b) DELFIA technique

Figure 9.2 Flowchart for competitive immunoassays, showing the procedure for (a) radioimmunoassay and (b) DELFIA.

determine the working range of the assay. Sensitivity can be adjusted by increasing or decreasing concentration of labelled analyte in the assay, and by manipulating antibody concentration. It is also possible to increase sensitivity by using disequilibrium techniques (see below).

If matrix effects are present, the curve will be displaced to the right, reducing sensitivity of the assay.

9.5.6 Matrix effects

These are usually seen as a decrease in binding of analyte to antibody, reducing sensitivity and giving falsely high estimates of concentration. Such effects are found particularly in urine, where pH can vary widely. The effects are sometimes worse with DELFIA systems, probably because urine contains many acidic compounds which could chelate europium. Sometimes simply diluting the samples with water can overcome matrix effects.

Protein binding can be a major problem with immunoassays for certain small pharmaceuticals. This can be reduced by:

- dilution of samples with 10% acetonitrile in buffer, which disrupts protein binding without disturbing the antibody binding;
- use of competitive displacement agents: There are at least two specific drug-binding sites on human plasma albumin. Drugs can be displaced by the use of an appropriate competitor, for example DL-tryptophan, octanoic acid, bilirubin, warfarin or salicylic acid.

9.6 Data handling

9.6.1 Standard curves

A typical standard curve from a competitive immunoassay is shown in Figure 9.3. It is sigmoidal in shape and usually has a linear concentration–binding portion of about two orders of magnitude.

9.6.2 Fitting

As a consequence of the sigmoidal nature of the standard curve, sophisticated curve-fitting and curve-interrogation software is required. Several computer software packages are available, and log-lin graph paper can also be used to plot the data by hand.

9.6.3 Precision profile

The precision profile is a graph of the calculated precision of the assay versus the concentration of the analyte in the assay. A typical profile is shown in Figure 9.4. It should be noted that the imprecision rises at each end of the concentration range. This can be understood by looking at the shape of the calibration line. At the ends of the range, the curve flattens off, so a small change in bound (e.g. counts) results in a larger change in measured concentration than in the centre of the range. Thus the imprecision for the assay is concentration-dependent and increases at the extremes of

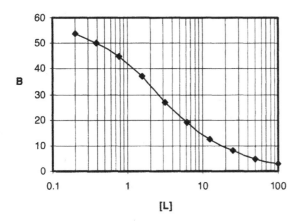

Figure 9.3 A typical standard curve from a competitive immunoassay.

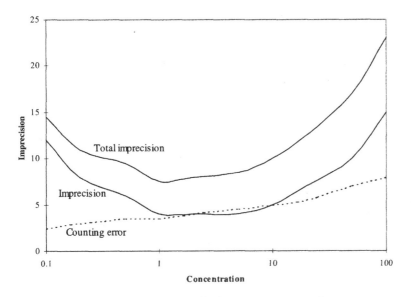

Figure 9.4 A typical precision profile for a competitive radioimmunoassay.

the sigmoidal calibration curve. This rise at one end will be worse in radioimmuno-assays since the counting imprecision rises as the counts decrease. This is not the case for DELFIA. Note that the linear range for competitive immunoassays occurs in the centre of the calibration curve, with linearity falling off at both upper and lower extremes of the curve. In general, competitive immunoassays are useable over about three orders of magnitude. This depends to a large extent on the individual character-istics of the antibody and label in use.

9.7 Advantages of immunoassay

9.7.1 Sensitivity

Immunoassay can achieve extremely low limits of quantification, down to tens of pg/ml or less and does not rely on any particular characteristic of the analyte, such as a chromophore or fluorophore or ability to ionise. Thus for those compounds requiring a very sensitive assay, unachievable otherwise, immunoassay should be considered. To increase sensitivity a disequilibrium assay can be used in which antibody is added to the assay prior to addition of label. After a period (1 hour at room temperature, or overnight at 4°C) the label is added and incubated for an equal further period. An increase in sensitivity of an order of magnitude is obtainable in this way.

9.7.2 Throughput

Immunoassays are capable of achieving very high throughput. For example, using DELFIA, four 96-well plates can easily be run each day by an analyst operating manually, giving a weekly throughput of 800 samples. Using automated liquid handling, double or triple this throughput could be attained.

9.7.3 Selectivity

Antibodies are extremely selective, but will usually be unable to distinguish between parent compound and closely related metabolites. This will depend on the epitope, which will in turn depend on where the molecule was coupled to the carrier protein. If the metabolism is on the epitope, then cross-reactivity is less likely. If a metabolite is expected to circulate at 10% of parent and it cross-reacts 10%, then the residual 1% contribution is unlikely to be an issue. If a late-forming metabolite circulates at higher levels than parent and cross reacts 30–100%, then this will be a major contribution. Some knowledge of the likely cross-reacting species is useful, since this can then steer the design of the hapten-conjugate to minimise chances of cross-reaction.

9.7.4 Ease

Immunoassays are some of the simplest methods to run once developed and validated. They are also cheap to contract out. For the 96-well plate format DELFIA systems, simply add sample, add antibody, add label, incubate, wash, count. The whole procedure can be carried out from start to finish in less than 4 hours.

9.7.5 Automation

Since the assays are so easy, it is not hard for a robotic liquid-handling system to carry out these assays. A Tecan, Packard or similar system can be used.

9.8 Disadvantages of immunoassays

9.8.1 Time: how long does it take?

The system requires design, synthesis and coupling of the hapten where small molecules are involved, immunisation, boosts, checking titres, synthesis of label, validation of assay. Sometime the analyte itself is a suitable hapten, with a reactive group that can be used for coupling to carrier protein – an amino or carboxylic acid group is ideal. A spacer (4–6 carbons) is usually necessary. A minimum of 4 months is likely. If an early enough start is made, the pre-clinical team in a pharmaceutical company can also use the assay for increased sensitivity.

9.8.2 Selectivity

The cross-reactivity profile for the assay must be checked by using the assay with whatever known metabolites (or other known related compounds) are available. If the metabolites can compete for the antibody binding site, the IC_{50} concentration (analyte concentration at which half the measured label binding is displaced) is determined and the ratio IC_{50} (analyte) $\times 100/IC_{50}$ (metabolite) is the percentage cross-reactivity. A further test that needs to be done as part of the validation procedure is 'parallelism'. This involves taking pooled study samples and assaying them undiluted and diluted 2, 4, 8, and 16 times with blank matrix. If there are cross-reacting metabolites of different cross-reactivity to the analyte, the measured concentrations will not be parallel with the dilution factor. If this is the case and metabolite cross-reactivity is a factor, then it must be eliminated. These are several ways of doing this. One is to use HPLC, collect fractions and assay appropriate dried fractions using immunoassay. Another is to use SPE to selectively remove the cross-reacting compounds. SPE is also a useful way of reducing matrix effects. When carried out in a 96-well format it does not add too much further time to the assay procedure.

9.8.3 Matrix effects

The matrix that a radioimmunoassay has to operate within is often of little importance, since the assays are so sensitive and the sample is diluted greatly. However, urine can often present a problem, especially with DELFIA. This may be because urine contains a number of polycarboxylic acids which have the potential to interfere by chelating the lanthanide. Urine samples are also likely to show a large variation in pH. See below for further discussion.

9.9 What can go wrong?

9.9.1 Matrix effects

It is a temptation to use water calibration curves when analysing urine. This is possible to do but it is important to be aware of the possibility of the urine matrix, in particular, interfering with the immunoassay, either by interfering with the antibody–antigen binding (perhaps by pH effect) or by interfering with the label in a DELFIA

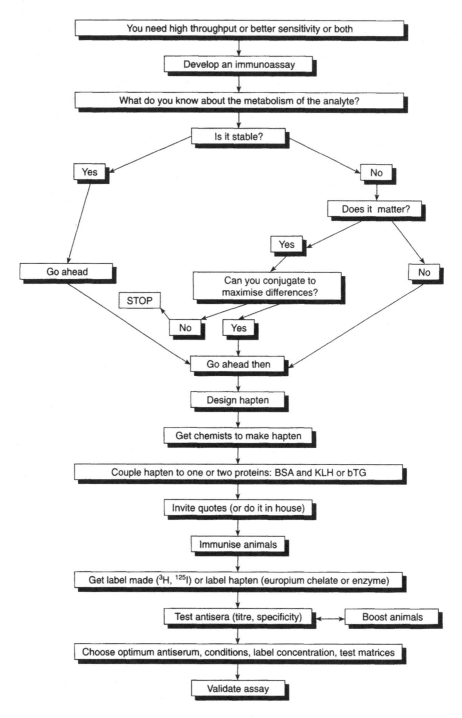

Figure 9.5 General scheme for the development of an immunoassay.

method. An assessment of a number of different samples of the matrix (as in any other method), both blank and spiked, is essential.

9.9.2 Concentration effects

The reader may be aware of certain high-profile problems with an immunoassay for detection of anti-HIV antibodies which became notorious in the media. This assay suffered from false negatives when extremely high levels of antigen were present. Such problems are possible, especially when non-competitive sandwich immunoassays are used with single incubation steps. The very high antigen concentration can bind all the available detector antibody, leaving essentially none to bind (via another epitope on the antigen) to the immobilised primary antibody. This is simply a consequence of the laws of mathematics, physics and mass action. Such events can be easily prevented either by running dilutions of all samples, or using several steps in the assay.

9.10   Immunoassay strategy

Figure 9.5 shows a general strategy for immunoassay development. Part of the validation process should be cross-validation of the immunoassay with a physical, specific method (preferably LC–MS–MS) for a number of clinical samples. If there is good correlation with no obvious bias, then the immunoassay will not be subject to interference from metabolites. Special patient groups should be considered also, such as those with renal or hepatic impairment, since these groups can have altered metabolite profiles. It is also important to carry out 'parallelism' checks on clinical samples in which pooled samples are diluted two-, four-, eight- and 16-fold. These diluted samples should give the correct undiluted concentration when multiplied by the corresponding factor. If this is not the case, either a matrix effect is responsible or metabolites are cross-reacting less than the parent compound.

9.11   Example

9.11.1 Sampatrilat

Figures 9.6–9.9 show the structure of sampatrilat, the labelling scheme that was used to develop the europium-labelled tracer, the calibration curve obtained and cross-validation by LC–MS–MS. This immunoassay was used to support the phase I clinical studies on this compound (Venn *et al.* 1998).

9.12   Affinity chromatography

Affinity chromatography is a technique that uses the specificity and binding affinity of antibodies to provide solid-phase or on-line extraction methods. Such systems are extremely specific and provide very good recoveries from plasma or urine. Large volumes can be used (50ml urine is not unusual) but the technique demands that antibodies be available. These are not always very robust and column re-use capacity is limited.

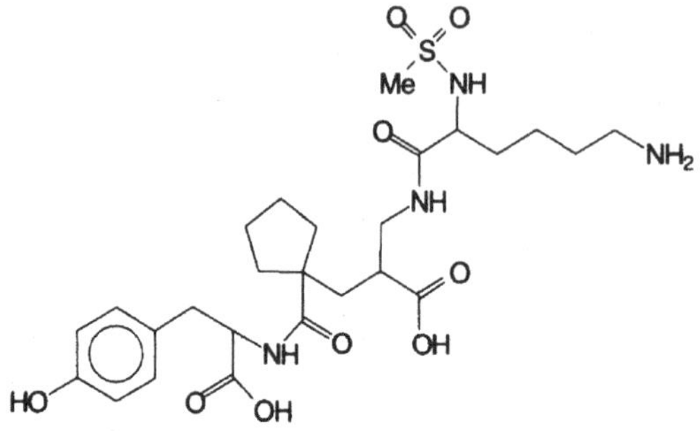

Figure 9.6 Structure of sampatrilat.

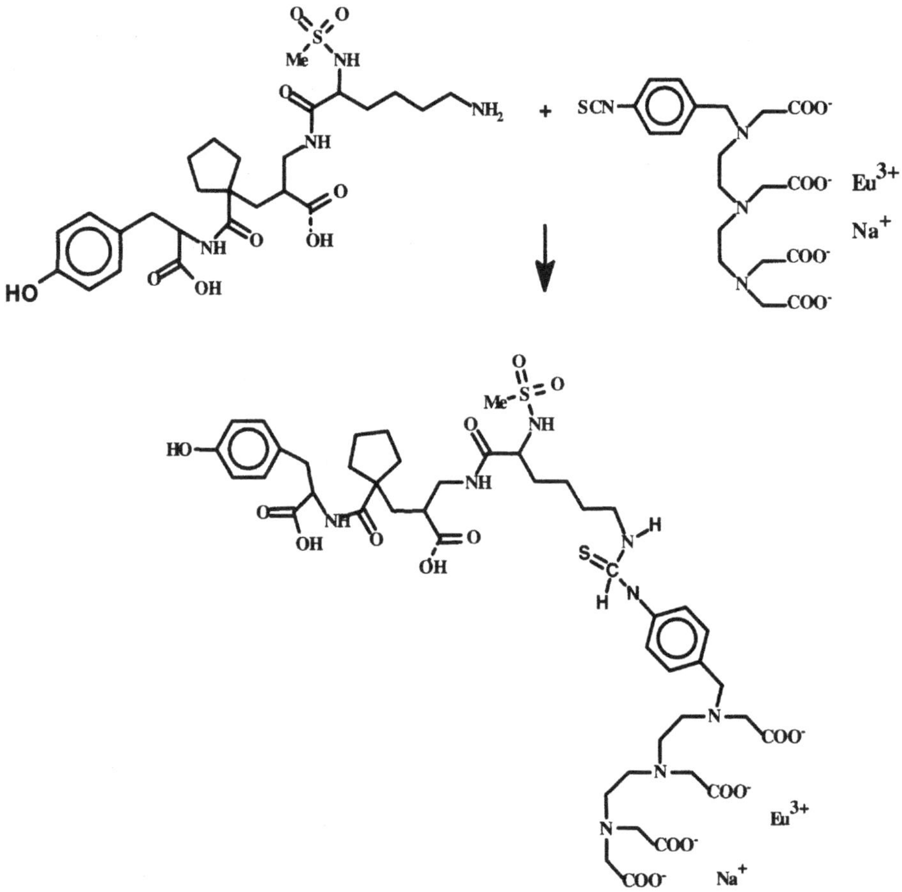

Figure 9.7 Labelling of sampatrilat.

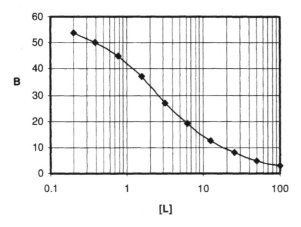

Figure 9.8 Typical calibration curve for sampatrilat.

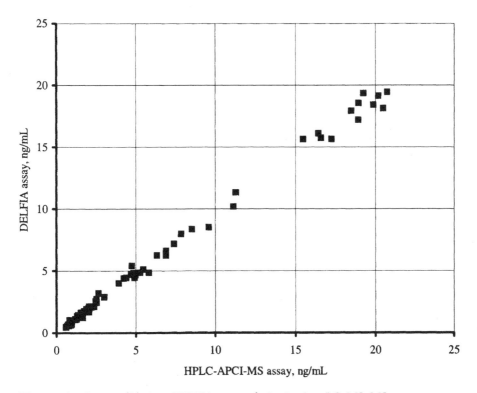

Figure 9.9 Cross-validation, DELFIA versus derivatisation–LC–MS–MS assay.

The component to be affinity purified is extracted from contaminants by temporarily allowing it to interact with an immobilised ligand; its affinity for the ligand permits the unwanted components to be removed and any interfering compounds to be washed from the immobilisation media; the component is then released from the ligand by manipulation of any one or more of a number of desorption methods,

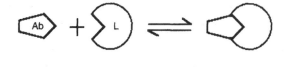

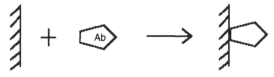

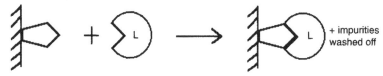

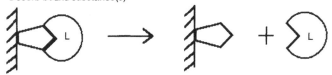

Figure 9.10 Diagram showing the principles of immunoaffinity chromatography.

such as pH, ionic strength, organic solvent concentration, or specific biochemical interaction.

Figure 9.10 shows the principle of affinity chromatography and the steps involved. The ligand is immobilised on a matrix (usually bead-formed, although this is not necessary and many other immobilisation forms have been used) and the matrix washed thoroughly to remove any unbound ligand. A solution containing the compound to be purified is allowed to equilibrate with the matrix-bound ligand: this is can be carried out in a beaker or tube, or in a chromatographic column. The matrix bearing covalently bound ligand and analyte bound to the ligand is then washed to remove unwanted contaminants, and the analyte or compound is then eluted from the matrix–ligand. The matrix can then be re-used to purify more compound.

9.12.1 Immobilisation techniques and media

It is important in affinity chromatography to ensure that all the ligand (in the clinical analytical situation, almost always antibody) has access to all the surface of the support matrix, thus ensuring maximum loading and a minimum of other chromatographic processes occurring. When soft gels are used, this presents no problem – Sepharoses normally have size exclusions in the 10^5–10^6 range. However, when silica or other supports are used, it is necessary to use wide-pore material, e.g. 300Å.

Soft gel supports

In the past affinity chromatography has relied almost entirely on the use of soft gels such as Sepharose 4B in low-pressure systems using peristaltic pumps or gravity (hydrodynamic pressure) to provide the flow, particularly in biochemistry laboratories. Some use has been made of controlled-pore glass, but the chemistry is generally not simple or accessible. Sepharose 4B and 4B-CL are excellent support matrices for low-pressure work, but are compressible and easily damaged. The chemistry used is often the cyanogen bromide method, in which the gel (an agarose or cross-linked agarose) is activated with cyanogen bromide in acetonitrile at pH 11 for a limited time; the gel is then washed and the ligand to be coupled (containing a primary amine) is added in aqueous solution. This mixture is then gently shaken overnight and then extensively washed. The ligand-bound gel is then ready for use. Providing that only low back-pressures are used, such gels can be used successfully for affinity chromatography in the analysis of clinical samples; for example, when column-switching is used and the eluent from the low-pressure gel column is passed onto a small concentration column, and from there to the high-pressure analytical column.

Rigid supports

Silica is often used for immobilising antibodies for high-performance liquid affinity chromatography (HPLAC). Silica, of course, has all the advantages of rigidity and the long experience in HPLC of silica and modified silicas, and will withstand very high pressures. Epoxy-silica (made by reaction of the selected silica with glycidoxypropyltrimethoxysilane) is the normal starting point for immobilisation; this can be used directly for coupling ligands containing primary amino groups. Diol-silica can also be synthesised from epoxy-silica, which can in turn be converted to tresyl-silica or aldehyde-silica. Each of these can be reacted with primary-amine-containing ligands to form immobilised antibody supports. Columns containing activated silica are available commercially; Pierce sell a tresyl-activated Selectispher-10™ 300Å column which is ready for coupling with the antibody of choice. Pierce also sell ready-coupled boronate, Cibacron Blue F3G-A, concanavalin A and protein A columns.

9.12.2 Elution techniques

Once the compound of interest has been adsorbed onto the solid support and the support washed free of any non-specifically bound contaminants, the analyte must be desorbed from the support and subjected to analytical methods, such as HPLC for quantification. It can be difficult to desorb the compound without resorting to extremes which could damage the antibody. Most elution methods rely on reducing the binding affinity to release the bound analyte, and this can damage the antibody irreversibly. The final choice of elution conditions will depend on the stability of the antibody when bound to a support, and the availability of specific analogues of the compound of interest with sufficient cross-reactivity, yet separable on HPLC from the analyte of interest (see below).

pH

Raising or lowering the pH of the eluent can effect a release of bound component. Care must be taken to return the pH to close to neutral quickly, and that the pH chosen does not damage the antibody. A pH of 2.5 for short periods is usually tolerated well by antibodies.

Ionic strength

Increasing the ionic strength of the eluent can effect desorption.

Temperature

Raising or lowering the temperature of the column has occasionally been used to disturb the equilibrium and release the bound component.

Polarity manipulation

This is probably the most common desorption method used in high-performance liquid affinity chromatography. Addition of methanol, ethanol or acetonitrile (10–60% concentrations) is usually sufficient to desorb the compounds from the affinity column. If the organic modifier concentration is low enough (10–20%), this will usually allow the analyte to be passed straight to an analytical column or to a concentration column. At higher concentrations, the eluent must be diluted with aqueous media prior to trace enrichment on a small pre-column.

Denaturing agents

Urea or guanidine can also be used with care to desorb the analyte from the affinity column, but at the risk of damaging and shortening the life of the immobilised antibody.

Biospecific desorption

This method involves the introduction of an excess of a competing similar analogue which displaces the bound component. The released component is then separated from the competing ligand by HPLC. This leaves the antibody with another specifically bound compound adhering to it, which must then be removed.

9.12.3 Re-use/reconditioning

Following desorbtion, the column or support needs to be returned to its original state, ready for the next sample. This would simply be done by returning to the conditions used for adsorption. If an excess of competing biospecific reagent was used, this would need first to be eluted from the column, with more aggressive means; for example, 70% methanol has been used.

9.12.4 *The interface between affinity chromatography and analysis*

Once affinity purified, the analyte(s) of interest must be quantified. This can be by HPLC with the full range of detection systems available, or GC, again with the full range of detectors available. Usually the analyte is not completely purified but elutes from the affinity absorbent with non-specifically bound components or related compounds. Thus a separation and quantification procedure is required. It is often the case that the purified analyte is desorbed from the affinity media in a large volume, and this requires a further concentration step. Thus, most on-line immunoaffinity procedures involve a secondary concentration column in the column-switching arrangement prior to the analytical column. It is, of course, possible to use affinity chromatography in a similar way to solid-phase extraction (SPE), using empty SPE cartridges to hold the affinity media, but often the volumes eluted are then larger than can be conveniently evaporated to dryness. Most examples of the technique in the literature use column-switching techniques to achieve the interface.

9.13   The future

The following paragraphs outline in the briefest detail some of the current directions research is following in the quest for faster, more sensitive analytical techniques.

9.13.1 *Phage libraries for antibodies*

In this technique the entire human library of antibodies is encoded in DNA transfected to bacteriophages – many copies of one single antibody per phage. The library is screened by immobilising the analyte, then incubating with the library. Unbound phages are washed off and bound (bearing cross-reacting phages) are then grown and the antibody extracted and used. The advantages are many: it is fast, no animals are used, unlimited quantities of antibody are available and the result is a single reagent similar to monoclonal antibody. The current disadvantages are that usually there is no maturation of response and thus a relatively low affinity. This can be an advantage for affinity chromatography.

9.13.2 *Monoclonal antibodies*

A technique using mouse spleen cells fused to myeloma cells can be used to generate monoclonal (pure single reagent) antibodies to analytes of interest. Different specificities for the same analyte can be generated. The advantages are that single reagent is available in unlimited quantities. The current disadvantages are cost, clonal drift and (usually) low affinity.

9.13.3 *Molecular imprinting*

This is a technique in which the shape of a template molecule (analyte or near analogue) is imprinted in a polymer by polymerisation of the appropriate monomers around the template. The polymer is ground to a small particle size and washed extensively to remove template. This imprinted polymer is then used, for example, in solid-phase extraction, for chiral LC, or as an antibody mimic for 'immunoassay'.

9.13.4   Non-competitive assays for small molecules

Currently only large molecules with two well-separated epitopes can be determined by non-competitive immunoassay. Techniques are becoming available to overcome this, usually using monoclonal anti-idiotypic antibodies (raised against the antibody itself) or metatypes. Such assays will be advantageous in terms of speed, sensitivity and dynamic range.

9.13.5   Use of low-specificity immunoassay for discovery compounds

Here, antibodies (possibly from a phage library) of low specificity are generated to the central core of a series of pharmaceutical compounds in the early discovery process. These are used to assay all the compounds in the series, allowing an increase in analytical and compound throughput.

9.13.6   Indwelling optical fibre probes

Antibodies covalently bound to an optical fibre alter the properties of laser light being transmitted down the fibre when they bind to the analyte of interest. The change is detected and quantified. It is conceivable that such a fibre optic could be inserted through a cannula into a vein, giving a real-time read-out of concentration for administered compounds.

9.14   Summary

Immunoassay is an important technique that can be used for the bioanalysis of small molecules. It can be extremely sensitive, offering lower limits of quantification than any other technique. However, it does have significant potential drawbacks. The time required to raise antibodies can be very long (up to a year) and it requires a suitable hapten to be available that can be coupled to a carrier protein in order to generate the antibodies. This may prove difficult. Immunoassays are not always as selective as might be desired and cross-reactivity with closely related molecules may be an issue. The advantages, however, often outweigh the disadvantages. For example, the running costs of an immunoassay are small, and usually very low sample volumes (in the order of 50µl) are required. The throughput can be very high, allowing several hundred samples to be assayed in 1 day. This contrasts very favourably with other HPLC-based assay methods. If large numbers of samples are to be assayed and a suitable hapten can be designed to minimise cross-reactivity, then immunoassay is a technique that should be investigated. When sample clean-up is a problem and interfering, co-eluting endogenous compounds adversely affect a conventional HPLC assay, or if greater sensitivity is required, then immunoaffinity chromatography is a powerful tool that can be exploited to the analyst's advantage.

9.15   Bibliography

Creaser, C.S., Feely, S.J., Houghton, E., Seymour, M. and Teale, P. (1996) On-line immuno-affinity chromatography–high-performance liquid chromatography–mass spectrometry for the determination of dexamethasone. *Analytical Communications*, 33, 5–8.

Farjam, A., De Jong, G.J., Frei, R.W. *et al.* (1988) Immunoaffinity pre-column for selective on-line sample pre-treatment in high-performance liquid chromatography determination of 19-nortestosterone. *Journal of Chromatography*, **452**, 419–433.

Farjam, A., Vreuls, J.J., Cuppen, W.J.G.M., Brinkman, U.A.Th. and de Jong, G.J. (1991) Direct introduction of large-volume urine samples into an on-line immunoaffinity sample pretreatment-capillary gas chromatography system. *Analytical Chemistry*, **63**, 2481–2487.

Glencross, R.G., Abeywardene, S.A., Corney, S.J. and Morris, H.S. (1981) The use of oestradiol-17β antiserum covalently coupled to Sepharose to extract oestradiol-17β from biological fluids. *Journal of Chromatography*, **223**, 193–197.

Haasnoot, W., Schilt, R., Hamers, A.R.M. *et al.* (1989) Determination of β-19-nortestosterone and its metabolite α-19-nortestosterone in biological samples at the sub parts per billion level by HPLC with on-line immunoaffinity sample pretreatment. *Journal of Chromatography*, **489**, 157–171.

Haasnoot, W., Ploum, M.E., Paulussen, R.J.A., Schilt, R. and Huf, F.A. (1990) Rapid determination of clenbuterol residues in urine by HPLC with on-line automated sample processing using immunoaffinity chromatography. *Journal of Chromatography*, **519**, 323–335.

Harlow, E. and Lane, D. (1999) *Using Antibodies – A Laboratory Manual*, Plainview, N.Y.: Cold Spring Harbor Laboratory.

Krause, W., Jakobs, U., Schulze, P.E., Nieuweboer, B. and Humpel, M. (1985) Development of antibody-mediated extraction followed by GC/MS (antibody/GC/MS) and its application to iloprost determination in plasma. *Prostaglandins, Leukotrienes and Medicine*, **17**, 167–182.

Larsson, P.-O., Glad, M., Hansson, L., Mansson, M.-O., Ohlson, S. and Mosbach, K. (1983) High-performance liquid affinity chromatography. *Advances in Chromatography*, **21**, 41–85.

Mazzeo, J.R. and Krull, I.S. (1988) Immobilized boronates for the isolation and separation of bioanalytes. *Biochromatography*, **4**, 124–130.

Takasaki, W., Asami, M., Muramatsu, S. *et al.* (1993) Stereoselective determination of the active metabolites of a new anti-inflammatory agent (CS-670) in human and rat plasma using antibody-mediated extraction and high-performance liquid chromatography. *Journal of Chromatography*, **613**, 67–77.

Turková, J. (1993) Bioaffinity chromatography. *Journalof Chromatography Library*, **55**.

Van Ginkel, L.A. (1991) Immunoaffinity chromatography, its applicability and limitations in multi-residue analysis of anabolizing and doping agent. *Journal of Chromatography*, **564**, 363–384.

Venn, R.F., Kaye, B., Macrae, P.V. and Saunders, K.C. (1998) Clinical analysis of sampatrilat, a combined renal endopeptidase and angiotensin-converting enzyme inhibitor. I: assay in plasma of human volunteers by atmospheric-pressure ionisation mass-spectrometry following derivatisation with BF_3-methanol. *Journal of Pharmaceutical and Biomedical Analysis*, **16**, 875–881.

Venn, R.F., Barnard, G., Kaye, B., Macrae, P.V. and Saunders, K.C. (1998) Clinical analysis of sampatrilat, a combined renal endopeptidase and angiotensin-converting enzyme inhibitor. II: assay in the plasma and urine of human volunteers by dissociation enhanced lanthanide fluorescence immunoassay (DELFIA). *Journal of Pharmaceutical and Biomedical Analysis*, **16**, 882–892.

10

AUTOMATION OF SAMPLE PREPARATION

Chris James

Dictionary definition, automatic: 'working of itself, without human actuation'.

10.1   Introduction

Although the term 'automation' is often used to refer to sample preparation when discussing bioanalysis, we should remember that analysing a biological sample typically involves the following processes:

- sample preparation, e.g. sorting, defrosting, aliquoting;
- preparation of calibration and QC samples;
- sample extraction;
- chromatography/analysis;
- data handling;
- reporting.

Some of these processes have been so successfully 'automated' that we take them for granted. In particular, automation of sample injection, and the acquisition, processing and reporting of chromatographic data are now routine using modern autoinjectors and computerised chromatography data systems. Only older bioanalysts will remember repeatedly injecting samples manually into Rheodyne™ valves for HPLC analysis, and then measuring the heights of the chromatographic peaks printed on a chart recorder traces with a ruler, or even cutting peaks out and weighing them to assess area.

Equipment to automate extraction procedures has been available for well over a decade, but automation of sample extraction has not yet reached as high a level of application as might be expected. In fact, virtually all bioanalytical laboratories still run some extractions manually, and in many laboratories all extractions are manual. Two reasons in particular may account for this. First, the most widely used sample preparation methods of SPE and protein precipitation are already quite efficient procedures, and may actually take less of an analyst's time than is first thought. Secondly, although the total number of samples analysed by many laboratories is large, the number for any single study varies from less than 100, to perhaps 1500 samples, leading to the requirement to frequently set up, and then change from one method to the next. The additional overhead to establish automated assays, and, in the past, the additional reliability problems caused by the automation instruments themselves, have therefore made the overall benefit of automation seem equivocal for many laboratories.

With the introduction of any new procedure or equipment the overall objective must be kept in mind. Automation is not a goal in itself, the aim is to have efficient analytical processes that make best use of staff and equipment, and allow laboratories to cope with the seemingly ever-increasing numbers of biological samples supplied by drug discovery and development programmes.

A combination of factors has led recently to an increasing interest and application of automation, which may eventually take us to the goal of fully automated assays for all bioanalytical methods. The most important of these is the dramatic increase in the use of LC–MS methods, which will probably replace HPLC as the predominant bioanalytical methodology in future years. Many LC–MS methods have very short run times (2–3 minutes) and are capable, in theory at least, of sample throughputs of several hundred per day. To fully utilise the capabilities of such instruments, there is clearly a need for automated sample preparation. The high selectivity of LC–MS–MS methods also reduces the need for complex and highly selective extraction methods, making automation easier. The majority of assays, even those requiring sub-ng/ml senitivities, can now often be performed with straightforward SPE or protein precipitation extraction methods.

These rapid, high throughput assays are also changing the requirements for automated systems. The speed of the automated extraction has become a key issue not only to keep pace with the analytical system, but as rapid batch extraction of samples off-line seems to be preferred to support LC–MS methods. This allows a simple (and highly reliable) autosampler to be connected to the analytical instruments, improving reliability and maximising the use of high-cost LC–MS instruments. The more complex and potentially less reliable extraction system can be run in the working day when staff are available to deal with any problems.

Although the following chapter discusses a number of commercially available instruments, this is not intended to be a full market survey or review; rather, details are given to illustrate various approaches to automation.

10.2 Approaches to automation

10.2.1 SPE

Although it is possible to automate virtually any laboratory procedure, much of the automation equipment introduced during the past few years has focused on the technique of solid-phase extraction (SPE), which has become the predominant method for many sample extractions, and is applicable to most types of analyte. It is also easier to automate SPE methods, which involve mostly liquid-handling steps, compared with liquid/liquid extractions that contain steps such as centrifugation and separation of solvent and aqueous layers.

For automated SPE methods, a combination of physical movements of the robot and the relatively low liquid flow rates that are needed for cartridge extractions, mean that each extraction can take up to 10–15 minutes, although with careful optimisation this time can be reduced to a few minutes per sample. However, sample throughput can really be increased dramatically with instruments that process SPE extractions in parallel, and instruments are now available that can extract a batch of about 100 samples in under 1 hour.

10.2.2 Protein precipitation methods

The technique of protein precipitation (PP) to prepare plasma samples for analysis, has always been seen as a simple and rapid procedure and consequently one with less to be gained from automation. However, as batch sizes increase from a typical 60–70 injections to 100 or even 200 samples, automation of even part of such simple procedures will become more beneficial. Semi-automated methods for PP in 96-well plates (see below) can also be expected to feature more strongly in future.

10.2.3 Multi-well plate technology

The introduction of SPE extraction 'cartridges' in a 96-well format has emphasised the potential uses of moulded multi-well plates for sample handling in bioanalysis. With batches of 100–200 samples becoming commonplace, handling 100 or more separate SPE cartridges, sample tubes, autosampler vials, etc. is not only time consuming, but is also a source of potential error as tubes and vials can easily be misplaced. Direct injection of samples from the plates by appropriate autosamplers (this is only possible with certain autosamplers) is necessary for the full benefit of this format. Due to the widespread use of multi-well plates in other fields of analysis, an extensive range of plates is available (e.g. from 12 to 1024 wells; plates that can be read spectroscopically; plates containing a wide variety of filtration membranes, etc.) from a number of manufacturers (e.g. Porvair, Polyfiltronics, Whatman). In addition, both manual (e.g. multi-channel pipettes) and automated equipment are available to aid processing of samples in this format. Such plates are increasingly (and perhaps belatedly) being adopted for bioanalytical assays.

For preparation of plasma samples by protein precipitation, 96-well plates offer two interesting and highly efficient possibilities. First, the precipitation can be performed in the plate, the complete plate centrifuged to spin down the precipitant, and then the supernatant injected directly with the height of the autosampler needle programmed not to disturb the precipitant. Secondly, a plate equipped with a filtration membrane can be used to separate the precipitant from the supernatant, with the supernatant being collected by applying a vacuum to draw it through the membrane into a collection plate.

10.2.4 Liquid-handling procedures

As with PP techniques, when typical batch sizes are 60–70 samples, there is often little benefit in automating simple liquid-handling steps, as these can usually be efficiently performed with hand-held repeating pipettes or dispensers. However, as batch sizes increase, not only do procedures such as aliquoting of samples, addition of reagents or internal standard, dilution, transfer of liquids between tubes, or transfer of an extract to an autosampler vial or plate become tedious and time-consuming, but the risk of error increases significantly. We can therefore expect greater use of liquid-handling robots for these 'simple' tasks as batch sizes increase, particularly as multi-probe instruments can perform such manipulations remarkably quickly.

10.2.5 Avoiding evaporation

Sometimes avoiding one particularly inconvenient part of a method can give a great improvement in the efficiency of the procedure overall. Evaporation of the final extract is often used to concentrate samples after SPE, but is difficult to automate, slow, and can risk loss of analyte due to binding to the tube, instability or volatility of the analyte. In the SPE chapter the use of disc cartridges, or cartridges with small packing weights, to reduce elution volume and hence avoid evaporation is described. Column switching methods (see below) may also be used to avoid evaporation after a SPE extraction. Here the eluted extract is diluted so that the analytes will be retained, and hence concentrated, on a short trace enrichment cartridge. This is then switched in line with the main analytical column to elute the extract for analysis. While both approaches are applicable after manual or automated SPE, they are particularly useful with automated systems where it is difficult to incorporate an efficient evaporation procedure.

10.3   Simple automation

As the overall goal is efficiency of the analytical process, not automation itself, it should be considered that simple changes in parts of the sample preparation procedure can make dramatic improvements in the process overall. These can be considered as an alternative to, or to be used in combination with, automation of other parts of the assay. One of the reasons that automation has not always produced the promised improvements is that with careful thought manual procedures can often be made relatively efficient and easy, and the human element both retains flexibility and is able to identify and rectify problems at an early stage.

Taking a little time for a simple analysis of your procedure is always worthwhile when considering improvements such as automation. A flow chart can be constructed with Post-it™ notes, recording time taken for various steps, and where problems regularly occur. This will usually reveal parts of the assay which are already easy and quick to perform, and those which are time consuming, tedious and/or prone to error. It may also reveal that the procedure to be automated does not consume as much of the working day as first thought. Table 10.1 gives a list of potential advantages and disadvantages of automation. Although some of these comments are somewhat obvious, it does emphasise that one should carefully consider exactly what will be achieved by 'automation'. Several simple improvements that can significantly improve efficiency and make life more pleasant for the analyst are given below.

The use of automatic repeating pipettes and dispensers, larger centrifuges that allow all samples to be spun in one batch, use of solid-phase extraction (SPE) instead of liquid/liquid extraction, more efficient equipment to evaporate samples (e.g. Zymark TurboVap™), can all provide significant increases in throughput. Liquid/liquid extraction procedures can also be improved by simple means: micro-volume solvent extractions using small sample and solvent volumes (< 1ml) can be performed very

Table 10.1 Advantages/disadvantages of automation

Advantages	Disadvantages
Better use of skilled staff	Longer development time
Improved staff attitude	High capital cost
Increased sample throughput	Increase space requirement
Reduced costs	Reduced flexibility
Improved repeatability	Slow extraction
Improved reproducibility	Trained staff to program and maintain robot
Safety (minimises contact with toxic	Reliability problems
reagents or biologically hazardous samples)	Inability to cope with the unexpected (e.g.
Allows staff to do other tasks	coagulated samples, variation in tube sizes)

efficiently, and freezing the aqueous layer to allow solvent to be poured off is a lot less tedious than transferring solvents with a pipette after extraction. Programmable autosamplers can be used to spike calibration curves and add internal standard to all samples at the start of the extraction. Changes that increase the selectivity of the chromatographic and/or detection system can dramatically reduce the need for a selective (and complex) extraction, e.g. a change from UV to fluorescence detection could allow a simple protein precipitation extraction instead of SPE; with LC–MS techniques much of the selectivity of the assay can be provided by the 'detector', and again it may be possible to replace a highly selective but complex extraction with a much simpler procedure.

10.4 Column switching

Column-switching methods can provide a high degree of automation by using a short pre-column repeatedly to perform a solid-phase extraction procedure. A sample of biofluid is injected onto the pre-column (typically 2–10mm long) which is packed with a bonded silica material. The sample is pumped through the pre-column with a weakly eluting solution (e.g. aqueous buffer for a reversed-phase system) which washes off more polar materials while the analyte is retained. At the same time, mobile phase for the analysis is being pumped through the analytical column via a separate channel in the switching valve. The valve is then 'switched', bringing the pre-column in line with the analytical column, and the analyte is eluted by the mobile phase onto the analytical column for analysis (Figure 10.1). This is the simplest column-switching set-up, and much more complex systems are possible, involving, for example, additional wash or regeneration steps for the pre-column, or 'heart cutting' a peak from one column to another to greatly increase the selectivity of the system.

Column-switching systems can be set up relatively cheaply, and require only the purchase of a switching valve, pump and suitable pre-column in addition to a standard HPLC system with autosampler.

The biggest disadvantage of this technique is the need to repeatedly pass a biological sample through the same 'extraction' pre-column. This not only risks blockages but increasing contamination of the pre-column can affect the analysis of samples later in the run (e.g. by reducing recovery of analyte). These problems can be reduced by the use of short extraction columns packed with large-particle-size bonded silica

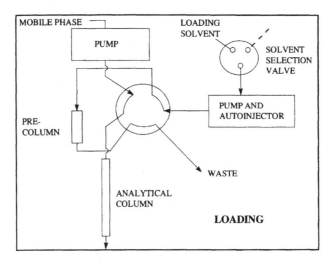

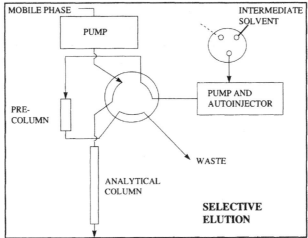

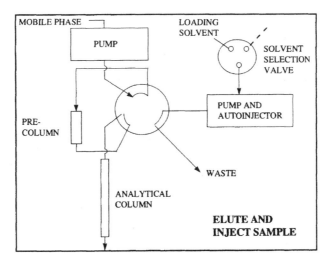

Figure 10.1 Diagram of a simple column-switching system.

Figure 10.2 The Prospekt.

(40μm) retained with stainless-steel meshes rather that frits. Up to 20ml of plasma can be passed through the columns before they need to be replaced. Use of pre-columns filled with 'restricted access' phases (e.g. Merck LiChrospher® RP-ADS, BioTrap™) are also reported. This type of material was originally developed as an analytical column into which plasma could be directly injected (Pinkerton *et al.* 1986). The chemistry of the bonded groups inside the pores is different to that at the surface of the material. Inside the pores, the silica has been bonded with a typical lipophilic (e.g. C8, C18) phase, but the outer surface of the silica is polar (e.g. diol) and compatible with proteinaceous materials. Small molecules enter the pores and are retained, while plasma proteins are excluded and can be washed readily from the column.

The development time of column-switching methods may be longer than that for off-line SPE methods, and choice of cartridge type and conditions perhaps more limited by the need for the 'extraction' cartridge to be in-line with the analytical column. However, once established, column-switching methods can provide a high degree of automation with modest equipment and running costs.

10.5   Prospekt™ and Merck OSP-2™

Two instruments, the Prospekt™ (Figure 10.2) and the Merck OSP-2™ (Figure 10.3), in effect, extract and analyse samples in the same way as a column-switching system, but with the major advantage that a new cartridge is used for each sample. The instruments are distinguished from other automation systems, as they contain a system to clamp short, disposable extraction cartridges which can be switched in line with the analytical column. Any problems of the sample concentration are avoided, but the analyte must be eluted from the extraction cartridge with the HPLC mobile phase. Extraction cartridges for the Prospekt™ are contained in a cassette of 10 cartridges, whereas the OSP-2™ uses short, metal pre-columns contained in a carousel.

10.6   Benchtop instruments – sequential sample processing

A number of benchtop instruments have been produced specially for SPE extraction. Those listed in this section process one sample at a time, and due to the speed of these extractions, are best suited for use in an on-line mode, as batch extraction of an

Figure 10.3   The Merck OSP-2.

Figure 10.4   The Zymark BenchMate.

analytical run of 100 or more samples would require processing overnight. The Waters Millilab™ and H-P Workstation™ have a similar function, but are not described below.

10.6.1   *Zymark BenchMate™*

The Zymark BenchMate™ (Figure 10.4) uses technology derived from the Zymate™ robot, including a similar robotic arm to move samples around the instrument. For a benchtop instrument, it is unique in offering vortex mixing and a built-in balance to confirm liquid-handling operations gravimetrically. Other instruments rely on bubbling gas through the samples, or repeated aspiration and dispensing to mix liquids. SPE is performed using standard 1ml or 3ml SPE cartridges, and a membrane filtration option is available.

10.6.2   *Gilson ASPEC XL™*

The Gilson ASPEC XL™ (Figure 10.5) is based on the familiar Gilson autosampler, and is designed primarily for SPE using standard SPE cartridges. Each cartridge is

Figure 10.5 The Gilson ASPEC XL.

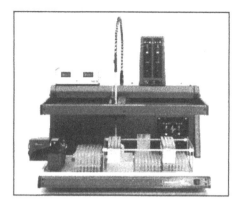

Figure 10.6 The Hamilton MicroLab.

fitted with a special cap with a small hole in it which allows the liquid-handling probe to push solutions through the cartridges. The ASPEC™ is relatively compact and can be located easily in a standard fume cupboard or biohazard cabinet for processing of hazardous samples. The ASPEC™ is something of an 'unsung hero' and appears to have been put to good use by most purchasers.

10.6.3 Hamilton MicroLab™

The Hamilton Microlab™ (Figure 10.6) is a liquid-handling system that has been adapted for SPE extraction. The unit has a single liquid-handling probe and solutions are pushed through the cartridges with positive pressure. The two special features of this unit are:

1 there is feedback control on the application of pressure to drive liquids through the cartridges, giving much better control of flow through different cartridges; and
2 the liquid probe has multiple channels leading to the tip, hence the system does not have to be flushed for each solvent change, making it significantly faster than some other instruments.

Figure 10.7 The Zymark RapidTrace.

10.7 Benchtop instruments – parallel sample processing

The current trend to support high throughput bioanalytical methods is with off-line automated SPE instruments, which process sample in parallel (between 4 and 10 samples simultaneously). This greatly improves sample throughput, allowing a typical batch of about 96 samples to be processed in 1 hour (exact times will, of course, depend on both the instrument used and details of the extraction procedure).

10.7.1 Zymark RapidTrace™

The RapidTrace™ (Figure 10.7) is a compact unit (approximately 10cm wide) that can perform solid-phase extraction. Each unit can process about 10 samples in 1 hour for a typical SPE method. Up to 10 units can be connected together and controlled by a single PC, giving the possibility of extracting 100 samples in under 1 hour. An interesting feature is the use of magnetically encoded sample racks to recall preprogrammed methods.

10.7.2 Gilson ASPEC 4™

This is modified version of the Gilson ASPEC XL™ (Figure 10.8), which has four probes to extract four samples in parallel. The instrument has no facility for on-line injection into HPLC, but can use either standard SPE cartridges or 96-well MicroLute™ plates. Extraction of 50 samples in 1 hour is reported.

10.7.3 Multiple probe liquid-handling robots

A number of liquid-handling robots are available for general laboratory use, some of which can be equipped with multiple liquid-handling probes (usually four or eight probes). Two of these in particular, Packard MultiPROBE™ (Figure 10.9) and Tecan Genesis™, have been adapted to perform SPE with 96-well plate format. As with other parallel processing instruments, rapid sample extraction is achieved. Various solutions are simultaneously drawn through all 96 wells of a SPE extraction plate under vacuum, which is controlled automatically. However, manual intervention is required to place the sample collection plate into the instrument prior to final elution of the

Figure 10.8 The Gilson ASPEC 4.

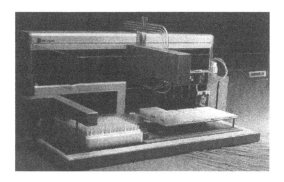

Figure 10.9 The Packard MultiPROBE.

extracts. The extracted samples are collected into a 96-well plate, from which extracts can be directly injected into an HPLC system.

The multiprobe instruments typically have the option to automatically collect, use and then discard disposable pipette tips for liquid processing. A pipette tip containing SPE material in a disc format is reportedly in development (SPEC PLUS PT™, Anasys™). These tips would potentially change such standard liquid-handling robots into automated SPE instruments without any special adaptation. This is an interesting possibility, particularly as two problems with the use of 96-well SPE plates, variation of vacuum between different wells and the manual intervention needed to insert collection plates, would be avoided. However, as only four samples would be processed simultaneously, using such SPE tips might be expected to be somewhat slower than with vacuum processing of 96-well plates.

10.8   Gilson ASTED™

The Gilson ASTED™ (automated sequential trace enrichment of dialysates; Figure 10.10) is a unique instrument which is worthy of discussion. It extracts biofluid samples by dialysis followed by trace enrichment of dialysate on a short column to

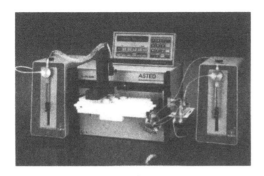

Figure 10.10   The Gilson ASTED.

concentrate the extract prior to analysis by HPLC (for a review, see Turnell and Cooper 1989). The sample (e.g. plasma) is diluted and passed through the donor channel of the dialysis unit. Simultaneously, buffer is passed through a recipient channel on the other side of the dialysis membrane and analyte diffuses across the dialysis membrane while plasma proteins are retained. The recipient fluid is passed through a trace enrichment cartridge (TEC) where the analyte is concentrated. The TEC is then switched in line with the analytical column to elute the analytes for analysis. For appropriate analytes the process works well, and once established the running costs of the system are very low, but the dialysis procedure is not suitable for all molecules (e.g. highly protein-bound drugs). This instrument also offers, in effect, an alternative extraction technique for compounds that prove difficult to analyse by other methods (e.g. analysis of highly polar compounds; Buick and Sheung 1993).

10.9   Full robotic systems

The market for 'complete' robotic systems (e.g. Zymark Zymate™, Hewlett Packard ORCA Robot™) has been dominated for over a decade by Zymark. Such robots are distinguished from the simpler instruments described above by offering the possibility to automate almost any laboratory procedures. The Zymark system (Figure 10.11) includes a robotic arm that moves samples between various work stations on the instrument. These robots can perform extraction methods such as SPE, liquid/liquid extraction, and protein precipitation, and direct injection of samples into LC systems, and also tasks such as vortex mixing, uncapping tubes, centrifugation, weighing and evaporation of solvents. Although generally the capital costs are high, and installation and method development times may be longer, such robots offer flexibility and capabilities that are not matched by other instruments. Combination of such robots with multiprobe liquid-handling instruments is possible to allow fully automated processing of several 96-well plates, without manual intervention.

10.10   When to automate?

The benefits of automation for a particular assay will depend on many factors, including the analyte, the nature of the assay, time available for method development, equipment and experience of staff available, etc. Clearly, the more samples that are

Figure 10.11 The Zymark Zymate.

available for analysis, the more likely it is that automation will be beneficial (Kushinsky 1991; Lingeman *et al.* 1991). However, do not assume that automation must be better. Poor instrument reliability, or analytical problems associated with the automation procedure (e.g. carryover of analyte in the instrument) can easily lengthen method development times, or negate the benefits of the automated procedure.

10.11   Example methods

The greater use of automated methods is reflected by a large and increasing number of publications in the literature. Some recent examples of published automated methods include those using column switching (Oosterkamp *et al.* 1996; Zell *et al.* 1997), the Varian Prospekt (Brandsteterova *et al.* 1994), the Gilson ASPEC (van Tellingen *et al.* 1994; Bosman *et al.* 1996), the Gilson ASTED (Turnell and Cooper 1987; Buick and Sheung 1993; Ceccato *et al.* 1997), the Zymark BenchMate (Ascalone *et al.* 1993; Hsieh *et al.* 1994), Zymark RapidTrace (Diamond *et al.* 1996) and Zymark robotic systems (Hempenius *et al.* 1994; Whigan and Schuster 1995; Dunne and Andrew 1996; Parket *et al.* 1996). Use of 96-well technology for SPE is also reported (Allanson *et al.* 1996; Kaye *et al.* 1996; Plumb *et al.* 1997).

10.12   Conclusions and future perspectives

1   Automation of sample preparation is increasing and can be expected to become routine for most biofluid assays in future.
2   Automation is not a panacea, and is not without its own costs and problems. Improved instrumentation is still required.
3   Currently, manual SPE can be regarded as 'competitive' with automated techniques, particularly if using 96-well format for SPE. Batches of up to 100 samples

can also be extracted efficiently with conventional SPE cartridges, if extracted simultaneously with appropriate vacuum manifolds.

4 The changing nature of bioanalytical methods (particularly the greater use of LC–MS) has led to the requirement for the extraction of large batches of samples. Rapid off-line batch extraction seems to be best suited for these methods, and instruments are now becoming available that can meet these requirements.

5 Rapid automatic extraction of large batches of samples may prove equally advantageous for standard HPLC assays, as HPLC autosamplers have proved to be more reliable than most automated extraction instruments over long, unattended runs.

6 All the above assume that the analytical instrument needs significant clean-up of samples before injection. If future analytical systems can tolerate direct introduction of plasma, none of these efforts may be required!

10.13 Bibliography

Allanson, J.P., Biddlecombe, R.A., Jones, A.E. and Pleasance, S. (1996) The use of automated solid phase extraction in the '96 well' format for high throughput bioanalysis using liquid chromatography coupled to tandem mass spectrometry. *Rapid Communications in Mass Spectrometry*, 10, 811–816.

Anon (1997) *Gilson guide to SPE automation*, Gilson, Inc., France.

Ascalone, V., Guinebault, P. and Rouchouse, A. (1993) Determination of mizolastine, a new antihistaminic drug, in human plasma by liquid–liquid extraction, solid-phase extraction and column-switching techniques in combination with high-performance liquid chromatography. *Journal of Chromatography*, 619, (2), 275–284.

Bosman, I.J., Lawant, A.L., Avegaart, S.R., Ensing, K. and de Zeeuw, R.A. (1996) Novel diffusion cell for *in vitro* transdermal permeation, compatible with automated dynamic sampling. *Journal of Pharmaceutical and Biomedical Analysis*, 14, (8–10), 1015–1023.

Brandsteterova, E., Romanova, D., Kralikova, D., Bozekova, L. and Kriska, M. (1994) Automatic solid-phase extraction and high-performance liquid chromatographic determination of quinidine in plasma. *Journal of Chromatography A*, 665, (1), 101–104.

Buick, A.R. and Sheung, C.T. (1993) Determination of 1-(beta-D-arabinofuranosyl)-5-(1-propynyl)-uracil and a metabolite, 5-propynyluracil, in plasma using ASTED (automated sequential trace enrichment of dialysates) combined, on-line, with high-performance liquid chromatography. *Journal of Chromatography*, 617, (1), 65–70.

Ceccato, A., Toussaint, B., Chiap, P., Hubert, P. and Crommen, J. (1997) Enantioselective determination of oxprenolol in human plasma using dialysis coupled on-line to reversed-phase chiral liquid chromatography. *Journal of Pharmaceutical and Biomedical Analysis*, 15, (9–10), 1365–1374.

Diamond, F.X., Vickery, W.E. and de Kanel, J. (1996) Extraction of benzoylecgonine (cocaine metabolite) and opiates (codeine and morphine) from urine samples using the Zymark RapidTrace. *Journal of Analytical Toxicology*, 20, (7), 587–591.

Dunne, M. and Andrew, P. (1996) Fully automated assay for the determination of sumatriptan in human serum using solid-phase extraction and high-performance liquid chromatography with electrochemical detection. *Journal of Pharmaceutical and Biomedical Analysis*, 14, (6), 721–726.

Hempenius, J., Hendriks, G., Hingstman, J., Mensink, C.K., Jonkman, J.H. and Lin, C.C. (1994) An automated analytical method for the determination of felbamate in human plasma by robotic sample preparation and reversed-phase high performance liquid chromatography. *Journal of Pharmaceutical and Biomedical Analysis*, 12, (11), 1443–1451.

Hsieh, J.Y., Lin, C., Matuszewski, B.K. and Dobrinska, M.R. (1994) Fully automated methods for the determination of hydrochlorothiazide in human plasma and urine. *Journal of Pharmaceutical and Biomedical Analysis*, **12**, (12), 1555–1562.

Hubert, P., Chiap, P., Evrard, B., Delattre, L. and Crommen, J. (1993) Fully automated determination of sulfamethazine in ovine plasma using solid-phase extraction on disposable cartridges and liquid chromatography. *Journal of Chromatography*, **622**, (1), 53–60.

Janiszewski, J., Schneider, P., Hoffmaster, K., Swyden, M., Wells, D. and Fouda, H. (1997) Automated sample preparation using membrane microtiter extraction for bioanalytical mass spectrometry. *Rapid Communications in Mass Spectrometry*, **11**, (9), 1033–1037.

Kaye, B., Herron, W.J., Macrae, P.V., Robinson, S., Stopher, D.A., Venn, R.F. and Wild, W. (1996) Rapid, solid phase extraction technique for the high-throughput assay of darifenacin in human plasma. *Analytical Chemistry*, **68**, (9), 1658–1660.

Kushinsky, S. (1991) Managing robotics in the bioanalytical/metabolic environment of a pharmaceutical company. *Journal of Automatic Chemistry*, **13**, 13–16.

Lingeman, H., McDowall, R.D. and Brinkman, U.A. (1991) Guidelines for bioanalysis using column liquid chromatography. *Trends in Analytical Chemistry*, **10**, 48–59.

Oosterkamp, A.J., Irth, H., Heintz, L., Marko-Varga, G., Tjaden, U.R. and van der Greef, J. (1996) Simultaneous determination of cross-reactive leukotrienes in biological matrices using on-line liquid chromatography immunochemical detection. *Analytical Chemistry*, **68**, (23), 4101–4106.

Parker, T.D. 3rd, Wright, D.S. and Rossi, D.T. (1996) Design and evaluation of an automated solid-phase extraction method development system for use with biological fluids. *Analytical Chemistry*, **68**, (14), 2437–2441.

Pinkerton, T.C., Perry, J.A. and Rateike, J.D. (1986) Separation of furosemide, phenylbutazone and oxyphenbutazone in plasma by direct injection onto internal surface reversed-phase columns with systematic optimisation of selectivity. *Journal of Chromatography*, **367**, (2), 412–418.

Plumb, R.S., Gray, R.D. and Jones, C.M. (1997) Use of reduced sorbent bed and disk membrane solid-phase extraction for the analysis of pharmaceutical compounds in biological fluids, with applications in the 96-well format. *Journal of Chromatography B: Biomedical Applications*, **694**, (1), 123–133.

Turnell, D.C. and Cooper, J.D.H. (1987) Automated sequential process for preparing samples for analysis by high-performance liquid chromatography. *Journal of Chromatography*, **395**, 613–621.

Turnell, D.C. and Cooper, J.D.H. (1989) Automation of liquid chromatographic techniques for biomedical analysis. *Journal of Chromatography*, **492**, 59–83.

van Tellingen, O., Pels, E.M., Henrar, R.E. *et al.* (1994) Fully automated high-performance liquid chromatographic method for the determination of carzelesin (U-80,244) and metabolites (U-76,073 and U-76,074) in human plasma. *Journal of Chromatography*, **652**, (1), 51–58.

Whigan, D.B. and Schuster, A.E. (1995) Manual and automated determination of 1-beta-D-arabinofuranosyl-E-5-(2-bromovinyl)uracil and its metabolite (E)-5-(2-bromovinyl)uracil in urine. *Journal of Chromatography B: Biomedical Applications*, **64**, (2), 357–363.

Zell, M., Husser, C. and Hopfgartner, G. (1997) Column-switching high-performance liquid chromatography combined with ionspray tandem mass spectrometry for the simultaneous determination of the platelet inhibitor Ro 44-3888 and its pro-drug and precursor metabolite in plasma. *Journal of Mass Spectrometry*, **32**, (1), 23–32.

11

FUNDAMENTAL ASPECTS OF MASS SPECTROMETRY: OVERVIEW OF TERMINOLOGY

Mira V. Doig

11.1 Introduction

The buzz words in mass spectrometry (MS) these days are APCI (atmospheric-pressure chemical ionisation), electrospray, tandem MS, LC–MS and LC–MS–MS, MALDI (matrix-assisted laser desorption ionisation) and ToF (time of flight analysis). To understand this terminology it is useful for a scientist to understand some basic fundamentals and historical aspects of MS, hence the contents of this chapter are ordered in a historical perspective.

The scheme in Figure 11.1 outlines the entire process of mass spectrometry (MS) and correlates this process with various components of a mass spectrometer. The sample, butane, represented by M, migrates by diffusion from the inlet into the ion source which is under vacuum at pressures varying from 10^{-2} to 10^{-5} torr. The butane is bombarded by a stream of electrons which, on collision with the atoms, knock electrons out of them. The ionised sample also absorbs excess energy, causing the molecule to break up, producing a range of fragment ions. These are accelerated towards the detector through the analyser where they are separated according to their mass to charge ratio (*m/z*). The data from the detector are analysed by a data system to produce a bar graph mass spectrum where the abscissa is *m/z* and the ordinate is the relative abundance.

11.2 Inlets

There are many types of inlet systems into a mass spectrometer and the type of inlet used is dependent on the volatility and purity of the compound to be analysed.

11.2.1 *Septum inlet*

Gases or compounds volatile at room temperature can be injected through a septum into a glass reservoir which is connected to the ionisation source by a narrow glass tube and a set of valves. The compound is then literally allowed to 'leak' into the vacuum system of the mass spectrometer. Similarly this system can be heated to 300°C to allow for the injection of less volatile liquids.

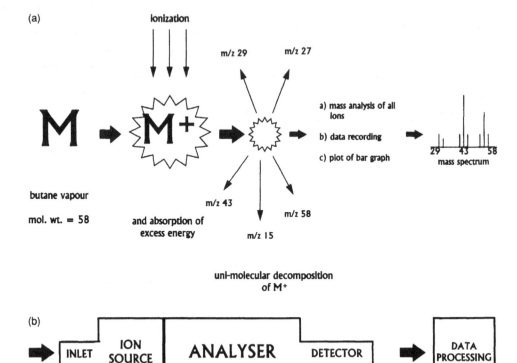

Figure 11.1 (a) Synoptic outline and (b) block diagram of a mass spectrometer.

11.2.2 Direct probe inlet

Compounds not sufficiently volatile can be inserted directly into the ion source by means of a probe passing through a vacuum lock. At the low pressures inside the ion source (10^{-6}–10^{-7} torr) and with the addition of heat, many compounds are sufficiently volatile to yield good mass spectra. The main disadvantage of this inlet is that the sample must be pure. If there are any contaminants in the sample, they must have a significantly different volatility from that of the compound of interest, so that they will evaporate from the probe at different times. A conventional inlet probe is shown in Figure 11.2. A direct exposure probe with the sample mounted directly on a convex rather than concave surface obviates the necessity for the sample to be volatile and allows for ions to be desorbed directly from its surface. This mode of sample introduction is commonly known as direct chemical ionisation (DCI). This ionisation process will be described in Section 11.3.3.

11.2.3 GC inlets

The effluents from GC columns consist of the carrier gas (mobile phase), which in MS is normally helium, and the compounds that have been separated by the partition

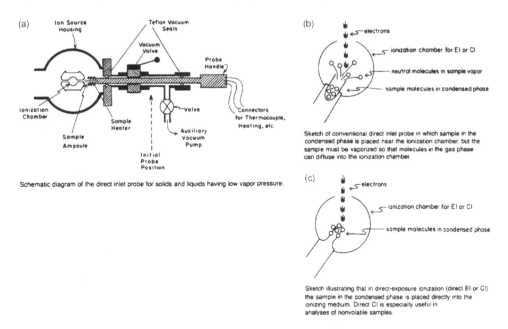

(a)

Ion Source Housing

Teflon Vacuum Seals

Vacuum Valve

Probe Handle

Ionization Chamber

Sample Heater

Sample Ampoule

Valve

Auxiliary Vacuum Pump

Connectors for Thermocouple, Heating, etc.

Initial Probe Position

Schematic diagram of the direct inlet probe for solids and liquids having low vapor pressure.

(b)

electrons

ionization chamber for EI or CI

neutral molecules in sample vapor

sample molecules in condensed phase

Sketch of conventional direct inlet probe in which sample in the condensed phase is placed near the ionization chamber, but the sample must be vaporized so that molecules in the gas phase can diffuse into the ionization chamber.

(c)

electrons

ionization chamber for EI or CI

sample molecules in condensed phase

Sketch illustrating that in direct-exposure ionization (direct EI or CI) the sample in the condensed phase is placed directly into the ionizing medium. Direct CI is especially useful in analyses of nonvolatile samples.

Figure 11.2   Diagram of a direct inlet probe interface (a), with sketches of (b) a conventional and (c) direct exposure probe tip.

process between the stationary phase and the mobile phase. These days the majority of GC inlets are heated tubes or boxes between the GC and the ion source through which a fused silica capillary column is fed. The temperature of the transfer box is important to maintain chromatographic resolution. Any cold spots between the GC and the mass spectrometer could broaden peaks, decrease sensitivity or, at worst, materials can condense on the cold spots and not be transferred to the mass spectrometer. Modern-day vacuum pumping systems can cope with maintaining the low pressures required in the source of the mass spectrometer even with a carrier gas flow rate from the gas chromatograph of 1–2ml/min. A more specialised mode of linking capillary GC to mass spectrometer is the open split interface. This is normally used if part of the GC effluent is not required in the mass spectrometer, if the sample concentration is too high or the user only wants to transfer part of the GC effluent during temperature programming. As the open split interface is rarely used in bioanalysis, reference to Throck Watson (1985) is recommended.

Similarly, packed GC columns are rarely used in bioanalysis these days so the various modes of linking packed GC columns to a mass spectrometer are not described but they are covered in Throck Watson (1985).

11.2.4   LC inlets

The problem with linking LC to MS is the volume occupied by the liquid phase when it expands into the vapour phase, e.g. 1ml of H_2O converted to vapour occupies 1200ml at stp (standard temperature and pressure) and 1ml of CH_3CN occupies 420ml at

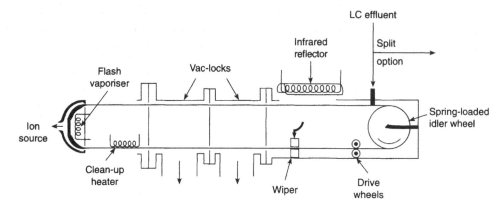

Figure 11.3 Diagram of a moving-belt interface for LC–MS.

stp. This illustrates the problem of linking conventional LC systems operating at 1–2 ml/min to ion sources operating at 10^{-6}–10^{-7} torr (10^{-4}–10^{-5}Pa) as there are no pumps that can mantain a vacuum when being bombarded with a 1000ml/min of gas. In addition, analytes normally have to be in the vapour phase for efficient ionisation.

Secondly, many of the compounds separated by HPLC, particularly reverse-phase HPLC, are so polar and involatile that they are not amenable to conventional ionisation techniques such as electron impact and chemical ionisation. As a result other ionisation processes have been developed; these will be covered in Section 11.3.

Direct liquid introduction (DLI)

As the name suggests, with this inlet, the HPLC is linked directly to a CI source of the mass spectrometer and the HPLC mobile phase acts as the chemical ionisation gas. As CI sources can only operate at reduced pressures of 10^{-3}–10^{-4} torr (0.01–0.13Pa) this interface is limited to flow rates of 30–40μl/min.

Moving-belt interface

This interface consists of an auxiliary vacuum chamber (Figure 11.3) through which a continuous train of events occurs. The belt transports the HPLC effluent under infrared heaters which evaporate the solvent. It then moves through the vacuum locks into the ion source of the MS where the non-volatiles are 'flash vaporised' into the ion source. The belt then passes under a clean-up heater, back through the vacuum locks under a wiper for its final clean and it is then ready to collect some more effluent from the HPLC. Although not used much now it was a useful way of introducing effluents from an HPLC into a mass spectrometer and to obtain library-searchable electron-impact mass spectra of the components in the effluent.

Particle-beam inlet

The particle-beam interface was developed as an alternative to the moving-belt interface. It, too, produces electron impact mass spectra, which is useful for people who

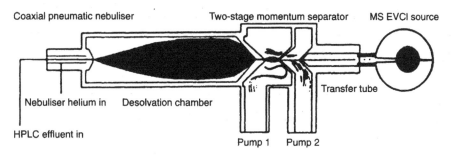

Figure 11.4 Diagram of a particle-beam LC–MS interface.

want to identify components in their HPLC effluent by comparison with a library and/or by its fragmentation pattern. Particle-beam interfaces are still used for tasks such as screening for drugs of abuse, screening for environmental pollutants and identifying breakdown products and impurities, since atmospheric-pressure ionisation is a softer ionisation technique which produces intense pseudomolecular ions with very little fragmentation.

The particle-beam interface is shown schematically in Figure 11.4 and operates as follows: the effluent from the HPLC enters the system through a coaxial pneumatic nebuliser which generates an aerosol. The aerosol passes through a desolvation chamber which is held at approximately 200 torr (26.6kPa) and 60°C. As the droplets are desolvated the more volatile components (such as the HPLC solvent) evaporate, leaving the less volatile components (e.g. analyte) as desolvated particles. At the end of the desolvation chamber, a mixture of helium gas, solvent vapour and desolvated analyte particles enters a two-stage momentum separator which consists of three parts: a nozzle and two skimmers. The vapour, gas and particles leave the nozzle at supersonic velocities. The heavier particles have a significantly higher momentum relative to the vapour and gas molecules and consequently pass through the momentum separator into the source. The lighter vapour and gas molecules have less momentum and are pumped away. This process results in analyte enrichment relative to the mobile phase and a pressure reduction from approximately 200 torr (26.6kPa) in the desolvation chamber to 5–10 torr (0.7–1.3kPa) in the first stage of the momentum separator to less than 0.5 torr (0.07kPa) in the source. Once the particles enter the MS source, they strike the heated source wall and are vaporised and ionised by electron impact ionisation (EI) or CI. The main limitations of this inlet are its limited sensitivity, quantification is highly dependent on the additives in the HPLC mobile phase and normally it has a limited linear response. This means that this interface is very rarely used for quantification but may be used for metabolite identification or verification.

Other inlets

There are other LC inlets but these direct systems require specialised ion source design and are therefore detailed in Section 11.3.

215

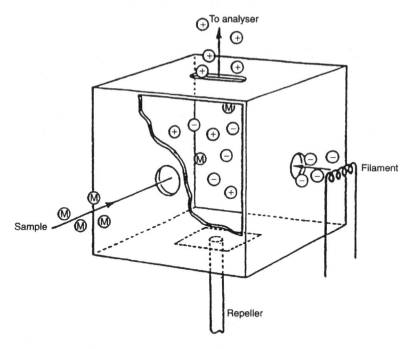

Figure 11.5 Schematic representation of an EI ion source.

11.3 Ion sources

11.3.1 Introduction

There are a number of different methods by which compounds can be ionised. Before 1981, the majority of samples were analysed by electron impact ionisation (EI) and chemical ionisation (CI). These techniques require that the sample is volatilised, i.e. in the gas phase, prior to ionisation. Provided that the sample molecules do not undergo thermal degradation on volatilisation, useful structural information can be obtained. However, these techniques are unsuitable for polar, thermally labile and involatile compounds (many drugs and their metabolites fall into this category) and derivatisation of sample molecules needs to be considered to improve their volatility and thermal stability. Techniques developed since 1981, i.e. fast atom bombardment (FAB), thermospray (TSP), ionspray (ISP), plasma and laser desorption have changed the strategy applied to the mass spectral analysis of polar, thermally labile, involatile molecules. The following sections explain the reasons why.

11.3.2 Electron impact ionisation

A schematic representation of an electron impact ionisation (EI) ion source is shown in Figure 11.5. The sample molecules, which must be in the vapour phase, are bombarded with a stream of high-energy electrons produced by electrically heating a rhenium or tungsten filament. Conventionally, an electron energy of 70eV is used,

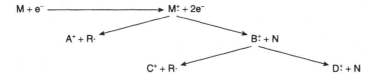

Figure 11.6   Electron impact ionisation of a molecule, M. R, radical; N, neutral.

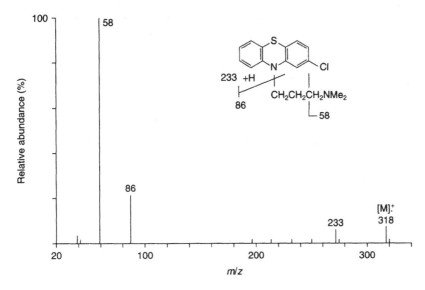

Figure 11.7   EI mass spectrum of chlorpromazine.

because it is sufficient to ionise the molecule and subsequently break bonds. Only about 0.1% of the molecules in the source are ionised.

In the ionisation process an electron is removed from the molecule, leaving a positively charged species with an odd electron, called the molecular ion, M^+. For most organic molecules the energy required to do this is about 10eV, hence excess energy is also transferred to the charged species during electron bombardment. This may be dissipated by the molecular ion undergoing fragmentation, by ejection of either a radical or a neutral species (Figure 11.6). If fragmentation is extensive, the molecular ion may not be present in the mass spectrum.

The EI mass spectrum of chlorpromazine is shown in Figure 11.7. A mass spectrum represents a snapshot of the abundances of the molecular and fragment ions plotted against their mass to charge ratio (m/z). The ion of highest abundance, m/z 58 in this instance, by convention is called the **base peak** and is given a arbitrary abundance of 100. The relative abundance of each peak is represented as a percentage of the base peak.

The mass spectrum of chlorpromazine shows a molecular ion of low abundance at m/z 318. The presence of an ion two mass units higher at m/z 320, which is approximately one-third of the abundance of the ion at m/z 318, is characteristic of the presence of chlorine in the molecule (chlorine-37 is the naturally occurring isotope of chlorine-35 and is present at about one-third of the abundance of chlorine-35). It

217

$$NH_3 + e^- \rightarrow NH_3^+ + 2e^-$$

Reagent gas

$$NH_3^+ + NH_3 \rightarrow NH_4^+ + NH_2\cdot$$

Reagent ion

either $M + NH_4^+ \rightarrow [M + H]^+ + NH_3$

or $\quad M + NH_4 \rightarrow [M + NH_4]^+$

Figure 11.8 Process of CI with reference to ammonia. M = sample molecules.

should be remembered that the molecular ion gives a molecular weight based on the sum of the masses of the lowest naturally occurring isotopes so for chlorpromazine these are $^{12}C_{17}$ $^{1}H_{19}$ $^{14}N_2$ ^{32}S ^{35}Cl, i.e. the monoisotopic mass of chlorpromazine is 318.01, hence the ion at m/z 318. This is different to the average molecular weight which takes into account the relative amounts of each naturally occurring isotope for an element, e.g. $^{12}C/^{13}C$, $^{1}H/^{2}H$, $^{14}N/^{15}N$, $^{32}S/^{33}S/^{34}S$, $^{35}Cl/^{37}Cl$. For chlorpromazine the average molecular weight is 318.87. The average molecular weight is the weight required for calculating the weight required for preparing mol/l or g/l solutions. It must not be used in mass spectrometry.

The mass spectrum shows some fragment ions; their structures are shown in Figure 11.7. The ion at m/z 233 still contains chlorine, as shown by the presence and relative abundance of the ion at m/z 235, whereas the ions at m/z 58 and 86 do not, since there are no ions present at m/z 60 and 88.

11.3.3 Chemical ionisation

During electron bombardment (EI) the excess amount of energy absorbed by the molecular ion may result in complete fragmentation, thus the ion of highest mass in the mass spectrum may not be representative of the molecular weight of the compound. Chemical ionisation (CI), first demonstrated by Munson and Field in 1966 (Munson 1977) provides a less energetic method of producing ions. It is known as a 'soft' ionisation technique.

A reagent gas (methane, isobutane and ammonia tend to be the most commonly used gases) is introduced into the ion source at a pressure of about 133Pa. Figure 11.8 illustrates the process of CI with reference to ammonia. First, some of the molecules of ammonia are ionised by electron bombardment, these then undergo collision with neutral molecules resulting in the formation of secondary or reagent ions. The latter collide with the gaseous sample molecules, leading to ionisation either by proton transfer or by addition, to give an adduct ion. In the gas phase the ammonium ion will readily protonate molecules more basic (e.g. aliphatic amines) than itself, whereas with molecules that are less basic, it may form ammonium adduct ions.

In contrast to EI mass spectra, which are characterised by a large number of fragment ions, CI mass spectra tend to be dominated by protonated molecules with few fragment ions (Figure 11.9). Thus the techniques provide complementary information. CI being used to obtain the molecular weight of a compound, EI being used for structural information.

Negative ion formation is more readily achieved in CI than in EI, although formation of a M^- is 100 times less favourable than M^+. The types of processes occurring:

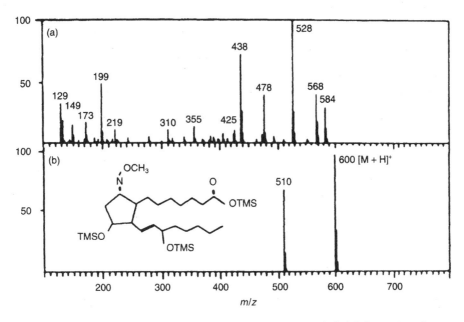

Figure 11.9 EI and CI mass spectra of a methoxime, trimethylsilyl derivative of prostaglandin E₁ (molecular weight 599Da). (a) EI mass spectrum; (b) ammonia CI mass spectrum.

$$AB + e^- \rightarrow AB^{\overline{\cdot}} \qquad \text{Resonance electron capture}$$

$$AB + e^- \rightarrow A^- + B\cdot \qquad \text{Dissociative electron capture}$$

$$AB + e^- \rightarrow A^- + B^+ + e^- \qquad \text{Ion pair formation}$$

are not 'true' chemical ionisation since they result from the presence of an excess of thermal electrons produced during the chemical ionisation process. Ion/molecule reactions, such as:

$$AB + [OH]^- \rightarrow [AB - H]^- + H_2O$$

$[OH]^-$ produced from N_2O/CH_4 mixture

are characteristic of chemical ionisation. Negative-ion CI, while providing complementary molecular weight information to positive-ion CI, can sometimes provide a very sensitive method for the quantitative determination of drugs (Chapter 12).

11.3.4 Atmospheric-pressure chemical ionisation

Atmospheric-pressure ionisation sources in mass spectrometry are not new. Their use with liquid chromatography was first reported by Horning *et al.* in 1974. More recently, interest has focused on the use of a heated nebuliser interface for combining HPLC with atmospheric-pressure chemical ionisation (APCI) mass spectrometry. The interface and ion source are shown schematically in Figure 11.10; the HPLC eluent (up to 1.5ml/min) flows through a narrow-bore stainless-steel tube, directly into the

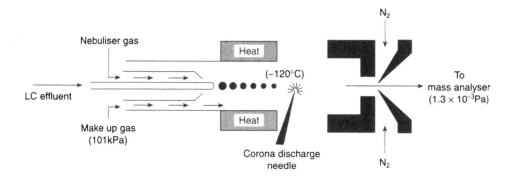

Figure 11.10 Schematic representation of a heated nebuliser interface and atmospheric-pressure chemical ionisation source.

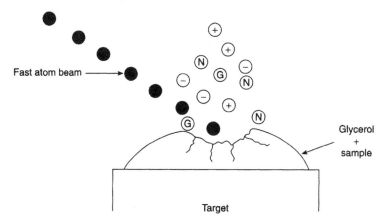

Figure 11.11 Representation of processes occurring during fast atom bombardment.

nebuliser. The eluent is nebulised by a coaxially introduced gas stream. The droplets are swept by the gas stream through a heated tube and solvent and solute are volatilised. Electrons produced in the region of the corona discharge needle ionise the volatilised solvent molecules. The latter act as a reagent gas for the chemical ionisation of the solute molecules. Ions are accelerated via a small orifice into the analyser of the mass spectrometer.

11.3.5 Fast atom bombardment

Fast atom bombardment (FAB) was introduced by Barber and co-workers in 1981. The technique is used extensively for the analysis of polar, thermally labile and involatile compounds. A schematic representation of the FAB process is shown in Figure 11.11. Typically, between 0.1 and 2mg of the sample is added (in solution) to 1–2/ml of a liquid matrix (a viscous solvent of low vapour pressure) on a metal target. The target is introduced into the FAB ion source where it is bombarded by a stream of fast (energetic) xenon atoms or caesium ions. The energy transferred to the solution on the target results in both ions and neutral species representative of both the matrix and the sample

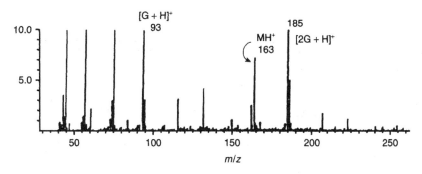

Figure 11.12 FAB mass spectrum of nicotine, showing ions derived from the sample and the glycerol matrix. G, glycerol; molecular weight, 92Da.

being sputtered from the target. Stable positive $[M + H]^+$ and negative $[M - H]^-$ ions are generated by this technique. The ions are sampled by the mass spectrometer.

FAB is a surface technique and the liquid matrix is important in ensuring that the surface is continuously replenished with the sample, so that mass spectra can be obtained over a period of time. Glycerol tends to be the most commonly used matrix, but several others, such as thioglycerol, and nitrobenzyl alcohol, are used. It should be remembered that FAB mass spectra can show ions derived from both the matrix and the sample (Figure 11.12).

This technique is also referred to as liquid secondary ion mass spectrometry (LSIMS), which reflects more accurately the use of either fast atoms or fast ions for ionisation.

11.3.6 Thermospray

Historically, the first most popular method available for combining liquid chromatography with mass spectrometry is the thermospray interface. This technique, introduced by Vestal and co-workers in 1980 (Blakley *et al.* 1980), is both a means of interfacing HPLC with mass spectrometry and a method of ionising polar, thermally labile and involatile molecules. A schematic diagram of a thermospray source is shown in Figure 11.13.

Thermospray works optimally with flow rates of 1–2ml/min and mobile phases of high aqueous content containing a volatile buffer such as ammonium acetate. The total eluent from the HPLC column is introduced into the mass spectrometer through a small-diameter stainless-steel tube, typically 100–150μm i.d. The capillary tube is heated, with the result that the liquid emerges as a superheated mist of droplets transported in a supersonic vapour jet. There are two types of ionisation processes believed to be occurring in thermospray, ion evaporation and gas-phase chemical ionisation.

Ion evaporation

This process is shown schematically in Figure 11.14. Ion evaporation is a process in which ions are emitted from charged droplets directly into the gas phase. The droplets produced in thermospray carry a surface charge which statistically may be positive or negative. The droplet itself contains ions derived from the eluent and solute.

221

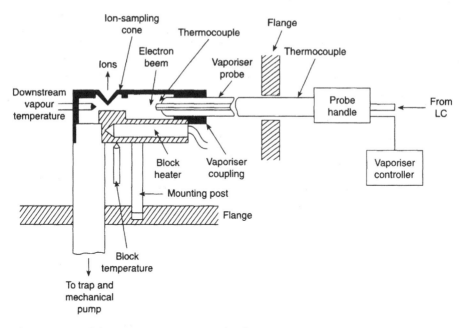

Figure 11.13 Schematic representation of a thermospray ionisation source.

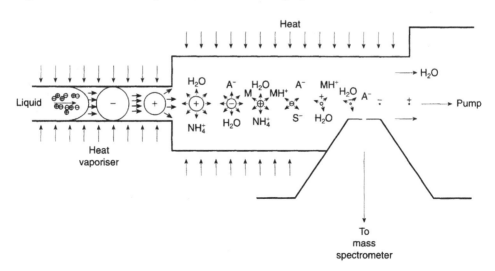

Figure 11.14 Proposed model for ion evaporation in thermospray.

As the droplets become smaller, desolvating under the influence of the vacuum and ion source temperature, the local electrical fields generated on the droplet surface increase. This effectively increases the field gradient across the charged droplets which eventually reaches a high-enough value for the ejection or evaporation of free ions from within the droplet. A proportion of the ions pass into the mass analyser via a sampling cone. A rotary pump attached to the ion source removes most of the vaporised eluent, thus maintaining the operating vacuum within the analyser of the mass spectrometer. The ion evaporation process is enhanced if the solute is ionic (hence the pK_a of the solute and the solution pH should be considered) and the solvent should contain a volatile electrolyte such as ammonium acetate.

Gas-phase ionisation

It has been suggested that thermospray ionisation may arise predominantly from a gas-phase process in which the reagent ions, NH_4^+ and CH_3COO^-, from the buffer, react with the sample molecules in a chemical ionisation process:

$$M + NH_4^+ \rightarrow [M + H]^+ + NH_3$$

or

$$M + CH_3COO^- \rightarrow [M - H]^- + CH_3COOH$$

Thermospray ion sources are also equipped with a filament or discharge electrode to promote chemical ionisation processes. Thermospray mass spectra are dominated by protonated or deprotonated molecular ions (Figure 11.15).

11.3.7 Electrospray

Important recent additions to commercially available ionisation techniques are electrospray and ionspray. A schematic diagram of an electrospray source is shown in Figure 11.16. Eluent, such as methanol/water containing the charged analyte, flows at a rate of 2–5µl/min through a needle (i.d. 0.1mm) held at a high field potential of typically 6kV. This disrupts the liquid, causing it to spread out as a plume of droplets which have a high surface charge. The process of ion evaporation described for thermospray is the dominant ionisation process occurring in electospray. However, unlike thermospray which generates both positively and negatively charged droplets, the surface charge of droplets produced in electrospray is dependent on the polarity of the applied electric field. Hence ions of only one polarity evaporate from the droplet.

This process takes place at atmospheric pressure and ambient temperature. The ions are drawn into the analyser of the mass spectrometer through a series of lenses, to focus the ion beam, and skimmers, to reduce pressure. Nitrogen gas flowing at 200–300ml/min into the source region prior to the lenses and skimmers serves two purposes, it prevents solvent vapour from entering the mass spectrometer and it breaks the hydrogen bonds in ion/solvent clusters which would otherwise complicate the mass spectrum. The first orifice/skimmer the ion beam goes through is commonly called the cone.

Ionspray, or pneumatically assisted electrospray, utilises a coaxially introduced gas stream, usually nitrogen, as well as a high electric field to create ionised droplets. Ionspray has the advantage over electrospray of being able to handle liquid flow rates up to 200µl/min. This allows direct connection of small-bore (2mm i.d.) HPLC columns, or split flow for connection to conventional bore (4.6mm i.d.) columns.

Electrospray/ionspray ionisation produces protonated or deprotonated molecular ions for polar, thermally labile and involatile compounds. Furthermore, these techniques can generate multiply charged ions from compounds such as proteins and

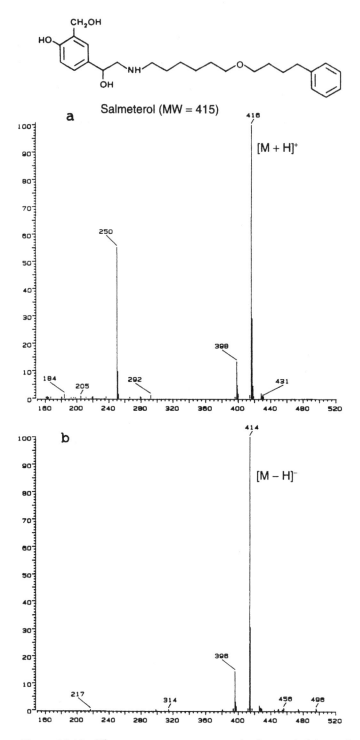

Figure 11.15 Thermospray mass spectra of salmeterol: (a) positive ion; (b) negative ion.

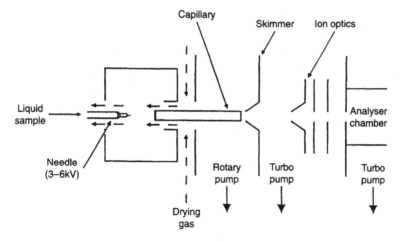

Figure 11.16 Schematic representation of an electrospray ionisation source.

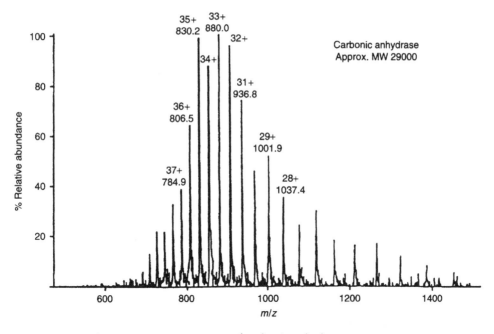

Figure 11.17 Electrospray mass spectrum of carbonic anhydrase.

peptides (Figure 11.17). The multiple charges produce ions whose m/z values fall within the range of a quadrupole mass spectrometer, since z is much greater than 1.

11.3.8 Other desorption techniques

Currently, there is considerable interest in the mass spectrometric analysis of large biomolecules (masses greater than 10 000Da). Desorption and ionisation of these molecules are achieved efficiently using highly energetic particles (100MeV energy) rather than the low-energy (keV) particles encountered in FAB. Two techniques which are very successful in this area are californium-252 plasma desorption and matrix-assisted laser desorption. As we are concentrating on techniques to analyse and measure small molecules, i.e. less than 1000Da, I have not reviewed these ionisation techniques. An excellent comprehensible and practical review of all the ionisation techniques used in mass spectrometry is covered in a recent addition to the RSC Analytical Spectroscopy Monographs (Ashcroft 1997).

11.4 Analysers

The common analysers used to separate the ions formed in the source of the MS are summarised in Table 11.1. The main physical and mathematical processes involved in each mode of ion separation are covered in the following sections.

11.4.1 Single-focusing magnetic instruments

The simplest form of this instrument is one that has a single magnet which generates a magnetic field to perform the mass analysis, as shown in Figure 11.18. The ions formed in the source are accelerated out of the source by the application of a voltage (V) between the source and the detector, hence the acceleration of an ion with a charge of magnitude (e) with (z) charges in an electrostatic field of voltage (V) imparts a kinetic energy of $\frac{1}{2}mv^2$ to the ion, where m is its mass and v its final velocity.

Table 11.1 Summary of analysers, i.e. modes of ion separation

Ion transport methods	
Electric/magnetic sector(s)	Deflection of ions in an electric/magnetic field. Up to the 1970s this was the workhorse for mass analysis. In combination, it can have high resolution/high sensitivity and a broad mass range. Operates best at 'low pressure', $< 10^{-7}$ torr. Initial problems with scanning speed. Large and expensive
Time of flight (ToF)	Simplest method, very high response and very high mass capability. Resurgence in protein, peptide field with MALDI
Quadrupole	Wide pressure tolerance for ion entry conditions (i.e. $> 10^{-5}$ torr). Cheap, compact, limited mass range and resolution capabilities
Hybrid	Combination of two or more of the above to increase versatility
Ion storage methods	
Fourier transform (ion cyclotron resonance spectroscopy)	Relative newcomer, capable of high resolution, good mass range. Operates at low pressure. Limited applications. More for ion/molecule reactions than bioanalysis
Ion trap	Newcomer. Used for GC and API techniques. Now developed to MS–MS and MS^n. Useful for metabolite identification. Many problems with quantification, especially 'needle in a haystack' applications such as bioanalysis

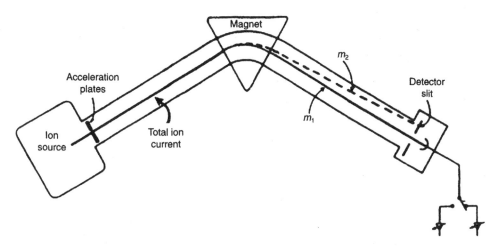

Figure 11.18   Schematic layout of a single-focusing magnetic sector mass spectrometer.

This relationship is expressed in equation 11.1.

$$zeV = \tfrac{1}{2}mv^2 \tag{11.1}$$

In order that ions of the same *m/z* value are detected at any one time by the detector, a collector slit is placed in front of the detector. The trajectory of each ion with a particular mass is altered by the passage through a magnetic field applied at right angles to the accelerating voltage. When this ion enters the magnetic field (*B*) it is subjected to a force *Bzev* which produces an acceleration toward the centre of the magnet mv^2/r, where *r* is the radius of the magnet. This relationship is expressed:

$$Bzev = \frac{mv^2}{r} \tag{11.2}$$

If equations 11.1 and 11.2 are combined to eliminate the velocity (*v*), then equation 11.3 is generated:

$$m/ze = \frac{r^2B^2}{2V} \tag{11.3}$$

As *r* is fixed and *e* is a constant, then mass analysis in this instrument is normally achieved by altering the magnetic field (*B*) or the accelerating voltage (*V*). However, because the quality of the mass spectrum in terms of ion abundance and ion focus is related to the magnitude of the accelerating potential, the mass analysis in this instrument is normally achieved by altering *B* rather than *V*, and equation 11.3 becomes:

$$m/z \propto \frac{B^2}{2V}. \tag{11.4}$$

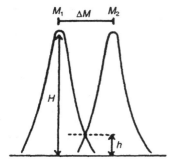
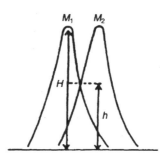

Figure 11.19 Graphical representation of unit resolving power with 10% valley (h/H) definition (left) and 50% valley definition (right).

A typical single-focusing magnetic sector mass spectrometer scans from m/z 10 to about m/z 750 and has a resolving power of about 1000 (10% valley, see Figure 11.19 for definition). Resolving power is defined by:

$$R = M/\Delta M \tag{11.5}$$

where M is the mean m/z value of two adjacent ions and ΔM is the difference between these ions. The allowable overlap between the two peaks is usually defined as 10% of the peak intensities, where the intensities of the two peaks are equal; for example, in Figure 11.19 $h/H \times 100 = 10\%$.

Care must be taken when discussing resolution of an MS since some manufacturers use a 50% valley rather than 10% when stating the resolving power of the instrument. The limitation of a single-focusing MS is that peak assignment can only be guaranteed to ±0.5 of a mass unit, commonly termed **nominal mass**. To achieve a higher resolution to achieve accuracy of a few parts per million, i.e. mass measured to 4–5 decimal places, multi-sector (i.e. 2–4 sector) instruments have been developed.

11.4.2 Double-focusing instruments

The equation for ion separation given in Section 11.4.1 makes no provision for the fact that ions of the same m/z value can have a range of kinetic energies due to the ionisation process and angular divergence due to charge repulsion. To remove this effect and improve the resolution of the instrument an electric sector can be placed between the ion source and magnet. This electric sector produces a radial electric field through which the ions are deflected. This deflection is independent of mass but proportional to energy, hence ions of different energies can emerge from the electric sector on different paths. These paths can be further analysed after passing through a magnet. For descriptions of the various geometries of these instruments and their application for accurate mass analysis see any traditional MS textbook (for example Millard 1978; Throck Wilson 1985). Their operation in tandem MS modes is discussed in Busch *et al.* (1988). The applicability of the increased resolution of m/z available from these double-focusing instruments will be illustrated in Chapter 12.

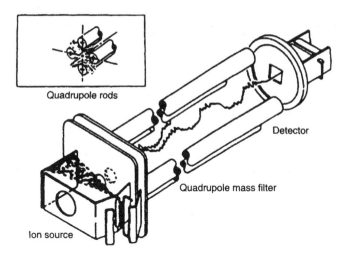

Quadrupole rods

Detector

Quadrupole mass filter

Ion source

Figure 11.20 Diagram of a quadrupole mass filter. The inset illustrates the ideal quadrupole hyperbolic electric field dispersion.

11.4.3 Quadrupole analysers

This non-magnetic analyser employs a combination of direct current (DC) and radio-frequency (RF) potentials to achieve mass analysis. Mechanically, the quadrupole consists of four parallel rods arranged symmetrically, as indicated in Figure 11.20. Ideally, these four rods should have the shape of a hyperbola in cross-section so that idealised hyperbolic fields can be produced according to quadrupole theory. However, in practice, cylindrical rods are often used to approximate the hyperbolic-field requirements. Opposite rods (i.e. those diagonally opposite) are connected together electrically and to RF and DC generators. Ions are extracted from the ion source and are accelerated (5–15V) into the quadrupole region between the four rods and along their longitudinal axis.

As the ions drift in the quadrupole region toward the detector they are influenced by the combined DC and oscillating RF fields. As the ions drift toward the detector along the Z axis, their motion in the X–Y plane is described by the Mathieu equation which defines a and q parameters as follows:

$$a_x = -a_y = 4zeU/mw^2r_0^2 \tag{11.6}$$

$$q_x = -q_y = 2zeV/mw^2r_0^2 \tag{11.7}$$

where r_0 is half the distance between opposite quadrupole rods, w is the alternating (RF) frequency, U is the magnitude of the DC field, and V is the amplitude of the RF field ($V \cos wt$ = instantaneous value of the RF field). Although the Mathieu equation is complex, it can be simplified to m/z is proportional to RF/DC. The parameters a and q define a stability diagram (Figure 11.21) that shows regions of (a, q) space in which ions have mathematically stable trajectories in the quadrupole field. The stable region is the patchwork area.

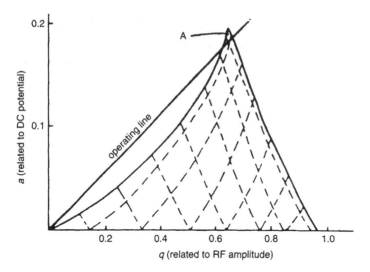

Figure 11.21 Stability diagram of *a* against *q* (equivalent to DC, RF), showing regions (patchwork area) that correspond to mathematically stable ion trajectories in a quadrupole mass spectrometer.

Two important points can be made while viewing Figure 11.21. For example, if the quadrupole were operated with a DC potential and an RF amplitude that defined the operating line in the stability diagram, only those ions with *m/z* values defined by the (*a, q*) space coordinates at the tip of the diagram would have stable trajectories. If the DC/RF ratio were adjusted to raise the operating line even higher, the resolving power of the instrument would increase, but the sensitivity would decrease, because fewer ions would follow a stable trajectory to reach the detector. In the RF-only mode, there is no DC field; thus $a = 0$, $U = 0$ in equation 11.6. Therefore, the operating line lies horizontally along the abscissa of the stability diagram, indicating that ions of all *m/z* values have stable trajectories and are transmitted through the quadrupole filter. This is known as 'RF-only' operation of a quadrupole.

As illustrated in Figure 11.20, the ion trajectories through the central space between the rods are complicated, and for any given level of DC/RF voltage, only ions of a specific *m/z* avoid collision with rods and successfully traverse the quadrupole filter to reach the detector, all other ions collide with the rods at some point. The entire mass spectrum is scanned as the voltages are swept from a pre-established minimum to a maximum value (approximately 400V DC and 2000V RF), but at a constant DC/RF ratio.

The quadrupole analysers are capable of scanning a mass spectrum in a few milliseconds. The mass range of the quadrupole is dependent on the output of the RF coil. Recent technological advances have made RF oscillators of improved output and stability so most of the commercial instruments can scan between *m/z* 10–1000 while some can extend to 4000. The resolution/resolving powers of most quadrupoles are identical to that of single-focusing magnetic sector instruments, i.e. nominal mass. The quadrupole is very well suited to selected ion monitoring (SIM) (quantification) since ions selected from any region of the mass spectrum can be monitored without altering the optimum conditions in the ion source or mass analyser. Furthermore, unlike changing the magnetic field, the parameters (superimposed RF and DC fields) can be altered rapidly, with stability and good response throughout the mass range.

11.4.4 Time of flight (ToF) analysers

The operating principle of a ToF MS involves, as the name suggests, measuring the time taken for an ion to travel from the ion source to the detector. After ionisation the ions are accelerated by an electric field (V) and gain kinetic energy $\frac{1}{2}mv^2$. As they have different masses but similar kinetic energies, they separate into groups according to velocity (and hence mass) as they traverse the region between the ion source and detector. The m/z value of an ion is determined by its time of arrival at the detector. Ions of low mass reach the detector before those of high mass. Solving equation 11.1 for v (velocity):

$$v = (2zeV/m)^{1/2}$$

Because it is impractical to measure ion velocity directly, the time of flight from the source to the detector, of distance L, is given by the equation:

$$\mathrm{ToF} = \frac{(m)^{1/2}L}{2zeV}.$$

This type of mass analyser is becoming popular in instruments that employ laser desorption techniques to obtain molecular weight information of large molecules, (i.e. > 5000Da molecular weight). Increased resolution and sensitivity have been achieved by the introduction of a reflective plate, known as a reflectron, within the flight tube (Ashcroft 1997).

11.4.5 Ion-trap mass analysers

The ion-trap mass spectrometer (ITMS) can be classified in the same category of mass analyser as a quadrupole. It is a three-dimensional quadrupole and similar equations for ion motion apply. The ion trap consists of a hyperbolic cross-section central ring electrode (a doughnut) and two hyperbolic cross-section end-cap electrodes. Figure 11.22 shows a cross-section of an ITMS. To obtain a mass spectrum, an RF voltage of variable amplitude and fixed frequency is applied to the central electrode and the end caps are grounded. As no DC voltages are used, the device can be thought of as a three-dimensional RF-only quadrupole. All ions above a certain m/z, determined by the RF amplitude, have trajectories that keep them trapped within the electrodes. As the RF amplitude is increased, ions of increasing mass sequentially become unstable in the z direction and are ejected from the trap through the end caps.

The operating sequence of the ITMS begins with the injection of a pulse of electrons into the ion trap to ionise a gaseous sample, or ions in the case of LC inlets using electrospray. The ring electrode is held at a low RF amplitude to trap all of the ions formed during this ionisation pulse. The RF amplitude is then scanned upward, and ions of increasing mass are sequentially ejected from the ion trap and detected with an electron multiplier. A key parameter in the operation of the ion trap is the

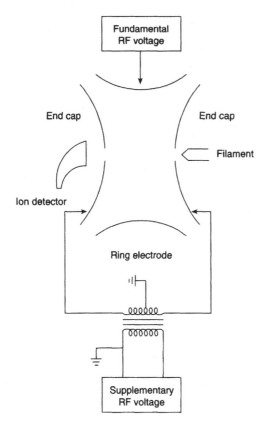

Figure 11.22 Cross-sectional diagram of an ion-trap mass spectrometer.

background gas pressure. Typically helium is added as a bath gas at a pressure of about 10^{-3} torr (0.1Pa). Collisions of the ions with the bath gas damp the trajectories of the ions toward the centre of the ion trap, providing better resolution and sensitivity. The scan range of a typical ITMS is m/z 600 with unit resolution.

This instrument was launched during 1983 as a low-cost mass analyser for GC–MS. It had some severe limitations in bioanalysis, the main one being a limitation in the number of ions ejected from the trap to the detector that could be processed by the data system before the signal saturated. When saturated, the data system switched off the ionisation voltage. As the background from bioanalytical samples is high, saturation of the signal was very easy to achieve during selected ion monitoring mode. This drawback of the instrumentation and its limited mass range have restricted its application in bioanalysis.

This analyser can also be operated in the MS–MS mode. To achieve this, a DC voltage (with the RF voltage) is applied to the ring electrode at an appropriate level so that ions of a specific m/z have a stable trajectory in the ion trap and all other m/z ions are ejected. At this point only the parent ion is left in the ion trap. The parent ion is then accelerated (excited) axially by application of a supplementary RF voltage to the end caps, causing the parent ion to undergo more energetic collisions with the background bath gas. The daughter ions formed by collision-induced dissociation are then detected by the conventional RF amplitude ramp of the ring electrode. At present the ITMS in MS–MS, or M to the nth (M^n) mode, is still in its infancy but is beginning to be useful in bioanalysis for identifying metabolites (Moseley *et al.* 1996).

Table 11.1 also mentions another ion storage method of mass analysis, fourier transform ion resonance [FT(I)MS]. This technique has not been applied routinely for bioanalysis and its application is unlikely in the near future. For explanation of this form of ion trapping and mass analysis, see Throck Watson (1985) and Busch *et al.* (1988).

11.5 Detectors

The ions from the mass analyser can be detected by a variety of devices, including photographic plates, faraday cups, electron multipliers, photomultipliers and photodiode arrays. The detector that is used in a majority of mass spectrometers, which satisfies the sensitivity, accuracy and response times necessary for bioanalytical mass spectrometry, is the electron multiplier. The other detectors are described in Throck Watson (1985).

11.5.1 Electron multipliers

The majority of the electron multipliers (EMs) in current instrumentation are the continuous dynode type, as shown in Figure 11.23. The ion beam from the mass analyser enters into the mouth of the cornucopia, hits the resistive coating (the continuous dynode) on the internal surface, which captures the ion beam and converts it into secondary electrons. The secondary electrons are attracted along the positive

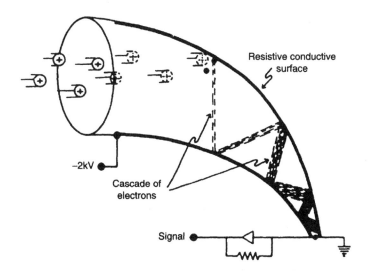

Figure 11.23 Diagram of a continuous dynode electron multiplier.

electrical gradient, deeper into the horn; each time these electrons collide with the wall, additional secondary electrons are expelled, providing amplification. These electrons are eventually measured by a pre-amplifier operating in analogue or pulse-counting mode and the signal transmitted to a computer for data processing.

11.5.2 Negative-ion detection

Modification of a magnetic MS to detect negative ions instead of positive ions is relatively simple. The polarity of the accelerating voltage and the magnetic current is reversed. The negative ions have sufficient kinetic energy, accelerating voltage typically 4kV, to enter the electron multiplier, even though the multiplier has a negative bias (charge) on its conversion dynode. In a quadrupole MS the ions have a relatively low kinetic energy, the accelerating voltage being 3–15V, hence the negative bias on the dynode of the electron multiplier would repel the negative ions and they would not be detected. In these instruments a special electrode that converts negative ions to positive ions is mounted near the entrance of a conventional electron multiplier so that the negative-ion beam is converted into positive ions by the first electrode then amplified by the electron multiplier.

11.6 Data acquisition and processing

With current instrumentation the computer linked to a MS performs three operations: instrument control, data acquisition and data processing. Ideally it should perform all these processes simultaneously, but in some instruments the instrument control and data acquisition are performed by one computer and the data processing is performed by a second unit.

11.6.1 Instrument control

The data system can monitor and control all important instrument parameters. The scope of this control and the range of parameters controlled can vary, but a majority of instruments have computerised control of the inlet systems. For example, the GC or HPLC control of the ion source and detector, which enables automated tuning for sensitivity and resolution and simplifies routine performance checking; switching between alternative modes of ionisation such as EI and CI; switching between positive and negative ionisation; and controlling the mass analyser to ease the operation during quantification (i.e. collecting data at a specified m/z value for tandem processes when using more sophisticated analyser systems such as double-focusing magnetic sectors, triple quadrupoles or hybrid instruments). A/D converters are used to input the required parameters from the MS into the data system, while instrument control is effected via D/A converters.

11.6.2 Data acquisition/preliminary data processing

The first job is to convert the time axis of a spectrum of ions detected by the MS into a mass axis. This is normally accomplished by introducing a sample of a compound whose mass spectrum is well known. The most common calibrants for EI operation is perfluorokerosene (PFK) which is suitable for m/z 30–800 and heptacosafluoro-tributylamine, sometimes referred to as FC43, which is suitable for m/z 40–600 and most often used for quadrupole MS. By matching and equating the acquired time/intensity file with the reference table of the calibrant, a mass/time conversion file, called a calibration table, is produced. This calibration table is then used to convert any subsequently acquired time/intensity files into mass/intensity files. Other calibrants used are fomblin oil (a polyfluoroether), m/z 3000; caesium fluids (m/z 3000, FAB); and a range of different molecular weight polyethylene glycols (PEGs).

The data obtained during a scan can be stored in two forms. In profile mode, all the digital samples are retained and the calculation of the centroids and sizes of the peaks are then calculated any time during or after acquisition. To acquire centroid data these calculations are performed in real time as the data are acquired. Then only the centroid size of each peak is stored and the original shape is lost. Compared with the centroid method, the profile method requires less computer time but more storage space. It also has the advantage that the shapes of the acquired peaks may be examined later for multiplicity, resolution and checking the operation of the MS. Most data systems allow you to select which mode of operation you prefer, although as computer storage has increased enormously over the past few years, profile ion (ion counting) is the most common mode of operation.

11.6.3 Secondary data processing/data presentation

A variety of data processing methods are used to extract specific analytical information from scanning or selected ion monitoring (SIM) experiments. The major application of selected ion monitoring is quantification, and the software requirements are typically procedures for calibration and peak height or area measurement. This will be covered in more detail in Chapter 12. When using the MS in scanning mode, especially when linked to a chromatograph, the normal strategy is to scan the required mass range as quickly as possible so that mass spectra with good signal-to-noise ratios can be obtained. Normally this means acquiring a new mass spectrum at least once every second. This has the advantage of not missing any useful information but has the disadvantage of acquiring a huge number of spectra. A variety of algorithms have been designed to help the user examine large numbers of mass spectra.

Total-ion current (TIC) chromatogram

The TIC comprises the summed ion abundances of each spectrum scanned, normalised to the largest of the values and plotted against time and scan number (Figure 11.24). Any spectrum of the chromatogram can be identified by its scan number and recalled for examination. For example the spectrum of the first peak, scan number 127, is shown in Figure 11.25 in the traditional bar graph format. The ordinate is the relative-intensity scan and the abcissa is the m/z scale. The scale of the ordinate is always 100% because the bar graph is always plotted as a percentage of the base peak which is the 100% value. In this example the base peak is m/z 315.

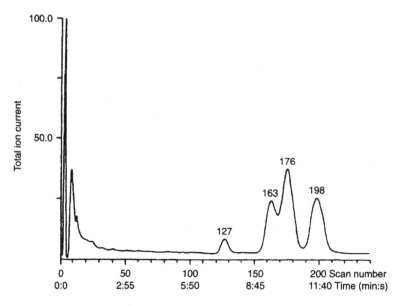

Figure 11.24 A TIC chromatogram for the analysis of sterol derivatives by GC–MS. Mass spectra were recorded every 3.5s; the peaks are identified by the number of the scan in which they maximise.

Background subtraction

Although the mass spectrum displayed in Figure 11.25a contains a lot of m/z values, to an experienced mass spectrometrist, it also contains a lot of ions that are constantly eluting from the GC. These can hinder interpretation of the structural identity of the compound that eluted during scan 127. This problem can be overcome by a numerically simple program that allows the operator to select a spectrum, or number of spectra, containing only background ions. This spectrum or average spectrum is then subtracted from the spectrum of interest, generating a background-subtracted mass spectrum as shown in Figure 11.25b. This process has completely changed the mass spectrum; the base peak is now m/z 129 instead of 315 and the largest m/z ion is now 458, which is actually the M^+ ion for this molecule. After the spectrum has been background subtracted, the operator can try to identify the compound from its fragmentation pattern or see if a similar mass spectrum is present in the mass spectral library stored on the data system. Systems now provide automatic background subtraction using algorithms linked to the TIC, and they can average the spectra. Averaging spectra is commonly termed 'smoothing'.

Library search routines

There is variety of proprietary retrieval systems. In the forward search routine, the unknown spectrum is compared with each of the entries in the library spectra. A match will always be achieved, but will be qualified by some sort of 'agreement index', i.e. number of common m/z values. In the reverse search, a given library entry is compared with the unknown for both m/z value and intensity. With this system a

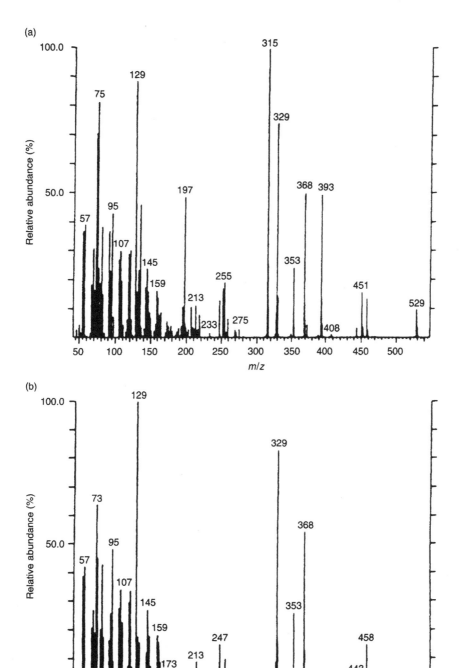

Figure 11.25 Mass spectra of scan number 127 before (a) and after (b) background subtraction.

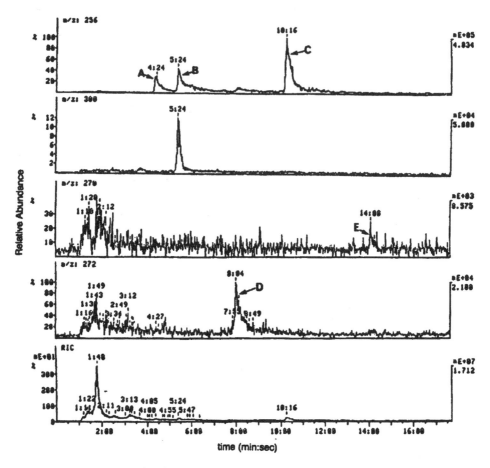

Figure 11.26 An example of the added information supplied by mass chromatograms. The chromatogram labelled RIC is equivalent to TIC.

match will only be achieved if the unknown contains peaks of similar intensity that are present in the reference spectrum. There are other, more sophisticated systems, such as weighting the m/z values to the intensity of the peaks and calculating a probability of the match being positive. Another allows the user to define search criteria, such as upper or lower limits of peak intensity, at m/z values that characterise the unknown.

In the pharmaceutical industry, when trying to identify metabolites of drugs, traditional library search routines are of very little use. Normally, processing and interpreting mass spectra are performed by applying MS interpretative skills such as isotope ratios and common fragmentation patterns (McLafferty 1992). This knowledge is then correlated with the structure of the parent drug, possible sites of metabolism and the likely physico-chemical properties of such metabolites. This will be illustrated in Chapter 12.

Mass chromatography

There are frequently times in bioanalytical data processing when the summed intensities of the ion abundances in the spectra do not vary enough to generate peaks and troughs in the TIC. When this occurs the operator has the ability of generating mass chromatograms, i.e. normalised plots of ion current at specified m/z values. In the example shown in Figure 11.26 the data were obtained by thermospray LC–MS of a urine sample. The m/z 256 ion was the protonated molecular ion for the administered drug, a novel anti-epiletic drug lamotrigine, and m/z 270 and 272 were the expected protonated molecular ions for a methylated and N-oxide metabolite. The complete story behind this example can be found in Doig and Clare (1991). It is an example of how thermospray LC–MS confirmed the metabolism of this drug in animals and man.

11.7   Bibliography

Ashcroft, A.E. (1997) *Ionisation Methods in Organic Mass Spectrometry*, Cambridge, UK: The Royal Society of Chemistry.

Barber, M., Bordoli, R.S., Sedgwick, R.D. and Tyler, A.N. (1981) Fast atom bombardment of solids (F.A.B.): a new ion source for mass spectrometry. *J. Chem. Soc., Chem. Commun.*, (7), 325–327.

Blakley, C.R., Carmody, J.J. and Vestal, M.L. (1980) A new soft ionization technique for mass spectrometry of complex molecules. *J. Am. Chem. Soc.* 102, (18), 5931–5933.

Busch, K.L., Glish, G.L. and McLuckey, S.A. (1988) *Mass Spectrometry/Mass Spectrometry: Techniques and Applications of Tandem Mass Spectrometry*, New York: VCH Publishers, Inc.

Doig, M.V. and Clare, R.A. (1991) Use of thermospray liquid chromatography–mass spectrometry to aid in the identification of urinary metabolites of a novel antiepileptic drug, Lamotrigine. *J. Chromatogr.*, 554, (1–2), 181–189.

Horning, E.C., Carroll, D.I., Dzidic, I., Haegele, K.D., Horning, M.G. and Stillwell, R.N. (1974) Atmospheric pressure ionization (API) mass spectrometry. Solvent-mediated ionization of samples introduced in solution and in a liquid chromatograph effluent stream. *J. Chromatogr. Sci.*, 12, (11), 725–729.

McLafferty, F.W. (1992) *Interpretation of Mass Spectra*, (4th edn), Mill Valley: University Science Books.

Millard, B.J. (1978) *Quantitative Mass Spectrometry*, London: Heyden.

Moseley, A., Tippin, T., Walsh, J. and Schwartz, J. (1996) Paper presented at the 43rd ASMS conference in Atlanta, May 21–26.

Munson, B. (1977) Chemical ionization mass spectrometry: ten years later. *Anal. Chem.*, 49, (9), 772A–778A.

Taylor, L., Singh, R., Chang, S.Y. and Johnson, R.L. (1995) *Rapid Communi. Mass Spectrom.*, 9, 902.

Throck Watson, J. (1985) *Introduction to Mass Spectrometry*, New York: Raven Press.

12

APPLICATIONS OF MASS SPECTROMETRY: QUANTITATIVE MASS SPECTROMETRY

Mira V. Doig

12.1 Quantification

The main reasons why MS started to be used as a chromatographic detector is because its compound versatility, sensitivity and specificity is better than that of any other detector. Initially mass spectrometers were large, complicated, expensive pieces of equipment, designed for a variety of applications, with many different types of inlet systems and ionisation modes. An example of such an instrument is a VG 7070 mass spectrometer. This instrument was popular during the 1970s to 1980s. It was a double-focusing instrument, i.e. had an electric sector (ESA) and magnet in series for mass analysis, and probe, septum and GC inlets as standard. It could perform electron impact (EI), chemical (CI) and fast atom bombardment ionisation techniques and could have other inlet systems fitted, such as the moving-belt LC–MS interface. Refer to Chapter 11 for more detail on these and other inlet systems.

12.1.1 Gas chromatography–mass spectrometry (GC–MS)

As technology advanced, with the development of flexible fused silica capillary GC columns and quadrupole mass analysers, one of the instrument manufacturers, Hewlett Packard, believed it was worth developing a bench-top mass spectrometer as a dedicated and fully automated GC detector. The advent of fused silica capillary GC columns instead of packed columns meant that the columns were flexible and the optimal flow rate of the mobile phase, helium, was 1–2ml/min instead of 20–30ml/min. The reduction in flow rate meant that a sophisticated interface was no longer required to interface the GC to the ionisation chamber (source) of the mass spectrometer, as conventional turbo, diffusion and/or rotary pumps could maintain the reduced pressure of between 10^{-3} and 10^{-5} torr (10^{-1}–10^{-3}Pa) that is required for efficient ionisation and operation of the mass spectrometer. The GC column in this instrument came out of the side of the GC, along a heated tube, straight into the ion source of the mass spectrometer. This instrument, launched in 1982, was called the HP 5970 mass selective detector (MSD). This, and other instruments based on similar technology, proved to be highly successful in bioanalysis. It is especially suited to automated quantitative analysis. It is relatively cheap, easy to use and reliable, hence during the 1980s the number of GC–MS methods developed and reported in the literature grew immensely. Since 1982, GC–MS systems based on this technology have developed so that most of the maufacturers of mass spectrometers produce such a system. Hewlett Packard now produces the HP 5973, which has ceramic quadrupoles that are so small that the analyser fits inside the oven of the GC.

The latest development in GC–MS/MS is the instrument launched in 1995 by Finnigan MAT, based on their new ion-trap system (ITMS) which can be operated either in the MS–MS mode or the MS to the nth mode, where n is less than 10. This instrument is known as the GCQ. Time will tell how useful it is for bioanalysis in the pharmaceutical industry; at present it is showing potential for scanning techniques but has ion suppression problems when performing quantification. This development, as with all GC–MS, is now overshadowed by the development of atmospheric-pressure ionisation (API). In Section 12.5 I have included an application area which has not been overtaken by API LC–MS.

12.1.2 Liquid chromatography–mass spectrometry (LC–MS)

Thermospray LC–MS was used for quantification (see Section 12.6). However, the heat produced to promote ionisation and the stability and reliability of this technique are such that it has now been overshadowed by API LC–MS as it not as reliable, versatile or as easy to use as the latter.

API led to an easier method of reliably interfacing LC with MS. The production of dedicated LC–MS and LC–MS–MS instruments, such as the Perkin Elmer SCIEX API III in 1989, changed bioanalysis completely, as it enabled the most versatile chromatographic technique to be coupled to the most versatile, specific and sensitive detector (Doig 1992). Initially most of the quantitative methods were based on APCI using the heated nebuliser, as this interface operates best with liquid flow rates of 1–2ml/min. As routine LC flow rates decrease, the number of quantitative methods based on electrospray and ion evaporation increase.

By 1996 every medium to large pharmaceutical company had at least one, if not more, API LC–MS–MS instrument dedicated to drug metabolism and quantitative bioanalysis. Many pharmaceutical companies have since decided that it is the technique of choice for all quantitative bioanalytical methods at any stage of drug development.

12.1.3 Quantitative API LC–MS and its contribution to the drug development process

Research/discovery

The introduction of combinatorial chemical synthesis has vastly increased the number of molecules to evaluate for activity and pharmacokinetic suitability. The specificity, versatility and speed of developing quantitative bioanalytical methods using API LC–MS is helping in this area. It enables 'cassette' dosing in animals, i.e. testing a range of analogues in one animal. Rapid screening for metabolic routes using *in vitro* techniques and sensitive methods allow pharmacokinetic profiling in single animals using tail or orbital blood sampling.

241

At this stage of drug development, fully validated methods are not required but specificity and speed of method development are important. As the results obtained at this stage can lead to promotion or rejection of further compound development, internal standardisation is vital, but homologues will suffice as tight (< 15%) precision and accuracy are not required.

Early development

At this stage of drug development, the bioanalytical requirements are specificity to measure drug and metabolites, validated methodology to keep the regulators satisified and sensitivity to keep the customer satisfied, either to minimalise sample volume requirements from animals or to keep the clinicians happy, especially for the all-important 'first administration to man'!

Late development

At this stage speed of analysis is important as the clinical studies generate large numbers of samples. This is where the ease of use, specificity and reliability using API LC–MS techniques for quantification become important. At present there is no other technique that can satisfy all these different requirements at every stage of drug development. At this stage automated sample preparation using micro SPE or Empore disc technology, with the use of autosamplers based on the 96-well plate format, means that hundreds of samples can be analysed automatically every day.

12.2   Internal standardisation

There are several reasons why internal standards should be used when developing quantitative methods using MS for detection. The first and second reasons are relevant for any chromatographic method. These are to correct for any losses during sample preparation and derivatisation, and to correct for any variability in injection volume. The third reason is because of the problem of obtaining reproducible ionisation and detection in a mass spectrometer. Ionisation efficiency and detector response can vary, depending on the condition (i.e. cleanliness) and the temperature and pressure in the ion source and mass analyser. The ideal internal standard is one that is an isotopically labelled analogue rather than a structural homologue. This is because ionisation efficiency can vary for structural homologues. The most common stable isotope incorporated into a chemical to make an internal standard is deuterium, rather than ^{13}C, ^{15}N or ^{18}O, because it is generally cheaper and easier to synthesize. It is important that the natural abundance of stable isotopes present in the unlabelled compound of interest does not interfere with the quantification of the internal standard. This means that the latter should contain sufficient stable isotopes to shift the molecular weight by at least 3 mass units. Then the m/z ions monitored are different for the compound and its internal standard even though their chemical structure is identical. It is also important for the internal standard, whether a stable isotopically labelled analogue or structural homologue, not to contain any analyte as an impurity. If it does, it will limit the lower limit of quantification (LOQ) and give a positive intercept on the calibration curve. For a detailed description of the calculation of the

Table 12.1 Masses and natural abundance of important stable isotopes

Element	Mass	Average abundance in atom%	Element	Mass	Average abundance in atom%
^{1}H	1.00783	99.985	^{29}Si	28.99649	4.70
^{2}H	2.01410	0.015	^{30}Si	29.97376	3.09
^{12}C	12.00000	98.89	^{32}S	31.97207	95.05
^{13}C	13.00335	1.11	^{33}S	32.97146	0.76
^{14}N	14.00307	99.63	^{34}S	33.96786	4.22
^{15}N	15.00011	0.37	^{36}S	35.96709	0.014
^{16}O	15.99491	99.759	^{35}Cl	34.96885	75.53
^{17}O	16.99914	0.037	^{37}Cl	36.96590	24.47
^{18}O	17.99916	0.204	^{79}Br	78.9183	50.54
^{28}Si	27.97693	92.21	^{81}Br	80.9163	49.46

natural abundance of stable isotopes and the use of stable isotopically labelled compounds as internal standards, see Doig (1992). Table 12.1 shows the natural abundance of some common elements used in bioanalysis.

12.3 Data acquisition

12.3.1 Selected ion versus mass chromatogram

The best technique for quantitative measurements is selected ion monitoring (SIM) as the instrument looks at only a few ions and can thus integrate ion currents for much longer than during a scanned spectrum. For example, with a scanned spectrum, assuming efficient ionisation and ultimate sensitivity and a GC peak width of 12 seconds, then the minimum concentration of sample to give a useful spectrum is about 20pg. For SIM, a minimum of 100 ions is needed to define the peak. This can be obtained from 1.6fg of sample. In practice, detection limits may not be as good as this because of background noise. For anyone who is interested in the theoretical formulae and calculations used to obtain these amounts please refer to Chapter 3 of Leis *et al.* (1990).

12.3.2 Mass analysis

To perform SIM using magnetic instruments (Section 11.4.1) the ions to be monitored can be focused by altering the acceleration voltage (V) or the magnetic field (B). In older instruments this was normally achieved by changing the accelerating voltage, as the hysteresis ('time constant') of the solid-core magnet was too slow to achieve the switching speed necessary to detect and define the peak shape of more than one ion, especially from a capillary GC column. Using the accelerating voltage has several disadvantages as the ion-source focusing and ion abundance varies with the acceleration voltage, i.e. the *m/z* being monitored. To keep this variation to a minimum, the dynamic range of the *m/z* values that required monitoring for quantification was limited to 30% of the highest mass. With newer instruments, laminated magnets allow rapid and accurate alterations of the magnetic field so this is no longer so important when selecting a mass spectrometer suitable for quantification.

Table 12.2 A few combinations of C, H, N and O that have a nominal molecular weight of 308

C	H	N	O	*Exact composite mass*
18	20	4	1	308.1637
18	12	–	5	308.0685
18	4	4	2	308.0334
19	2	1	4	307.9984
19	20	2	2	308.1525
19	36	2	1	308.2827
20	12	4	–	308.1062
20	24	2	1	308.1888

As mentioned in Section 11.4.3, quadrupole instruments are ideal for rapid SIM operation over any range of m/z values within the scan range of the instrument. For a typical SIM method, the required m/z values and dwell times are entered for each compound. The data are then stored in the method file together with the other data required for automation of the quantification (such as acquisition start and stop times, GC conditions, etc.). Then, with automated systems such as the HP 5973, the analysis can begin.

12.3.3 Calculation of the mass of the selected ion

Although quadrupole instruments have nominal mass resolution, the m/z values of the ions you are monitoring should be calculated to at least two places of decimals so that the window set up (which is normally the m/z value ± 0.3 mass units) has the m/z of interest in the centre of the SIM 'window'. When performing this calculation remember that mass spectrometrists calculate the mass based on the isotope present rather than the average atomic mass used by chemists. The average mass takes into account the relative amount of the naturally occurring isotope for an element. For example, a mass specrometrist will use the mass for chlorine as listed in Table 12.1, which is 34.96885, whereas the atomic weight of chlorine to a chemist is 35.453, which takes into consideration the natural abundance of the ^{35}Cl and ^{37}Cl isotopes. When using double-focusing or hybrid instruments utilising magnetic focusing, the specificity of SIM can be increased by using high-resolution mass measurement. This decreases the width of the peak window (e.g. m/z 386.2146 ± 0.0005). Consequently this leads to less interference from compounds that might elute at a similar retention time to the compound of interest, and hence decreases LOQ or increases the specificity of the method. To illustrate how high-resolution SIM can improve specificity and LOQ, Table 12.2 gives an example of eight different elemental compositions that would give an m/z value of 308.

Another method that is becoming increasingly popular for quantification because it increases specificity is the use of MS–MS techniques. This is because the cost of hybrid and triple quadrupole instruments has decreased and the number of efficient soft ionisation techniques used in combined chromatographic MS interfaces has increased. The method of using MS–MS for quantification is usually achieved by setting up the first quadrupole, Q1 to transmit the molecular ion into the second quadrupole Q2. This 'RF only' quadrupole or the 'collisional activation' cell is where the molecular ion undergoes collision-induced dissociation (CID) to generate daughter ions. The third quadrupole, Q3, is then set up to allow a specific daughter ion to be transmitted to the detector. This mode of SIM using MS–MS is known as multiple reaction monitoring (MRM). Tandem MS is described in more detail in Section 13.3.

12.3.4 Data storage and processing

Once all the details are known, such as the m/z value to be monitored, retention time, etc. the data are stored in the MS data system and automated sample injection and data acquisition can be performed. The data generated using SIM and MRM are similar to those from any chromatographic detector, i.e. the voltage from the detector is measured for set periods of time and stored digitally. The digital data are stored as area slices then integrated. The peak heights and areas are calculated as required, ratioed and finally processed to produce calibration curves, values to unknown samples, etc. using the same principles as any for other chromatographic integrator (Dyson 1990).

12.4 Developing a quantitative method

This is the suggested order of points to consider when developing a quantitative method. If instrumentation access is available, then API LC–MS–MS approaches are now usually the first approach.

1   Select the most appropriate mode of chromatography and ionisation technique. This will be dependent on such items as sample structure/overall charge, volatility, stability, ease of extraction, sensitivity required, etc. Talk to your MS expert first as many departments and companies have a standard approach to sample preparation and chromatography. At the Advanced Bioanalytic Services Laboratories this is normally solid-phase extraction followed by LC on a 5μm column (3cm × 4.6mm) using a simple mobile phase such as acetonitrile : water with the appropriate volatile buffer/pH modifications. The mobile-phase additions are dependent on whether APCI or electrospray is used and the potential of ionising the compound(s) of interest. The potential for ionising the compound depends on its structure. For example, nitrogen-containing compounds normally ionise easily under positive-ion conditions, while acids may give more intense ions under negative-ion conditions.

2   Obtain a suitable internal standard. Preferably a stable isotopically labelled one which increases the pseudo molecular ion by at least 3Da.

3   Obtain a full scan of the compound of interest and the internal standard in positive- and negative-ion mode. The most intense ions in the spectra are normally the parent (precursor) ions, e.g. $(M + H)^+$ and $(M - H)^-$ ions. Full-scan spectra on an LC–MS–MS instrument are obtained by only Q1 acting as a mass filter, i.e. having RF and DC to the Q1 quadrupole while Q2 and Q3 operate in transmission mode (i.e. RF-only mode). Vary the cone voltage to optimise the intensity of the precursor ion and vary the collision voltage to optimise the intensity of the product ion. Examine the blank material for any interference. If

the interference looks minimal, use LC–MS, especially if sensitivity is an issue, since fragmenting the precursor ion either in Q2 in MS–MS or by increasing the cone voltage in LC–MS, although increasing your specificity, can decrease your sensitivity. Most people with LC–MS–MS instruments will always fragment the precursor ion and examine its product ion spectrum. This is achieved by setting Q1 to transmit the precursor ion, introducing a collision gas such as argon into Q2 and scanning Q3 to obtain the product ion spectrum. Then select the most intense product ion for transmission through Q3. Then program the MS–MS to perform multiple reaction monitoring (MRM) for the quantitative method, i.e. setting Q1 to transmit the precursor ion and Q3 to transmit the selected product ion. This mode of operation for quantification is the most common. Other modes can be used, especially early on in drug development when looking for a range of molecules that all fragment to an identical product ion (e.g. sulphates) or measuring all the compounds that have a specific neutral loss, such as glucuronides.

4 Prepare samples spiked with the test substance(s) over the concentration range of interest. Examine the calibration curve. Some compounds have a limited range over which they will produce a linear response, especially using some APCI methods which struggle to generate a linear response over a range of greater than a factor of 2, i.e. 1 to 100.

These points will be highlighted further below. GC–MS and thermospray examples have been left in this chapter to help illustrate that GC–MS and thermospray LC–MS have been useful for quantification but the times spent on method development and sample analysis were a lot longer than those required for API LC–MS. One major change to quantitative analysis when using API LC–MS is that run times are typically 3–5 minutes. This means that traditional autoinjectors with maximum sample holders of 60–100 were a severe limitation to the efficiency of unattended operation, especially overnight. Technology has overcome this problem by the introduction of autosamplers based on injector trays that carry multiple 96-well plates. These and the developments of speedier sample processing are described in more detail in Chapter 10.

12.5 Analysis of prostanoids by GC–MS

Prostanoid research and development, since the late 1960s, has been dependent on GC–MS for qualitative and quantitative measurements. This method has been used to characterize and measure the metabolites of arachidonic acid (5,8,11,14-tetraeicosenoic acid). Arachidonic acid is metabolised by two pathways, the cyclo-oxygenase pathway to the prostaglandins (PGs) and thromboxanes (TXs) (Figure 12.1a) and the lipoxygenase enzymes to hydroxy-eicosatetraenoic acids (HETES) and leukotrienes (LTs) (Figure 12.1b). These metabolites have different pharmacological and biochemical properties (Table 12.3) and because of these differences it is important to have methodology available that can quantify each metabolite. Initially, in the early 1980s, the method of extraction was acidification and solvent extraction (Figure 12.2), and the easiest method to produce stable derivatives was to form the methyl ester, methoxime, trimethylsilylether derivatives (Figure 12.3). This enabled the cyclo-oxygenase and lipoxygenase products to be measured concurrently from the same sample using capillary GC–MS with EI ionization. Using this methodology,

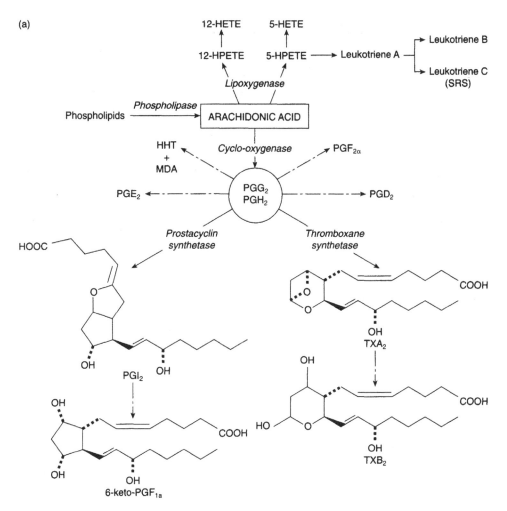

Figure 12.1 Pathways of arachidonic acid metabolism, (a) showing the structures of some of the cyclo-oxygenase products.

LOQs of 1–5ng/ml for each of these analytes was easily attained. Unfortunately, as can be seen in Figures 12.1a and 12.1b, the structures of these compounds are not easily ionised under API conditions, hence they are still assayed by GC–MS and GC–MS–MS today.

To attain sensitivity at these concentrations and to allow SIM on the highest m/z ion possible, the electron energy used for EI was reduced to 20eV instead of the normal 70eV. As time progressed the pharmacologists and biochemists wanted the LOQ lowered to pg/ml concentrations. To achieve this the derivatisation techniques and MS conditions were changed to decrease the fragmentation and improve the percentage of the total ion current that was carried in these ions. This improvement is illustrated well in Figure 11.9 which compares the EI and positive ammonia CI spectra of the methyl ester, methoxime, trimethylsilyl ether derivative of PGE_1. With experimentation the best method to enhance the sensitivity for these compounds

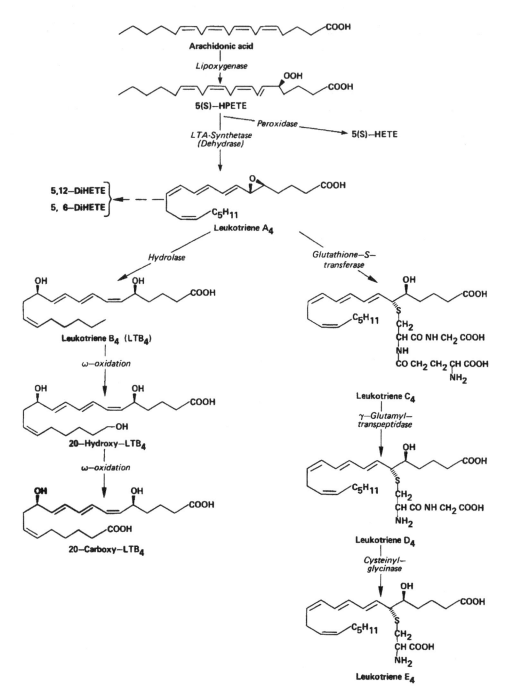

Figure 12.1 Pathways of arachidonic acid metabolism, (b) via the 5-lipoxygenase enzyme pathway.

Table 12.3 Summary of the biological activities of PGs, TXs and LTs

---

PGI$_2$ = prostacyclin = Flolan®
    vasodilator, leads to lowering of blood pressure
    inhibitor of platelet aggregation
    inhibitor of gastric secretion.
Therapeutic use: pulmonary oedema, anti-ulcer, lowering blood pressure, chronic heart failure

TXA$_2$ = thromboxane A$_2$
    opposite of PGI$_2$
Therapeutic use: block thromboxane sythetase enzyme

PGE$_2$ at higher concentrations equivalent to PGI$_2$ plus a hyperalgesic and pyrogenic agent

PGF$_2\alpha$ vasodilator, hyperalgesic, contraction of pregnant uterus
Therapeutic use: induction of abortion and labour

PGD$_2\alpha$ at high concs equivalent to PGI$_2$

LTB$_4$ aggregates white blood cells, chemokinetic

LTC$_4$ stimulates histamine release, which constricts the lung airways
Therapeutic use: blocks pathway for treatment of asthma

LTD$_4$ equivalent to LTC$_4$

All are thought to be proinflammatory

---

was found to be the use of negative-ion chemical ionisation using methane as the CI gas, and the pentafluorobenzyl (PFB) ester, trimethylsilylether, methoxime derivative. A full description of this method can be found in Millard (1978).

12.6 An example of thermospray LC–MS

This example is an automated method for the measurement of sumatriptan (structure I in Figure 12.4) in plasma. This drug is used for the acute treatment of migraine. The full method is described by Oxford and Lant (1989). The final internal standard was deuterated [^{2}H$_3$]sumatriptan (structure III, Figure 12.4) after the homologue (structure II Figure 12.4) was found to be unsuitable. The 1ml plasma samples were spiked with internal standard, loaded onto prewashed (1.8ml MeOH : 1.8ml H$_2$O) AASP C-2 cartridges, washed with water (1.8ml) followed by water : methanol (7 : 3, 1.8ml) and placed onto the AASP module for automated elution onto a HP 1090 HPLC system. This used a 50mm × 4.6mm i.d. 3µm Spherisorb ODS-2 column with a mobile phase of methanol : 0.1M ammonium acetate (60 : 40) at a flow-rate of 1ml/min. The run time between injections on the AASP was 4 minutes. The SIM ions monitored were *m/z* 296 and 299. These are the [M + H]$^+$ ions for sumatriptan and its internal standard. The optimal TSP ionisation conditions for this compound used the NH$_4^+$ and CH$_3$COO$^-$ ions generated from the buffer in the mobile phase in the positive-ion mode to ionise the drug and internal standard in the gas phase. To minimise the fragmentation, i.e. maintain a high percentage of ion current in the protonated molecular ions, the source temperature had to be kept at 250°C and the temperature at the tip of the thermospray probe was 148–152°C. The calibration curve was linear

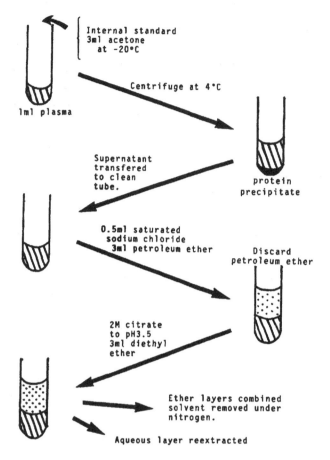

Figure 12.2 The multi-stage solvent extraction procedure for removing prostaglandins and leukotrienes from plasma.

($r = 0.9995$) over the range 2–50ng/ml using peak height ratios of m/z 296 : m/z 299 versus ng/ml of sumatriptan in plasma. The intra- and inter-assay variations are shown in Tables 12.4 and 12.5. The authors stated that the use of a stable isotopically labelled internal standard was essential to obtain acceptable intra- and inter-assay variation.

12.7   Examples of API LC–MS

The first method published by a pharmaceutical company was a method developed to assay a renin inhibitor, CP-80,794, in human serum. The full method is described by Fouda *et al.* (1991). The 1ml serum samples were spiked with 50ng of 'carrier' CP-81,489 and 1ng of internal standard, a homologue CP-88,587. All three structures shown in Figure 12.5.

The sample was basified and extracted into 5ml of *n*-butyl chloride, added to 2ml of acetonitrile, evaporated, resuspended into 60μl of methanol followed by 40μl of

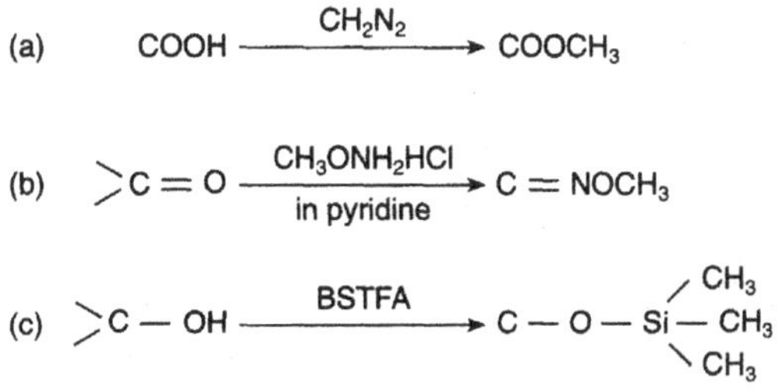

(a) $COOH \xrightarrow{CH_2N_2} COOCH_3$

(b) $\text{>}C=O \xrightarrow[\text{in pyridine}]{CH_3ONH_2HCl} C=NOCH_3$

(c) $\text{>}C-OH \xrightarrow{BSTFA} C-O-Si\begin{smallmatrix}\nearrow CH_3 \\ - CH_3 \\ \searrow CH_3\end{smallmatrix}$

Figure 12.3 Robust method for producing stable derivatives of eicosanoids.

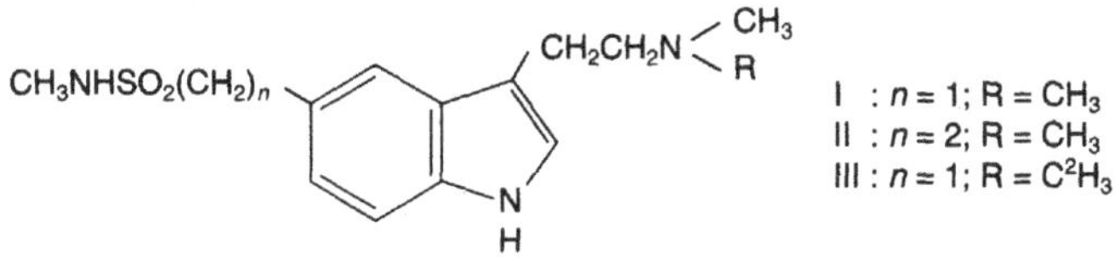

$CH_3NHSO_2(CH_2)_n$ — [indole structure with $CH_2CH_2N\text{<}\begin{smallmatrix}CH_3\\R\end{smallmatrix}$]

I : $n = 1$; $R = CH_3$
II : $n = 2$; $R = CH_3$
III : $n = 1$; $R = C^2H_3$

Figure 12.4 The structures of sumatriptan (I), a homologue (II) and the deuterated internal standard (III).

Table 12.4 Intra-assay variability with [$^2H_3$]sumatriptan (50ng/ml of plasma) as the internal standard

Nominal concentration (ng/ml)	Concentration found (mean ± SD, n = 6) (ng/ml)	Accuracy[a]	Coefficient of variation CV (%)
0	Not detectable	–	–
2	1.8 ± 0.2	10.0	11.1
5	4.7 ± 0.2	6.0	4.3
10	9.7 ± 0.3	3.0	3.1
20	19.9 ± 0.5	0.5	2.5
30	29.1 ± 0.6	3.0	2.1
50	48.5 ± 1.5	3.0	3.1

[a] Defined as the percentage difference between the mean concentration found and the theoretical concentration.

Table 12.5 Inter-assay variability with [$^2H_3$]sumatriptan (50ng/ml of plasma) as internal standard

Nominal concentration (ng/ml)	Concentration found (ng/ml)			Mean	Accuracy[a]	CV (%)
	Assay 1	Assay 2	Assay 3			
3.9	4.1	3.8	3.6	3.9	0	10.2
	4.4	b	3.4			
20.6	21.0	20.4	19.7	20.0	2.9	3.6
	20.5	19.0	19.6			
41.2	40.3	37.0	40.5	38.7	6.1	3.9
	39.1	37.7	36.8			

[a] Defined as the percentage difference between the mean concentration found and the theoretical concentration.
[b] Sample lost during analysis.

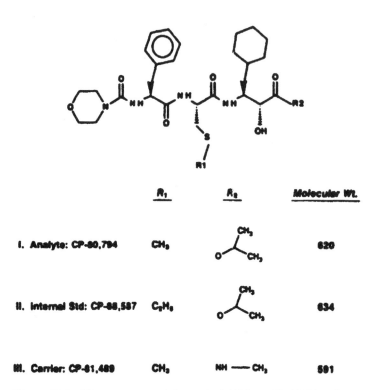

| | R₁ | R₂ | Molecular Wt. |

I. Analyte: CP-80,794 CH₃ (O-isopropyl group) 620

II. Internal Std: CP-88,587 C₂H₅ (O-isopropyl group) 634

III. Carrier: CP-81,489 CH₃ NH—CH₃ 591

Figure 12.5 The structures of the renin inhibitor, CP-80,794, the internal standard and carrier.

water. Injections (80μl) were made onto the LC–MS system. The LC system was a C-18 Nova Pak 150mm × 3.9mm column with a mobile phase of methanol : water (80 : 20) at a flow rate of 1ml/min. The run time was 4 minutes and ionisation was achieved using the corona discharge at –2.5mA (i.e. negative-ion APCI) using the heated nebuliser interface. The SIM ions monitored were the deprotonated molecular ions [M – H]⁻ at *m/z* 619 and 633. Using peak height ratio versus drug concentration in plasma to achieve quantification over the range 0.05–10ng/ml, two calibration curves had to be used, one for the range 0.05–1ng/ml and one in the range 1.0–10ng/ml. This limitation of the linear dynamic range has been found by others using APCI or electrospray on the SCIEX API LC–MS and appears to be a problem with either the ionisation efficiencies or the pulse-counting technique used by this instrument for the detection of ions. The intra-assay precision and accuracy for this assay over the range 0.05–10ng/ml were 5.0–12.5 and 96–119%, respectively. Factors of note from this assay are that it is far more sensitive than TSP LC–MS and it has a short run time. To maintain high precision and enhanced extraction efficiency, a high concentration (50ng) of a carrier molecule is required. This method has been used for many thousands of samples. To quote the company using this method '20 000 in 1 year doing analysis 4 days a week, 10 hours a day'. The only service required on the MS during this time was the cleaning or replacing of the corona discharge needle, an operation that takes about 1 minute. This was performed after every 50 serum samples. This example was one of the first demonstrating the potential of API LC–MS as a routine quantitative tool for increasing the efficiency of drug development within the pharmaceutical industry. Since 1991 other applications have appeared in the literature (Gilbert *et al.* 1992; Kaye *et al.* 1992; Fraser *et al.* 1994).

Kaye *et al.* (1992) describe the measurement of an α_1-adrenoceptor agonist, abanoquil, in human blood over the range 10–500pg/ml using APCI LC–MS–MS. Gilbert *et al.* (1992) describe an APCI LC–MS–MS method for the determination of a new cholecystokinin receptor antagonist, L-365,260 (i.e. a drug to control gastro-intestinal function) over the range 1–200ng/ml of plasma. Fraser *et al.* (1994) describe a method for quantifying a non-systemically available ACAT inhibitor, 447C88 (i.e. a drug that inhibits the uptake of cholesterol from the gut), with a limit of detection of 0.5ng/ml using 0.5ml of plasma. This reference also illustrates how an APCI LC–MS–MS method can be just as sensitive but far simpler and more efficient than a GC–MS method. This comparison could be performed because the GC–MS method was developed prior to the purchase of an APCI mass spectrometer.

As mentioned earlier, the selection of applications is personal. There are many applications in the literature and if the appropriate key words are selected they can be found by any literature search.

12.8   The future

The main techniques used for quantification in bioanalysis using MS as a detector are LC–MS using either APCI or electrospray. Initially APCI with the heated nebuliser interface appeared to be more useful and reliable for bioanalysis but as more people use narrow-bore HPLC columns and lower flow rates so electrospray or PE's turbo ionspray is becoming more popular. API LC–MS has now had a dramatic effect on quantitative bioanalysis and has become the method of choice in many pharmaceutical companies. There are still some compounds that are difficult to obtain sensitive methods for, such as some of the purines and pyrimidines, prostaglandins and leukotrienes and steroids. In addition, phosphates and *N*-oxide metabolites still tend to fragment rather than producing intense pseudo-molecular ions.

Innovations to watch for in the future are:

- Automation based on batch solid-phase sample preparation using a 96-well plate format (Allanson *et al.* 1996; Kaye *et al.* 1996).
- The introduction of ion-trap technology to perform LC–MS–MS to LC–MS to the *n*th where $n < 10$. This technology is available from Bruker, Varian and Finnigan MAT but as yet has not been proven in producing routine, reliable, quantitative bioanalytical assays. If, or when, it does, the LC–MS–MS instrument manufacturers producing instruments based on traditional quadrupole will find it hard to compete, as ion traps are a lot cheaper to manufacture.
- New sources that can handle involatile buffers and dirty samples without a decrease in performance.
- Even more sensitivity as more sensitive analysers are developed, e.g. QTOF instruments.

12.9 Bibliography

Allanson, J.P., Biddlecombe, R.A., Jones, A.E. and Pleasance, S. (1996) *Rapid Comm. MS*, **10**, 811–816.

Doig, M.V. (1992) Mass spectrometry with atmospheric pressure ionization, in E. Reid and I.D. Wilson (eds), *Methodological Surveys in Biochemistry and Analysis*, Vol. 22, (Bioanal. Approaches Drugs, incl. Anti-Asthmatics Metab.), pp. 307–310, Cambridge, UK: Royal Society of Chemistry.

Dyson, N. (1990) *Chromatographic Integration Methods*, London: Royal Society of Chemistry.

Fouda, H., Nocerini, M., Schneider, R. and Gedutis, C. (1991) *J. Am. Soc. M.S.*, **2**, 164–167.

Fraser, I.J., Clare, R.A. and Pleasance, S. (1994) Determination of the ACAT inhibitor 447C88 in plasma using LC–API–MS–MS, in E. Reid and I.D. Wilson (eds), *Methodological Surveys in Biochemistry and Analysis*, Vol. 23 (Biofluid and Tissue Analysis for Drugs, Including Hypolipidaemics), pp. 113–120, Cambridge, UK: Royal Society of Chemistry.

Gilbert, M.D., Hand, E.L., Yuan, A.S., Olah, T.V. and Covey, T.R. (1992) *Biological MS*, **21**, 63–68.

Kaye, B., Clarke, W.H., Cussans, N.J., Macrae, P.V. and Stopher, D.A. (1992) *Biological MS*, **21**, 585–589.

Kaye, B., Heron, W.J., Macrae, P.V., Robinson, D.A., Venn, R.F. and Wild, W. (1996) *Anal. Chem.*, **68**, (9), 1658–1660.

Leis, H.J., Welz, W. and Malle, E. (1990) *Chromatography and Analysis*, October, 5–7.

Millard, B.J. (1978) *Quantitative Mass Spectrometry*, London: Heyden.

Oxford, J. and Lant, M.S. (1989) *J. Chromatog.*, **496**, 137–146.

Rose, M.E. and Johnstone, R.A.W. (1982) *Mass Spectrometry For Chemists and Biochemists*, Cambridge: Cambridge University Press.

13

MASS SPECTROMETRIC
IDENTIFICATION OF METABOLITES

Janet Oxford and Soraya Monté

13.1 Objectives

The objectives of this chapter are to introduce the concepts of tandem mass spectrometry (MS–MS) for the identification of metabolites, to consider the role of isotopically labelled compounds in identifying metabolites and to consider some practical aspects and limitations of the mass spectrometric techniques used for metabolite identification.

13.2 Introduction

Most drugs undergo a biotransformation in the body which modifies their structure in such a way that they can be readily eliminated. The identification of such pathways is important for a number of reasons, including the following:

- In drug discovery, to ascertain positions of metabolic instability which may have led to reduction in the duration of action or potency of a drug.
- A metabolite may be pharmacologically active and patent protection may be required.
- Determination of metabolic pathways in laboratory animals is necessary to predict possible biotransformation in humans.
- Early information on the metabolism in humans is needed so that the species selected for long-term toxicity testing are those that metabolise the drug in a similar manner.

The identification of metabolites in biological fluids is extremely challenging. Metabolites are often present at low concentrations, typically parts per million or less in a biological fluid. Although NMR can sometimes provide unequivocal structural identification on *c.* 20µg of an isolated metabolite, mass spectrometry is often the first spectroscopic technique of choice for metabolite identification. It is inherently more sensitive than NMR, mass spectral data being obtainable for a metabolite often on less than 1µg sample. Furthermore, chromatographic techniques combined on-line with mass spectrometry give a rapid method for the characterisation of low amounts of a metabolite with minimal sample clean-up.

The important items of information that can be obtained from a mass spectrum are:

- the molecular weight of the metabolite; and
- fragment ions, which can give structural information on the site of metabolism.

The 'soft' ionisation techniques of thermospray (TSP), electrospray (ESP), heated nebuliser atmospheric-pressure chemical ionisation (APCI) and fast atom bombardment (FAB) generally yield the first and most important piece of information, the molecular weight of the metabolite. However, by the nature of these 'soft' techniques very little energy is imparted to the molecular species and consequently there is no driving force for it to fragment. Hence, structural information necessary for assignment of the site of metabolism is often not available. Tandem mass spectrometry (MS–MS) has now become a routine technique which enables structural information to be obtained from ions produced by 'soft' ionisation methods.

The concepts of MS–MS are outlined in the next section. For a more detailed account, the reader is referred to Busch *et al.* (1988).

13.3   Tandem mass spectrometry (MS–MS)

13.3.1   Theory

Tandem mass spectrometry has come to mean the mass spectrum of a mass spectrum, hence, MS–MS. This concept is illustrated schematically in Figure 13.1. Essentially in MS–MS, mass analysers are coupled together via an interface known as a collision cell. Ions of a selected mass (**precursor ions**), e.g. m/z 202 in Figure 13.1, are transmitted by the first mass analyser into the collision cell. Here, they collide with the neutral atoms of an inert gas; generally argon is used. In this process, known as collision-induced dissociation (CID), a fraction of the translational energy inherent in the precursor ion is converted into internal energy. The internal energy is lost by the precursor ion decomposing or fragmenting into **product ions** (Figure 13.2). The product ions are transmitted through a second mass analyser.

Throughout this chapter the nomenclature 'precursor' and 'product' ions will be used. These terms are current nomenclature replacing 'parent' and 'daughter' ion, respectively.

13.3.2   Instrumentation

Several different instruments are available for carrying out MS–MS experiments. The most commonly used instrument for drug metabolism studies is the triple quadrupole, shown schematically in Figure 13.3. Ions transmitted by the first quadrupole enter a collision cell, which in the early instruments was a quadrupole, but hexapoles or octapoles are commonly used now. The collision cell is operated in RF mode, with no DC component, which means that no mass analysis takes place (see Chapter 11), instead all ions produced in the collision cell are transmitted into the third quadrupole for mass analysis.

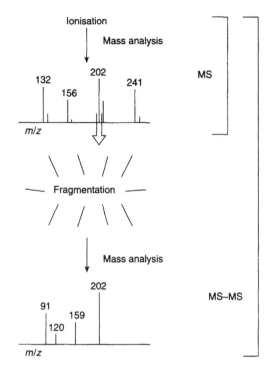

Figure 13.1 Schematic representation of MS–MS.

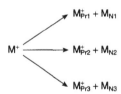

Precursor ion ⟶ Product ion + Neutral species

Figure 13.2 Scheme showing decomposition of the precursor ion. M^+, mass of precursor ion; $M^+_{Pr_n}$, mass of product ion; M_{N_n}, mass of neutral species.

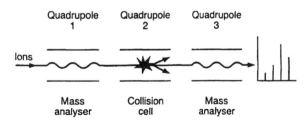

Figure 13.3 Schematic diagram of a triple quadrupole mass spectrometer.

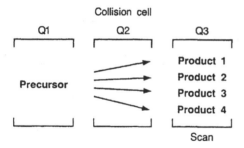

Figure 13.4 Schematic representation of the production of a product ion scan.

13.3.3 MS–MS scans and their application to metabolite identification

There are three methods of scanning the mass analysers, to give:

- a product ion mass spectrum;
- a precursor ion mass spectrum; and
- a neutral loss mass spectrum.

These scans are complementary, each giving different types of information.

Product ion mass spectrum

This is the mass spectrum produced by fragmentation of a given precursor ion. The method of producing a product ion mass spectrum is shown schematically in Figure 13.4. The first mass analyser, Q1, transmits ions of one m/z value into the collision cell, where they fragment. Product ions are mass analysed by scanning the second mass analyser Q3. The product ion mass spectrum is recorded.

This type of scan enables us to obtain structural information from protonated molecules produced by a 'soft' ionisation technique such as thermospray (TSP). This is demonstrated by the example shown in Figure 13.5. The TSP mass spectrum of ondansetron, a drug given for the treatment of emesis during cancer chemotherapy and radiotherapy, shows predominantly the protonated molecule at m/z 294 (Figure 13.5a). Product ions derived from the protonated molecule, m/z 294, are characteristic of the carbazalone nucleus. A metabolite of ondansetron is the N-demethylated analogue. The TSP mass spectrum of this compound shows only a protonated molecule at m/z 280. The product ion mass spectrum, however, shows which methyl group has gone (Figure 13.5c).

MS–MS can also be used effectively for the analysis of mixtures. Metabolites, even when isolated, are seldom pure. Invariably the isolated metabolite contains co-chromatographing or co-extracted components. These may arise from the original matrix or are introduced during sample isolation. Figure 13.6 illustrates that if a mixture of AB and CD is present in the ion source then the spectrum obtained is also representative of a mixture of AB and CD.

In an MS–MS instrument, Q1 could be set to transmit ions of a selected m/z value, in Figure 13.6 this is the molecular ion of AB. This ion undergoes collisionally

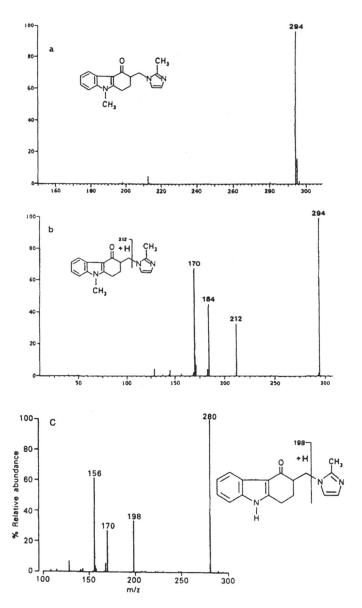

Figure 13.5 (a) The TSP mass spectrum of ondansetron; product ion mass spectra of (b) *m/z* 294 derived from ondansetron and (c) *m/z* 280 derived from its *N*-demethylated analogue.

induced dissociation and the resulting fragment ions are analysed by Q3 to produce a product ion mass spectrum. This mass spectrum is representative of AB.

An example of the use of product ion scanning for mixture analysis is shown in Figure 13.7 (Straub 1986). Guanethidine, an antihypertensive drug, undergoes metabolism *in vivo* to an *N*-oxide. An FAB mass spectrum of a methanolic SepPak™ fraction obtained from rat urine (0–4h; 10mg (base)/kg given orally) is shown in Figure 13.7a. Ions derived from the glycerol matrix are observed at *m/z* 93, 115, 185

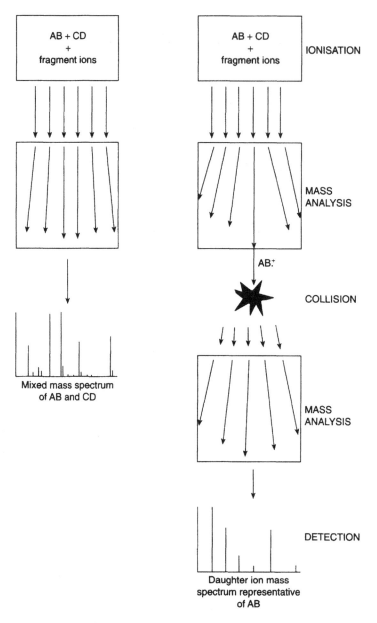

Figure 13.6 Schematic representation of the way MS–MS can be used to obtain a mass spectrum of a single component in a multi-component mixture. (Reproduced from Straub 1986.)

and 207. The product ion mass spectrum (Figure 13.7b) of *m/z* 199 confirms that guanethidine is present in the sample. Despite the absence of any obvious signal at the mass expected for an N-oxide metabolite, the product ion mass spectrum (Figure 13.7c) of *m/z* 215 confirms the presence of an oxygenated species, with oxidation occurring on the azocinyl group.

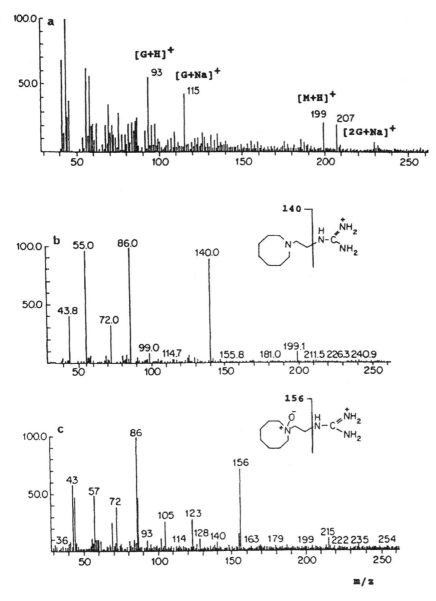

Figure 13.7 (a) FAB mass spectrum derived from an extract of rat urine obtained following oral administration of 10mg guanethidine/kg. (b) Product ion mass spectrum of *m/z* 199 derived from guanethidine. (c) Product ion mass spectrum of *m/z* 215 derived from the N-oxide metabolite. G = glycerol.

Precursor ion mass spectrum

This is the mass spectrum of a number of different precursor ions which, on fragmentation, produce a single given product ion. The method of producing a precursor ion mass spectrum is shown schematically in Figure 13.8. All the ions produced in the

261

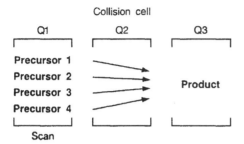

Figure 13.8 Schematic representation of the production of a precursor ion scan.

ion source are mass analysed by Q1 and then fragmented in the collision cell. So while Q1 is scanned, Q3 transmits a selected product ion which is a representative fragment ion of a series of precursor ions.

The use of this type of scan is shown by the following example taken from studies on the human metabolism of MDL257, a bronchodilator (Coutant *et al.* 1987). The volunteers each received an oral dose of 400mg MDL257, urine samples collected after dosing were concentrated and a portion submitted for mass spectrometric analysis using electron impact ionisation (EI). Metabolites generally contain a substructure representative of the drug. Hence the product ion mass spectra of the metabolites should contain product ions common to the parent compound. The product ion mass spectrum of *m/z* 217, M⁺ for MDL257, is shown in Figure 13.9a. In this example, any modification to the piperidine ring should give a metabolite yielding one or both of the product ions *m/z* 135 and 162. Figure 13.9b shows the precursor ion mass spectrum of *m/z* 162, showing molecular ions for MDL257-related metabolites. A component of molecular weight 249 is indicated. The product ion mass spectrum of *m/z* 249 is shown in Figure 13.9c. By comparison with an authentic standard, the structure of the metabolite was confirmed as a pentanoic acid derivative.

Neutral loss mass spectrum

The neutral loss mass spectrum is the mass spectrum of ions undergoing a common neutral loss, and the method of producing a neutral loss scan is shown schematically in Figure 13.10. With reference to Figure 13.2, the relationship between the mass of a precursor ion (M⁺), a product ion M_P^+ and a neutral loss M_N is

$$M^+ = M_P^+ + M_N$$

or

$$M_N = M^+ - M_P^+$$

Thus a neutral loss mass spectrum is obtained by scanning both Q1 and Q3 with a fixed mass difference between them. The mass difference is equal to the mass of the neutral loss.

An example of the use of this type of mass spectrum is shown in Figure 13.11. Under MS–MS conditions some glucuronic acid conjugates can lose a neutral species

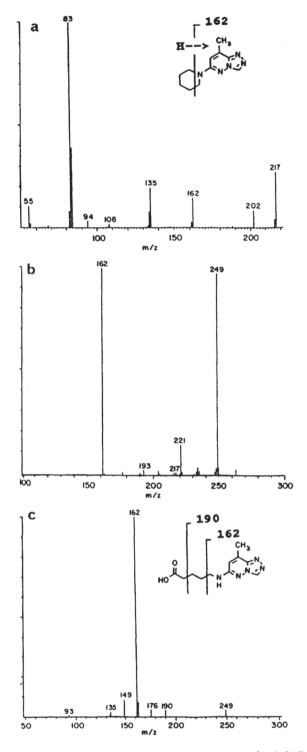

Figure 13.9 (a) Product ion mass spectrum of *m/z* 217 (M^{+}) derived from MDL257.
(b) Precursor ion mass spectrum of *m/z* 162 derived from analysis of urine extract.
(c) Product ion mass spectrum of *m/z* 249.

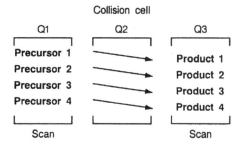

Collision cell

Q1	Q2	Q3
Precursor 1		Product 1
Precursor 2		Product 2
Precursor 3		Product 3
Precursor 4		Product 4
Scan		Scan

Precursor ion = product ion + constant neutral loss

Figure 13.10 Schematic representation of the production of a neutral loss scan.

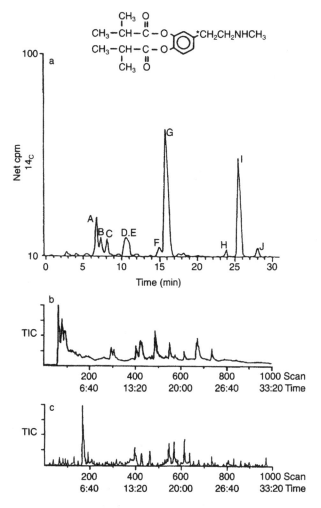

Figure 13.11 (a) HPLC radiochromatogram of a sample of monkey urine containing [^{14}C]ibopamine metabolites; (b) total ion current chromatogram and (c) chromatogram for a neutral loss of 176 mass units obtained by TSP LC–MS and TSP LC–MS–MS analysis of the same sample.

of mass 176 (i.e. glucuronic acid minus water). In the TSP LC–MS–MS analysis of the metabolites of [^{14}C]ibopamine, a positive inotropic agent, a 'neutral loss' scan was used to detect the presence of glucuronic acid conjugates (Rudewicz and Straub 1986).

Summary

A 'soft' ionisation technique such as TSP, APCI, electrospray (ESP) and FAB in combination with MS–MS provides an extremely powerful analytical tool for obtaining both molecular weight and structural information on polar drug metabolites. The combination of scans that are available provide screening techniques that are appropriate to the identification of drug metabolites. This approach has been termed 'metabolite mapping' (Straub *et al.* 1987).

13.4   Isotopically labelled compounds in metabolite identification

In Chapter 11 it was shown how the presence of chlorine in a molecule gave rise to a characteristic doublet of ions arising from the naturally occurring isotopes chlorine-35 and chlorine-37. Bromine-containing compounds are also easily recognised from their mass spectra because bromine-79 and bromine-81 exist naturally in the approximate relative abundance of 1 : 1. The mass spectral patterns produced by these types of compounds are very useful in helping to distinguish those ions in the mass spectrum which are drug related as opposed to those derived from endogenous compounds. Unfortunately not many of the compounds handled in drug metabolism studies contain these elements. The same effect can be created artificially if sufficient ^{14}C-labelled material is present, or a stable isotopically labelled analogue is added in a pre-defined ratio with the parent compound to the dosing solution or incubation substrate.

As drugs become more potent, so the relative amount of ^{14}C-labelled material added to 'cold' drug in the dose increases. Hence, this can provide a characteristic mass spectral pattern. The total ion current chromatogram obtained for an isolated metabolite of ondansetron shows two peaks. Only the second peak is derived from drug-related material, as evidenced by the mass spectrum (Figure 13.12). This shows the presence of two protonated molecules, m/z 310 and 312, the former being derived from the ^{12}C-labelled component and the second from the ^{14}C-labelled component. The relative abundance of these ions is approximately 3 : 1, which was consistent with the proportion of ^{14}C-labelled drug in the dosing solution.

The use of stable isotopes in metabolism studies has increased. Examples of the use of deuterium, carbon-13, oxygen-18 and nitrogen-15 are commonly encountered. Incorporation of deuterium atoms has the advantage that it is a relatively cheap synthetic process. Furthermore, it is often easy to incorporate more than one deuterium atom. However, care must be taken to ensure that the deuterium atom is in a metabolically stable position since an isotope effect may result, due to the different reaction rates between cleavages of C–H and C–D bonds.

The cost of synthesising ^{13}C-labelled analogues has been reduced in recent years. In addition, a ^{13}C-labelled analogue may be synthesised as a 'cold' run prior to synthesis of the ^{14}C-labelled material, so it is worth checking with in-house isotope chemists (if available) to see whether this has been done. An example of the use of this technique is shown in Figure 13.13.

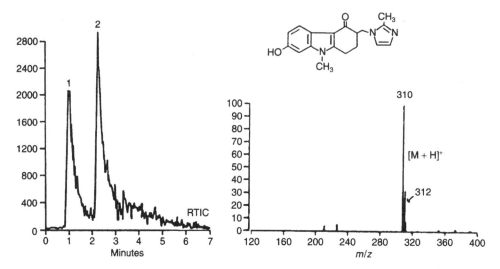

Figure 13.12 TSP LC–MS analysis of a HPLC fraction containing a hydroxylated metabolite of ondansetron. The metabolite was isolated from urine collected from a rat orally dosed with 15mg ondansetron/kg containing 25% [^{14}C]ondansetron.

13.5 Practical aspects for the identification of metabolites by mass spectrometry

13.5.1 Introduction

Chapter 11 and this chapter cover the different mass spectrometric techniques that can be used for metabolite identification. These techniques can be used to analyse metabolites of a wide range of polarities (Figure 13.14). It may take a combination of these techniques to solve a particular problem and one strategy for doing this is outlined in Figure 13.15.

The simplest approach is to analyse the untreated biological sample, obtaining the molecular weight by LC–MS, then use LC–MS–MS to derive some structural information. If this is not appropriate, possibly because of low concentrations of metabolites or unsuitability of the mass spectrometric technique, then alternatives to be considered are sample concentration (liquid/liquid extraction, solid-phase extraction, freeze-drying and reconstitution) or isolation of the individual metabolites, followed by re-analysis.

Successful identification of metabolites also depends on a consideration of sample preparation and of the operating characteristics of the different mass spectrometric techniques. There are a number of pitfalls to try to avoid. First, avoid using poly-ethylene glycol (PEG) as the dosing vehicle. It gives an excellent response in TSP, ESP and FAB generally to the detriment of any metabolites present. In fact it is used as a mass spectrometric calibration standard. Secondly, when storing or isolating samples it is worth considering what equipment is being used. Plasticisers leached from con-tainers, tubing, etc. give characteristic mass spectra which can affect mass spectral interpretation.

The following sections detail some of the operating characteristics of the different mass spectrometric techniques.

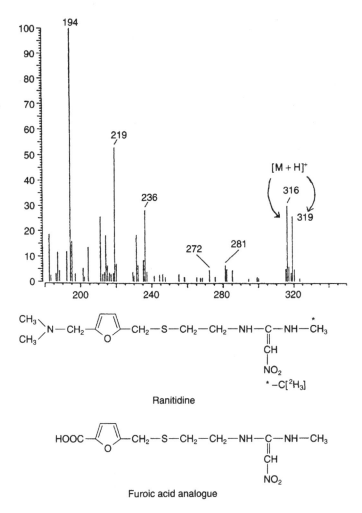

Figure 13.13 CI mass spectrum (DLI LC–MS) of the methyl ester of the furoic acid metabolite of ranitidine. The sample was isolated from human urine collected after oral aministration of ranitidine : [²H]₃ranitidine (relative amounts 1 : 1).

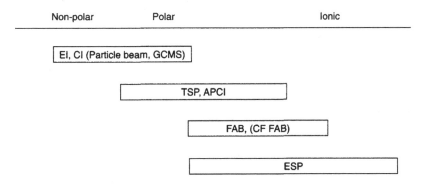

Figure 13.14 Applicability ranges of ionisation techniques as a function of solute polarity.

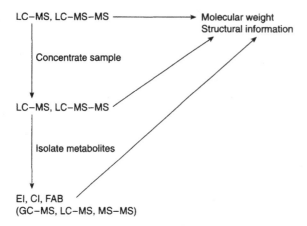

Figure 13.15   A proposed strategy for identifying metabolites by mass spectrometry.

13.5.2   *Electron impact ionisation (EI) and chemical ionisation (CI)*

These techniques have the following general characteristics:

- Isolated sample is required to minimise ion source contamination (probe analysis).
- On-line techniques are GC–MS, particle beam LC–MS, heated nebuliser LC–MS/APCI.
- Gas-phase ionisation.
- Thermally labile samples can degrade.
- Potential sample amounts are approximately 1µg (direct probe), or 100ng (direct exposure probe), 10ng (GC–MS), 100ng–10µg (particle beam), 10–100ng (heated nebuliser APCI).
- They are generally unsuitable for polar, ionic, thermally labile and involatile metabolites (although APCI can handle some of these).
- Mass spectrum: EI gives $M^{\ddagger}$ (possibly) and fragment ions; CI gives $[M+H]^{+}$, $[M-H]^{-}$ and M^{-}.
- Accurate mass measurement is possible.

Direct-probe analysis

The techniques of EI and CI require that the sample is volatilised prior to ionisation. If the sample is amenable to this type of analysis without thermal degradation, then useful structural information may be obtained on low amounts (1µg) of material. For those samples undergoing thermal degradation during volatilisation, there is a wide variety of chemical derivatives that can be considered so that the usual types of polar functionalities (e.g. amines, hydroxy groups) can be converted to compounds with a reduced propensity for hydrogen bonding. This does have a spin-off in that the analysis of derivatives can provide information on the number and type of functional groups present in the metabolite. For direct-probe analysis, the metabolite should be isolated in order to, first, minimise ion source contamination and, secondly, yield an

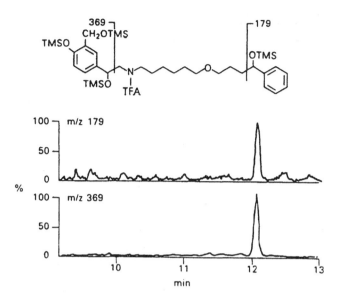

Figure 13.16 EI GC–MS of the trimethylsilyl, trifluoroacetyl derivative of the human faecal metabolite of salmeterol. TMS, trimethylsilyl; TFA, trifluoroacetyl.

identifiable mass spectrum which is not complicated by the presence of ions derived from other components.

The mass spectral information that may be obtained using these techniques is the molecular weight and some structural information that can sometimes define the site of metabolism. In addition, the use of accurate mass measurement (on a magnetic sector mass spectrometer) can provide additional supportive information on the elemental composition of an ion.

Gas chromatography–mass spectrometry (GC–MS)

The analysis of complex mixtures via capillary column GC–MS enhances the flexibility of using EI and CI for metabolite identification. Once again, sample volatility is vital and chemical derivatisation should be considered. An added bonus to making use of chemical derivatisation is that the mass spectrum of the derivative may contain structurally diagnostic ions, as shown by the example in Figure 13.16. Mass spectral data are obtainable on low nanogram amounts of material.

Particle-beam liquid chromatography–mass spectrometry (LC–MS)

The particle beam interface enables EI and CI mass spectra to be obtained from components in a HPLC eluent. Both reverse-phase (only volatile buffers are permitted) and normal-phase systems at flow rates up to 0.5ml/min can be used. In the interface (Chapter 11) the volatilised solvent molecules are pumped away and the heavier sample molecules are transferred through the interface into a conventional EI/CI ion source. The molecules are volatilised on impact with the heated surfaces in the ion source and are then ionised. Thus the particle-beam interface is restricted in

269

its application to samples that do not undergo thermal degradation. For metabolite identification, therefore, it has limited application and limited sensitivity (100ng–10μg, dependent on structure).

Heated nebuliser interface/atmospheric-pressure chemical ionisation

The heated nebuliser LC interface was developed to be used in combination with an atmospheric-pressure chemical ionisation (APCI) source. Both reverse-phase (volatile and non-volatile buffers are permitted according to the manufacturers) and normal-phase systems at flow rates up to 1.5ml/min are tolerated, although with normal-phase systems, care is needed with solvents with low flash points. The rapid desolvation and vaporisation of the mobile phase minimise thermal decomposition of the en-trained sample molecules (Chapter 11). Hence, the technique is suitable for many phase 1 and some phase 2 metabolites (phase 1 metabolism refers to metabolic modi-fication of the parent drug to introduce chemically reactive functional groups such as -OH, -NH$_2$ and -COOH; phase 2 metabolism refers to the conjugative reactions, e.g. glucuronidation and sulphation). Mass spectral data can be obtained on 10–100ng amounts of sample. The mass spectra are simple, generally showing the presence of protonated ([M + H]$^+$) or deprotonated ([M – H]$^-$) molecules.

13.5.3   Fast atom bombardment

Fast atom bombardment (FAB) ionisation has the following general characteristics:

- It is suitable for involatile, polar, thermally labile and ionic metabolites.
- The metabolite must be soluble in the FAB matrix.
- Detection limits vary with compound, 100ng – 2μg.
- Mass spectra obtained show [M + H]$^+$ or [M – H]$^-$.
- Ions from the matrix will contribute to the mass spectrum.
- Ion suppression can occur with mixtures.
- On-line systems are available with continuous-flow FAB.

The advent of FAB along with TSP meant that polar, involatile, thermally labile metabolites, in particular conjugates, could be analysed directly by mass spectrometry without the need for chemical derivatisation. The metabolite(s) (generally 0.1–2μg) are added in solution to a small amount (1–3μl) of a viscous liquid matrix, generally glycerol, contained on a metal target, prior to obtaining an FAB mass spectrum.

FAB is a 'soft' ionisation technique, producing predominantly protonated ([M + H]$^+$) and deprotonated ([M – H]$^-$) molecules. However, the interpretation of FAB mass spectra is not always straightforward. Ions derived from the sample matrix may be present in the mass spectrum of the metabolite (Chapter 11). To some extent their abundance can be minimised by subtraction, but it is not possible to remove them completely because it is not feasible to obtain a mass spectrum of the matrix which has ions of exactly the same abundance. The mass spectrum of the metabolite may also contain ions derived from co-extracted material. Here, the use of isotopically labelled compounds or MS–MS may improve specificity (Figure 13.6).

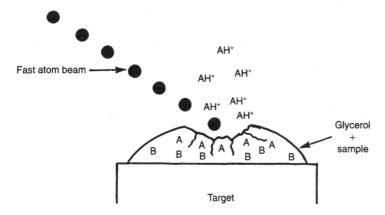

Figure 13.17 Dispersion of solutes of differing hydrophobicities throughout an FAB matrix. A is more hydrophobic than B.

Metabolite contamination with sodium and potassium salts seems to be the rule rather than the exception. Their presence is manifest in the mass spectrum by the presence of adduct ions, i.e. $[M + Na]^+$ and $[M + K]^+$. The presence of these ions can sometimes help in assigning the molecular weight. However, gross contamination with salts can lead to complete suppression of the analyte signal. Solid-phase extraction may be needed to remove the salts.

Components in complex mixtures disperse throughout the FAB matrix. The more hydrophobic samples tend to be present at the surface of the matrix; this effect is shown schematically in Figure 13.17. Since FAB is a surface technique, the mass spectrum will be representative of the more hydrophobic molecules closest to the surface. This effect is particularly noticeable if bile acid salts are present (Figure 13.18). Further sample purification is probably the only way to overcome this problem.

FAB has been adapted to deal with a flowing rather than a static matrix in 'continuous-flow FAB' (CF FAB) (Caprioli 1990). This technique is also referred to as dynamic FAB or frit FAB. Two different types of probe are in common use, one in which the eluent is conducted to a stainless-steel frit at the end of a probe and the other in which eluent flows out over what is essentially the drilled-out tip of a conventional FAB probe. Low amounts of glycerol (2–5%) are either incorporated into the eluent before chromatography or added post-column. Only very low flow rates (typically 5–10µl/min) are allowed, so for optimum operation a capillary column is coupled to the CF FAB probe. With large-bore columns, flow splitting is essential. CF FAB offers potentially a tenfold increase in signal to noise over static FAB because of the reduction in the abundance of matrix-derived ions.

13.5.4 Thermospray liquid chromatography–mass sectrometry (TSP LC–MS)

Thermospray LC–MS techniques have the following general characteristics:

- Reverse phase eluent is generally used.
- A volatile buffer is required.

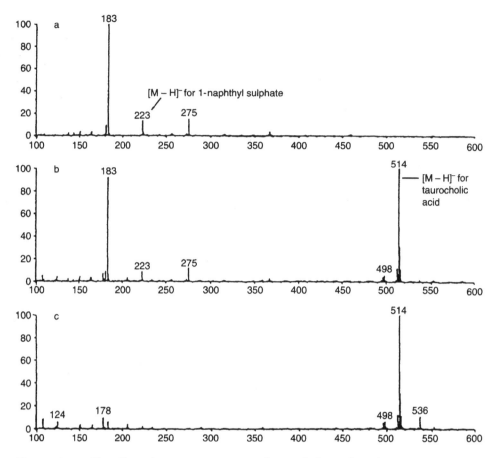

Figure 13.18 The effect of increasing amounts of taurocholic acid on the FAB spectrum of α-naphthyl sulphate in glycerol. (a) 1µg α-naphthyl sulphate; (b) 1µg α-naphthyl sulphate + 1µg taurocholic acid; (c) 1µg α-naphthyl sulphate + 10µg taurocholic acid.

- Normal-phase HPLC eluents are acceptable, but filament or discharge electrodes are then required for ionisation.
- Flow rates are 1–2ml/min.
- Isocratic or gradient elution can be used.
- Sensitivity is dependent on eluent composition, the ion source and the interface temperatures.
- The response is compound dependent (10ng–10µg).
- Thermal degradation can occur.
- Mass spectra obtained contain $[M + H]^+$, $[M - H]^-$ or M^-.

Until recently, thermospray has been the most widely used LC–MS technique for metabolite identification. TSP mass spectra can often yield molecular weight information for phase 1 and some phase 2 metabolites and then additional structural information is obtained using MS–MS.

Table 13.1 Suitable solvents for thermospray (from Mellon 1991)

Polar	Water, methanol, ethanol, acetonitrile
Non-polar	Hexane, methylene chloride
Buffers	Ammonium acetate, ammonium formate, acetic acid, trifluoroacetic acid, ammonium hydroxide, triethylamine, diethylamine, volatile ion-pair reagents

Ideally, chromatographic systems used for the characterisation of metabolites should be developed with a mass spectrometric end-point in mind. So the eluent should contain a volatile buffer. A list of suitable solvents and buffers is given in Table 13.1.

Ammonium acetate at a concentration of 0.05–0.1M is generally used. Involatile buffers and ion-pair reagents are not really compatible since their use leads to blockage of the thermospray interface and ion-source contamination. TSP works optimally at flow rates of 1–2ml/min. Both isocratic and gradient elution can be accommodated. Ideally the aqueous content of the eluent should be greater than 50% for optimum thermospray performance; however, with eluents containing a high percentage of an organic co-solvent a discharge electrode or filament can be used to promote chemical ionisation. It is worth bearing in mind though, that some sacrifice of the 'perfect chromatographic system' may be needed to achieve optimum performance in thermospray.

HPLC columns are tolerant of dirty samples so analysis of an untreated biological fluid (i.e. bile or urine) or an *in vitro* incubation mixture is feasible. Often a switching valve is used to divert the eluent from the first 2–3min post-injection to waste. This prevents deposition of polar, involatile material which would then block the interface, but it has the disadvantage that polar metabolites, such as conjugates, may be lost.

Sensitivity in TSP tends to be very dependent on compound type, solvent composition, TSP probe and ion-source temperature, and ionisation mode (i.e. positive or negative ions; filament or discharge switched on). Detection limits in TSP can be as good as 10ng or as poor as 10μg. With the ammonium ion present as the electrolyte, compounds containing a basic substituent tend to give the best response since they are readily protonated in the gas phase by NH_4^+ (cf. ammonia chemical ionisation, Section 11.3.3). The variability in response can create a problem if a compound is deaminated on metabolism. Sumatriptan (Figure 13.19), a drug used in the treatment of migraine, is extensively metabolised and the major metabolite results from oxidative deamination. Although a thermospray mass spectrum can be obtained from 10ng of sumatriptan, no response is obtained from 1μg of the carboxylic acid metabolite.

13.5.5 Electrospray liquid chromatography–mass spectrometry

In LC–MS in the electrospray (ESP) ionisation mode, the following are generally true:

- Reverse-phase eluent is used.
- Volatile buffer and/or acids or bases can be used. Trifluoroacetic acid is often added to promote ionisation in the positive-ion mode.

273

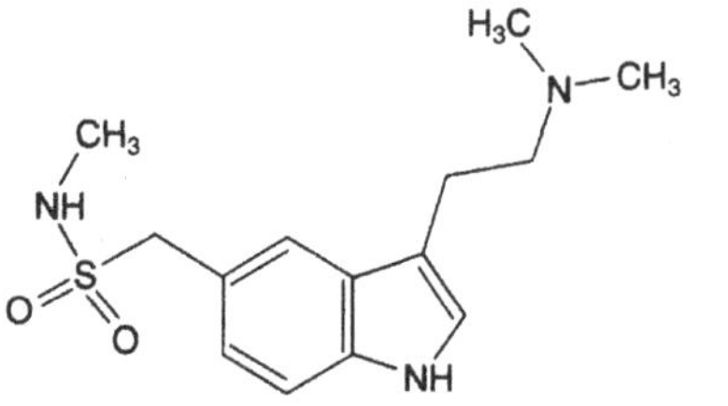

Figure 13.19 Structure of sumatriptan (I) and its carboxylic acid metabolite (II).

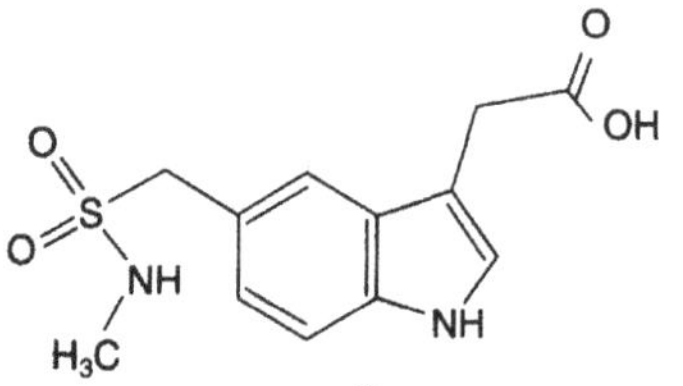

- Flow rates can be up to 1ml/min.
- Isocratic or gradient elution is used.
- Any metabolites should be ionised in solution (solution pH is important).
- A good response is obtained for polar, ionic, phase 1 and phase 2 metabolites that are not amenable to TSP.
- Sample amount can be down to 100pg.
- The mass spectrum obtained shows $[M + H]^+$, $[M - H]^-$ or M^-.
- Ionisation efficiency varies with solvent composition and type of analyte.

The ESP interface is the most recent LC–MS interface to gain widespread acceptance as a routine and robust interface for the analysis of metabolites. This technique is still continuing to be developed. Of particular importance is the ability to obtain mass spectral data on polar and ionic compounds, such as conjugates, which are not readily handled by TSP.

A good response is obtained in ESP for compounds that are ionised in solution. Reasons for poor response in ESP may be ion suppression caused by the presence of co-eluting compounds or the pH of the eluent, which reduces the proportion of molecules in the ionised state. Typical ESP mass spectra generally contain abundant protonated or deprotonated molecules and few fragment ions. For example, the carboxylic acid metabolite of sumatriptan (Figure 13.19) yielded the ESP mass spectrum shown in Figure 13.20. Phase 2 metabolites also give a good response in ESP.

13.5.6 Ion-trap mass spectrometry coupled to external atmospheric-pressure ionisation sources

This approach has recently been used to identify three isomeric monohydroxylated *in vitro* metabolites of bupropion (Figure 13.21), an antidepressant (Taylor *et al.* 1995). Initial LC–MS experiments were carried out with the ion-trap instrument using

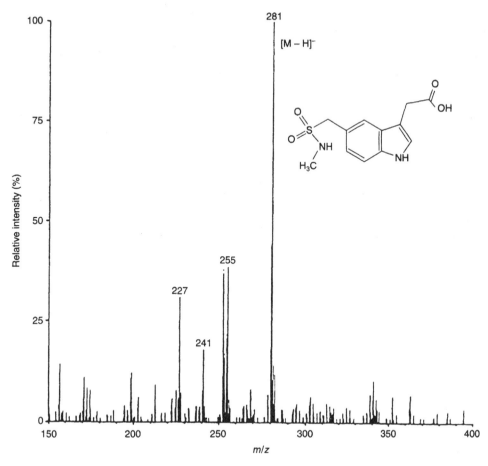

Figure 13.20 ESP mass spectrum of 10ng of the carboxylic acid metabolite (Figure 13.19, structure II) of sumatriptan.

bupropion, which established a detection limit for the $[M + H]^+$ ion of approximately 12pg (injected).

Using the ion trap, three monohydroxylated positional isomers were identified from their full-scan mass spectra followed by their product ion mass spectra, using a series of resonance excitation steps and MS^n. In this technique, the ions are trapped and then further fragmentation induced, providing a second mass spectrum. One of these fragment ions is then selected, trapped and again, further fragmentation induced. Thus a series of product ions can be produced, yielding much structural information. For further information, see Taylor *et al.* (1995).

13.5.7 *Summary*

For a number of years TSP and FAB have been the predominant ionisation techniques for providing molecular weight information for phase 1 and phase 2 metabolites. The API techniques of ESP and APCI are now more widely available for the trace

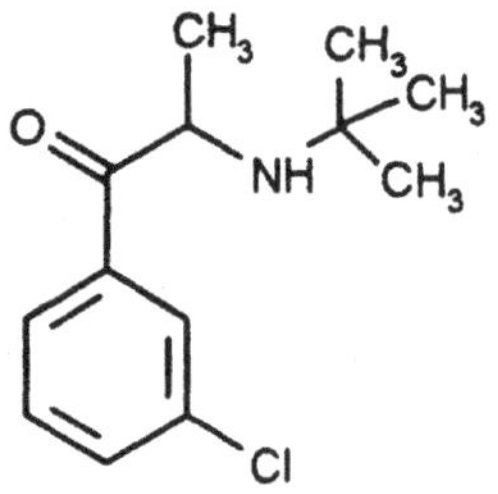

Figure 13.21 Structure of the antidepressant, bupropion.

analysis of polar ionic drugs and metabolites, extending the polarity range of compounds that can be handled successfully using on-line LC–MS and LC–MS–MS techniques.

13.5.8 Overall comments

Before embarking on a series of experiments to identify metabolites, there are a few points worth considering:

- Use the highest practicable dose or concentration for *in vivo* and *in vitro* studies to maximise the concentration of metabolites for identification purposes.
- Whenever possible use 'isotopic markers' to aid in distinguishing those ions derived from drug-related material from those ions derived from endogenous biological components and/or impurities introduced during sample work-up.
- If the concentrations of metabolites are high enough, biological samples could be screened first using NMR to evaluate the extent of metabolism and possibly the nature of the metabolites.
- Talk to your mass spectrometrist. It is important that a close working relationship is established between the drug metabolist and the mass spectrometrist so that each understands the practical considerations and limitations of their respective techniques.

13.6 Bibliography

Busch, K.L., Glish, G.L. and McLuckey, S.A. (1988) *Mass Spectrometry/Mass Spectrometry: Techniques and Applications of Tandem Mass Spectrometry*, New York: VCH Publishers.

Caprioli, R.M. (ed.) (1990) *Continuous Flow Fast Atom Bombardment Mass Spectrometry*, Chichester: John Wiley and Sons.

Coutant, L.E., Barbuch, R.J., Satonin, D.X. and Cregge, R.-L. (1987) Identification in man of urinary metabolites of a new bronchodilator, MDL 257, using triple stage mass spectrometry mass spectrometry. *Biomed. Environ. Mass Spectrom.*, **14**, 325.

Mellon, R. (1991) Liquid chromatography mass spectrometry. *VG Monographs in Mass Spectrometry*, **2**, (1).

Rudewicz, P. and Straub, K.M. (1986) Rapid structural elucidation of catecholamine conjugates with tandem mass spectrometry. *Anal. Chem.*, **58**, 2928.

Straub, K.M. (1986) Metabolic mapping of drugs: rapid screening for polar metabolites using FAB/MS/MS, in Gaskell, S. (ed.), *Mass Spectrometry in Biomedical Research*, Chichester: John Wiley and Sons, pp. 115–136.

Straub, K.M., Rudewicz, P. and Garvie, C. (1987) Metabolic mapping of drugs: rapid screening techniques for xenobiotic metabolites with MS/MS techniques. *Xenobiotica*, 17, 413.

Taylor, L., Singh, R., Chang, S.Y., Johnson, R.L. and Schwartz, I. (1995) The identification of *in vitro* metabolites of bupropion using ion trap mass spectrometry. *Rapid Comm. Mass Spectrom.*, 2, 902.

14

NUCLEAR MAGNETIC RESONANCE IN DRUG METABOLISM

Phil Gilbert

14.1 Introduction

It is now more than half a century since the first NMR signals were detected independently by the groups of Bloch and Purcell. In that time NMR has been developed into the most important structural analysis tool available in organic chemistry. The applications of NMR range far and wide, from determination of protein structures in solution, and the study of oils in the petrochemical industry, to the sophisticated NMR imaging methods in clinical studies. Until recently, the use of NMR in drug development groups has been relatively unusual. This is probably because NMR is historically found in drug discovery areas, providing a service to organic chemists in the pharmaceutical industry. However, metabolists and toxicologists are increasingly realising the value of this resource, and are using it to great effect alongside the other spectroscopic techniques such as mass spectrometry.

14.2 Basic theory of the NMR phenomenon

The fundamental property exploited in NMR is a property of the atomic nucleus, called *spin*. This is usually given the symbol I. The value of I can be 0, $1/2$, 1, ... etc., in units of $h/2\pi$ where h is Planck's constant. This value is determined by the mass and atomic number (Table 14.1).

From Table 14.1 we can soon work out that many common nuclei, e.g. ^{12}C, ^{16}O, have zero spin, so no NMR signals. The most important nuclei for our purposes are ^{1}H (proton), ^{13}C, ^{19}F, ^{31}P, ^{15}N and ^{3}H (tritium). All of these have a value for I of $1/2$, and the rest of the discussion will be restricted to spin-$1/2$ nuclei. It is important to

Table 14.1 Value of I

Mass number	Atomic number	Value of I
Odd	Even	1/2, 3/2, 5/2 ...
Odd	Odd	1/2, 3/2, 5/2 ...
Even	Even	0
Even	Odd	1, 2, 3 ...

remember, however, that other values for I are possible as you may sometimes come across spin-1 nuclei such as ^{2}H (deuterium) and ^{14}N.

If we place our nuclei into a magnetic field, they can align themselves in $2I + 1$ different ways. For spin-½ nuclei, then, two ways are allowed. We can imagine the nuclei behaving as tiny bar magnets, orienting themselves either with the magnetic field (low energy) or against the magnetic field (high energy). Thus, we have two possible states, and the energy between them is given by:

$$\Delta E = \frac{h\gamma B_o}{2\pi}.$$

In this equation, h = Planck's constant, B_o is the applied field and γ is the gyromagnetic ratio, which is a constant for each nucleus. Nuclei can cross between the two energy states with absorption or emission of energy. As in other spectroscopic techniques this energy is in the form of electromagnetic radiation the frequency of which, v, is given by the equation:

$$\Delta E = hv.$$

Combining this with the previous equation leads to the expression:

$$v = \frac{\gamma B_o}{2\pi}.$$

With the sort of magnets available to us, v comes into the radiofrequency range, between 60 and 800MHz, depending on B_o. It is usual in NMR to describe magnets by their ^{1}H frequency – as 'we've got a 400MHz system'. At equilibrium the relative populations of the two allowed energy states, α and β are described by the Boltzmann distribution:

$$\frac{N_\beta}{N_\alpha} = \exp(-\Delta E/kT).$$

The difference $N_\beta - N_\alpha$ is very small, about 1 in 10^5, so N_β/N_α is close to 1. This leads to the inherent insensitivity of the NMR experiment. However, the old adage applies, if you want a bigger bang, get a bigger hammer, and N_β/N_α increases with increasing B_o, or magnet size.

To observe the NMR signal, we can use the same methods as IR or UV spectroscopy, varying the frequency, v, and looking for an absorption of energy. In practice, however, this method (CW or continuous-wave) is very slow and insensitive. Modern spectrometers work on the fourier transform (FT) principle. In this, all the nuclei are excited at once by a pulse of radiofrequency energy. The system is then allowed to relax, giving a complex decay pattern (the FID, free induction decay). The FID can be converted back to a conventional frequency display by the mathematical process known as fourier transform, which is done using a dedicated computer. The pulse can be repeated, and FIDs co-added to increase sensitivity. Any modern spectrometer likely to be used in metabolism work will operate in this way.

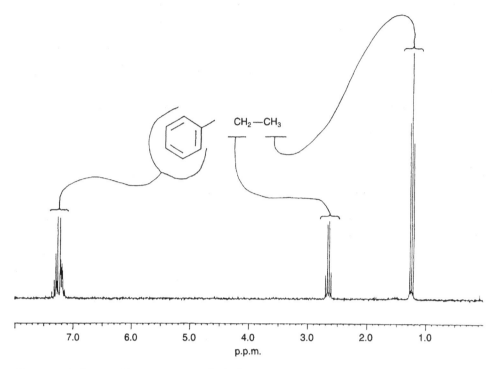

Figure 14.1 Proton NMR spectrum of ethylbenzene.

14.3 Parameters of the NMR spectrum

14.3.1 Chemical shift

None of the foregoing would be of much use if all you got was a single line telling you that your sample had, say, protons in it. There are easier and cheaper ways of learning this! Fortunately, each chemically distinct nucleus in a molecule is in a slightly different magnetic environment. Each nucleus is shielded from the applied magnetic field by the electrons around it, and the electron density will be different for each different chemical environment. Hence, in a molecule we see a range of different absorption frequencies. This is the phenomenon of the chemical shift, δ. If B_o is the applied field, and B_{local} is the actual field seen at the nucleus, then the chemical shift, δ, is defined as:

$$\delta = \frac{B_o - B_{local}}{B_o} \times 10^6 \text{p.p.m.}$$

Since this is a dimensionless quantity, it is necessary to use an internal standard. For this purpose, tetramethylsilane, $Si(CH_3)_4$ (TMS) is defined as 0p.p.m. for ^1H and ^{13}C NMR. Since v is proportional to B, we can rewrite the chemical shift definition as:

$$\delta = \frac{v_{\text{sample}} - v_{\text{TMS}}(\text{Hz})}{\text{operating frequency in MHz}}$$

Hence, at 100MHz, 1p.p.m. will be 100Hz and at 400MHz it will be 400Hz.

Spectra are shown with TMS at the right-hand, **high-field** or **upfield** end of the spectrum. The range of chemical shifts varies from nucleus to nucleus. Most proton shifts fall between 0 and 15p.p.m. ^{13}C shifts cover about 200p.p.m., and for some exotic nuclei the range can be thousands of p.p.m.

If (at last!) we look at a spectrum (Figure 14.1) we can see the effect of chemical shift. In ethylbenzene there are three distinct groups of proton signals at different chemical shifts, corresponding to the methyl, CH_2 and benzene ring of the nucleus. Each signal has its own chemical shift, characteristic of its molecular environment (Figure 14.2). Inspection of the ethylbenzene spectrum shows us that the signals are not single lines but have fine structure. This leads us on to the next of the NMR parameters, spin–spin coupling.

14.3.2 Spin–spin coupling

We have already seen how the position of an NMR line is determined by the actual magnetic field at the nucleus, and how this value is different from the applied field because of electronic shielding. A further factor in the actual value is the effect of nearest neighbours. Consider two protons, H_x and H_y, on adjacent carbon atoms:

$$\begin{array}{cc} \mathbf{H_x} & \mathbf{H_y} \\ | & | \\ \text{---C---C---} \\ | & | \end{array}$$

Imagine you are standing on nucleus H_x; the magnetic field you experience is, as before, made up of the applied field B_0, less an amount due to electronic shielding. However, your nearest neighbour has nuclear spin, and is acting like a bar magnet, so generating its own contribution to B_{local}. Because it can be aligned either with or against the field, H_y's contribution to the magnetic field at H_x has two discrete values. This causes H_x's signal to be split into two lines (Figure 14.3). This phenomenon is called coupling and the distance between these lines is called the coupling constant, symbol J, measured in units of Hz.

This value is not dependent on the field strength, so as this increases spectra become less liable to overlap and easier to interpret. In general, the effect of proton–proton coupling is only seen over three bonds, which helps to prevent spectra becoming impossibly complex (Figure 14.4).

As a further example, consider the spectrum of ethanol (Figure 14.5). The highest field signal belongs to the methyl group and appears as a triplet. The methylene group appears at $\delta = 3.7$p.p.m. as a quartet, and the OH as a singlet at 2.8. The three protons on the methyl group are identical, hence they show *no visible coupling to*

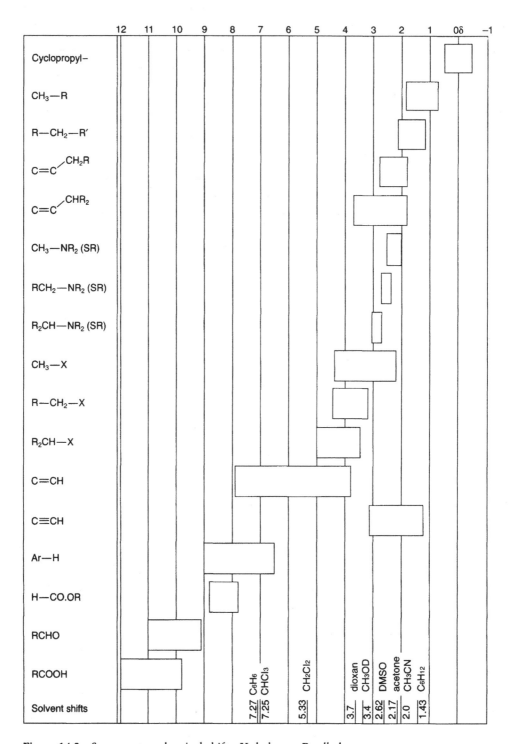

Figure 14.2 Some proton chemical shifts. X, halogen; R, alkyl.

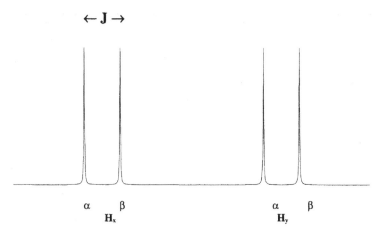

Figure 14.3 The effect of proton–proton coupling on the NMR spectrum.

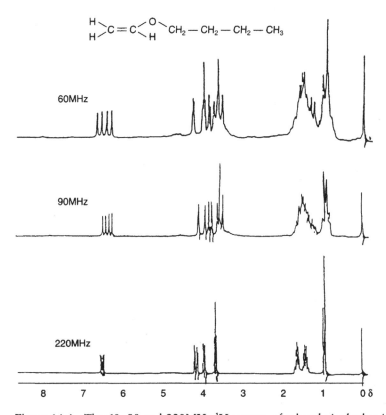

Figure 14.4 The 60, 90 and 220MHz ^{1}H spectra of *n*-butyl vinyl ether in CDCl$_3$.

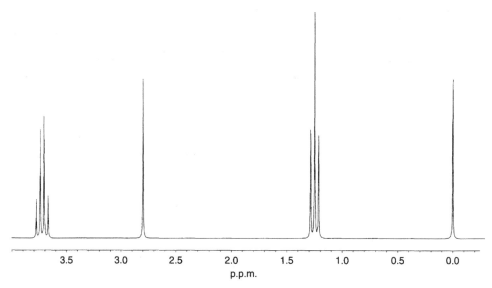

Figure 14.5 Proton NMR spectrum of ethanol.

one another. However, the spins of the two protons on the adjacent methylene carbon can exist in a number of combinations:

↑   ↑

↑   ↓

↓   ↑

↓   ↓

Clearly, the middle two have the same energy so give only one line. Hence the methyl group 'sees' a CH_2 with three possible states, one of which is twice as likely as the others. The result of this is a $1 : 2 : 1$ triplet. Similar reasoning gives a $1 : 3 : 3 : 1$ quartet for the methylene group. In general, if one nucleus couples to n other nuclei, then its signal consists of $n + 1$ lines. The intensity of these is given by the coefficients of Pascal's triangle, i.e. $1 : 2 : 1$, $1 : 3 : 3 : 1$, $1 : 4 : 6 : 4 : 1$, etc. Notice in the ethanol example that no coupling is seen between the CH_2 and the OH protons. This is because the OH protons are exchanging rapidly with one another, and we are observing them in a time-averaged environment. This effect is general with OH, SH and amine NH signals. In D_2O or CD_3OD, often used in studies of drug metabolites, these signals will not be visible at all, as they exchange with the solvent. As a final example of this important idea, consider aromatic systems.

The protons attached to a benzene ring each couple to the others with characteristic J values (Figure 14.6). These values can be used to show the position of the substituents around the ring, for example following metabolic oxidation of an aromatic drug. As an example of an aromatic coupling pattern, consider the spectrum of 3-nitrophenol (Figure 14.7). The four proton signals are clearly resolved from one another, and we can use the coupling values to assign the spectrum. The signal at

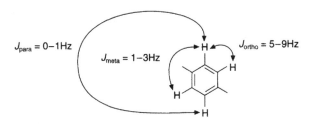

Figure 14.6 J values for couplings around a benzene ring.

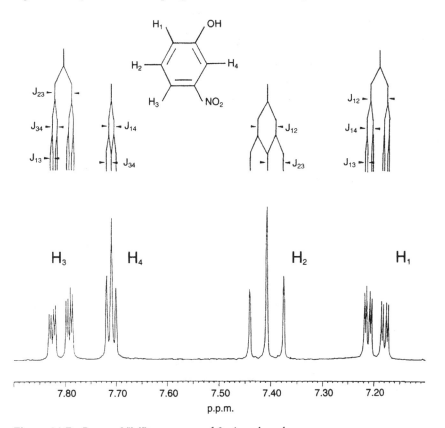

Figure 14.7 Proton NMR spectrum of 3-nitrophenol.

$\delta = 7.71$ comprises a triplet. This arises from two equal couplings, whose value we can measure as 2.4Hz, telling us they are meta couplings. No other splittings are visible, so we can assign this signal to H_4. The para coupling to H_2 is not resolved, or has a value of 0. A similar argument shows that the triplet at $\delta = 7.41$ has two equal ortho couplings of 8.3Hz. This is therefore assigned to H_2. The remaining signals, due to H_1 and H_3, both show similar patterns; we can assign the signal at lowest field to H_3 because of its chemical shift.

The splitting of the H_3 signal is due to an 8.3Hz ortho coupling to H_2, a 2.4Hz meta coupling to H_4 and a further meta coupling to H_1, this time of 1.0Hz value. This combination of couplings leads to the signal for H_3 being split into eight lines.

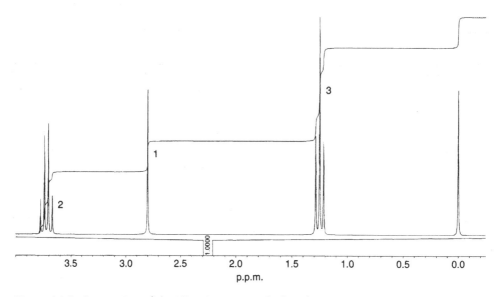

Figure 14.8 Integration of the NMR spectrum of ethanol.

14.3.3 Intensity

There is a third important parameter to consider – intensity. The area under an NMR peak is proportional to the number of nuclei producing it. Thus, if we reconsider our spectrum of ethanol and integrate (Figure 14.8), we see that the signals are in a ratio of 3 : 2 : 1. Clearly, we could also calculate absolute quantities in the presence of an internal standard. Care must be taken when making quantitative measurements, especially under FT conditions. It should also be remembered that the accuracy of NMR is only fair as a quantitative method.

14.4 Practical considerations

14.4.1 Types of spectrometer

We have already seen that the NMR experiment is inherently insensitive. Coupled with the small amounts of sample usually available, this means using the largest spectrometer you can lay your hands on! These days, this will invariably work on the fourier transform (FT) principle. High-resolution NMR looks at materials in solution. Since FTNMR involves exciting all signals at once, we must ensure that signals due to solvents, buffers, etc. are kept to a minimum. For example, H_2O is about 110M in protons. If we are seeking submilligram quantities of a metabolite in aqueous solution, the large water signal can make the small signals of interest impossible to detect. This is because they must be digitised for the computer, and the ratio of the largest to the smallest peak must not exceed the capacity of the instrument's digitiser. Although recent instruments *can* give excellent results in H_2O, it is usual to use 'heavy' solvents, with the hydrogen replaced by deuterium. For our purposes the most useful solvents are D_2O, CD_3OD and $CDCl_3$.

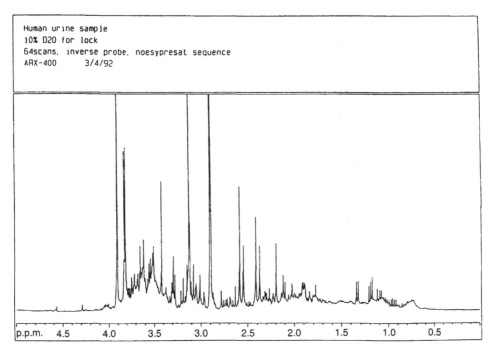

Figure 14.9   NMR spectrum of a human urine sample.

14.4.2   Sample preparation

The simplest form of sample preparation for NMR is none at all! With some modern instruments it is quite possible to look at samples of neat biofluids by NMR. An example is shown in Figure 14.9, which is a 400MHz spectrum of control human urine. The sample was simply filtered into the NMR tube. Solvent suppression is now so effective with the latest instruments that the 110M water signal at $\delta = 4.8$ has been effectively wiped out. Normally, however, it is necessary to extract and concentrate the sample. Any of the methods already outlined in this book could be tried. However, most success usually comes from the use of HPLC or solid-phase extraction methods. Commercial systems are now available for direct coupling of HPLC to NMR. For the principles of HPLC–NMR coupling, a description of its problems and advantages and some applications, see Woolf (1999) and Webb (1999). The acquisition of NMR data remains essentially the same whether the sample is introduced to the magnet via HPLC or a conventional tube. However, few labs as yet have access to direct-coupling technology, so most samples will come from isolated fractions. It is important that protonated solvents and buffers are removed carefully. In particular, freeze-drying ammonium acetate is much less effective than you might think! Solid-phase extraction can give rise to very rapid answers in conjunction with NMR detection, and examples are shown in the next section. One can also use SPE columns to quickly switch from protonated to deuterated solvents without freeze-drying or blowing down – an advantage for volatiles. Preparative TLC methods for NMR samples can be applied; however, this method invariably introduces substantial quantities of impurities and is best treated with caution.

14.5 NMR applications in drug development

NMR has been used within the pharmaceutical industry for many years. Traditionally, it has been an important tool for synthetic chemists, but not in development areas. It has often been dismissed as too insensitive and demanding of clean samples. These are not now serious drawbacks, as I hope to illustrate with these examples of identification of urinary metabolites, using different approaches.

14.5.1 No sample preparation

CAS 493 has the structure:

and its NMR spectrum is shown in Figure 14.10a. We can compare this with a spectrum of urine from a dog dosed at 25mg/kg with the drug (Figure 14.10b). In this case, samples are neat, filtered urines. In the dosed sample, signals from two drug metabolites are clearly visible (marked **v**). No peak can be seen due to the aldehyde proton at $\delta = 9.5$, so we could postulate one of the metabolites as the carboxylic acid. Close examination of the spectrum shows a further signal at $\delta = 5.5$ (marked $\star$). This is characteristic of the anomeric proton of the β-glucuronide group. Treatment of the sample with deuterated sodium hydroxide results in the two sets of metabolite signals collapsing to one (Figure 14.10c), whose identity as the carboxylic acid metabolite can be proved by spiking with authentic material. We have thus identified two metabolites, the carboxylic acid and its glucuronide ester conjugate, with no sample clean-up at all, using less than 1ml of urine, and in only a few hours. Clearly this is a powerful method. This example is, of course, particularly favourable in the dose, excretion route and NMR parameters. However, it is a quick and non-destructive experiment, well worth using as a first look for a particular problem.

14.5.2 Solid-phase extraction sample preparation

A much more likely scenario finds us with too little drug in the urine to see directly, and the spectrum heavily overlapped by the endogenous components of urine. In this case, use of solid-phase extraction cartridges can be a very rapid way of obtaining enough sample for NMR. For example, in Figure 14.11b, we see a complex spectrum of urine from a human volunteer dosed with 400mg of HWA 448, the structure and

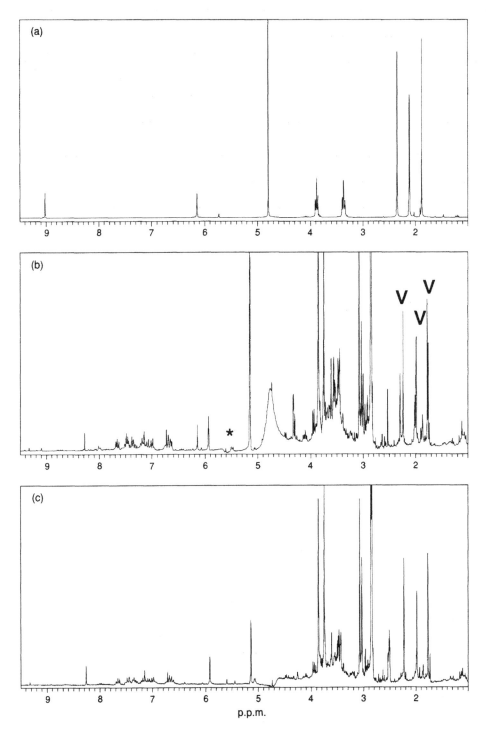

Figure 14.10 NMR spectra of (a) CAS 493; (b) urine from a dog dosed with CAS 493; (c) the dosed sample treated with deuterated sodium hydroxide.

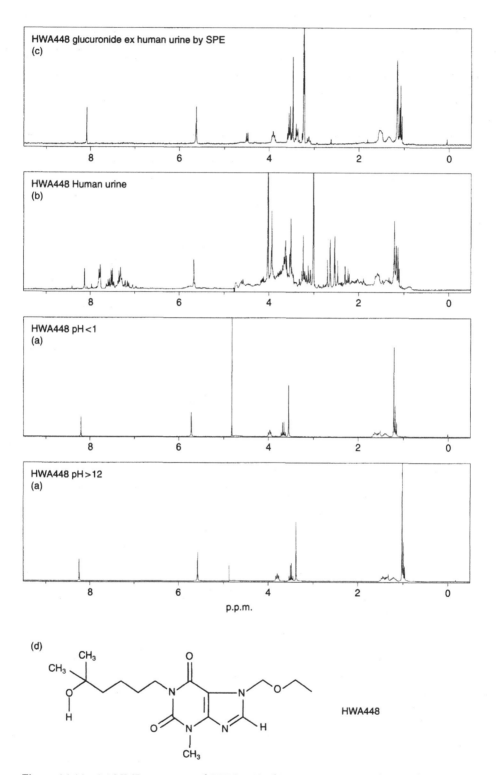

Figure 14.11 (a) NMR spectrum of HWA 448; (b) NMR spectrum of urine from a human volunteer dosed with HWA 448; (c) NMR spectrum of HWA 448 glucuronide isolated from human urine; (d) the structure of HWA 448.

spectrum of which are shown in Figures 14.11d and 14.11a. Although some signals due to the drug are visible, it is not possible to deduce the structure of any metabolite. However, if a 5ml sample of urine is passed directly onto a 3ml C18 column, all the drug-related material is retained, even at neutral pH. A subsequent wash with 20% methanol/water clears nearly all the endogenous material. A further wash of 40% methanol/water gives a very clean extract, easily identifiable as the O-glucuronide of HWA 448 (Figure 14.11c). If you are in a real hurry, the SPE columns can be washed using deuterated methanol/water, and the samples put directly into the NMR tube!

This approach again works well at moderate to high dose levels. However, use of the very large columns now available allows concentration from considerable quantities of urine. A further advantage of this technique for NMR is that it de-salts the sample, as some highly sensitive modern probes can be a little awkward with samples of high dielectric constant.

14.5.3   HPLC fractions

If you have to use NMR facilities outside your own department, then it is very likely that your samples will be fractions isolated by HPLC. You'll probably be competing for instrument time with synthetic chemists, so it's a good idea to prepare your sample as effectively as possible. In particular, ensure that protonated solvents and buffers are completely removed. It is all too common to see samples marked 'single peak by HPLC' in which the spectrum is overwhelmed by a huge acetate signal. This said, good-quality HPLC plus NMR will often give the answer. The amount of material you need to isolate depends on how clean you can get it, and how big a spectrometer you can use, but at the relatively modest field of 250MHz, I estimate 100µg of clean material as a minimum amount. An example of good HPLC is shown in Figure 14.12, where we can see an unusual metabolite of HWA 138. Notice the signal near 0p.p.m. – this is due to silicone oil from a faulty HPLC pump and caused this sample to be almost uninterpretable by mass spectrometry.

14.5.4   Fluorinated compounds

The presence of fluorine in a new drug molecule is a gift to the NMR spectroscopist. The nucleus ^{19}F has spin-$\frac{1}{2}$, is almost 100% abundant and gives a sensitivity 88% of proton NMR. The background signal in biological fluids is essentially zero. Hence, we can observe ^{19}F signals with no sample pre-treatment. ^{19}F has a huge chemical shift range and even distant modifications to a molecule cause observable changes to the ^{19}F spectrum. As an example, Figure 14.13 shows the ^{19}F spectrum of a neat, filtered dog urine sample from an animal dosed with Imirestat:

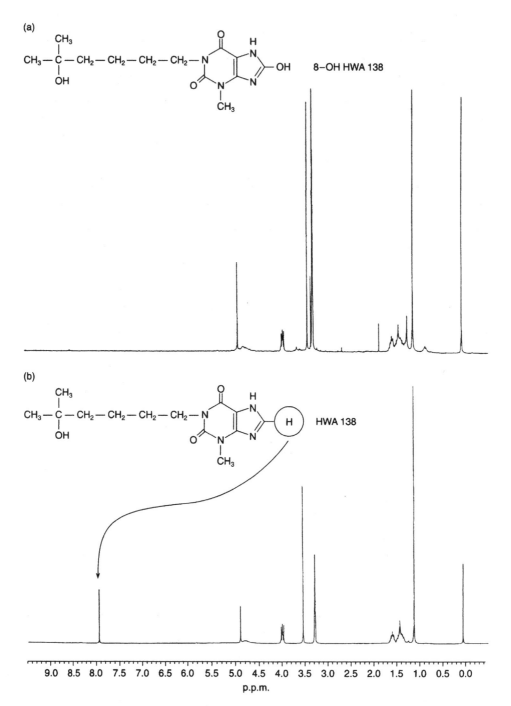

Figure 14.12 (a) NMR spectrum of 8-hydroxy HWA 138; (b) NMR spectrum of HWA 138 (the signal near 0 p.p.m. is due to silicone oil).

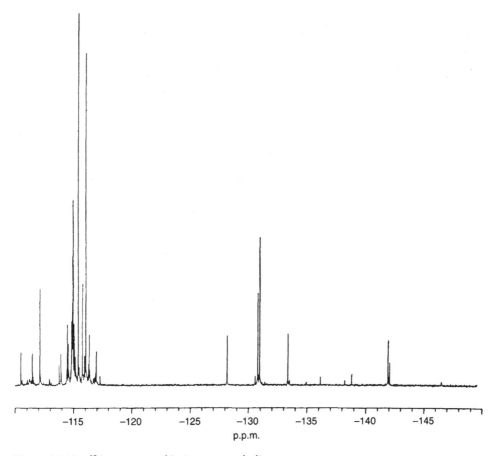

Figure 14.13 ^{19}F spectrum of Imirestat metabolites.

Over 50 individual signals are visible, showing that the metabolism of this compound is complex. Using ^{19}F NMR as a 'detector' for HPLC work identified nearly 20 metabolites of this compound, including some unexpected oddities.

14.5.5 Stable isotope labelling

We have been mostly concerned with ^{1}H and ^{19}F nuclei, because they give high sensitivity and are useful for the small quantities of metabolites generally on hand. However, the ^{13}C nucleus can provide a great deal of structural information. At natural abundance only 1% of carbon atoms are ^{13}C, making it difficult to detect sub-milligram quantities. However, it is quite possible to label molecules with ^{13}C, in the same way that we regularly use ^{14}C for radiolabelling. This then provides us with a 'magnetic marker' which can give structural information in the same way as ^{19}F. This method, of course, involves special synthesis, but can be very useful in resolving problems with heterocyclic compounds, or isomer problems that are untraceable by mass spectrometry.

14.6   Plasma metabolites

So far most of the discussion has centred round identification of urinary metabolites. In general, the low sensitivity of NMR precludes direct observation of plasma metabolites. However, in favourable cases it can be possible, especially using ^{19}F. Protein binding will cause NMR signals to broaden and disappear, so some degree of sample preparation will be required.

14.7   Biochemical changes

Sometimes the pattern of endogenous components in a spectrum of urine can be drastically altered by administration of a test compound (e.g. Figure 14.14). Information like this sometime comes as a bonus during metabolite identification, and it is worth remembering that much biochemical information can be gleaned.

14.8   Summary

- NMR spectroscopy is a very powerful tool in drug metabolism. Its use should *always* be considered when identifying metabolites alongside the better-known mass spectrometry techniques.
- It is relatively insensitive, but has a very high information content.
- It is non-destructive of samples, and a 'transparent' method – what you see is what you get!
- Although NMR is perhaps conceptually difficult it is practically simple – it may be possible to get away with very little sample preparation.
- For fluorinated drugs in urine or plasma, NMR may well be the method of choice.
- Finally, NMR techniques are rapidly advancing, sensitivity is improving all the time. There is no doubt that nuclear magnetic resonance now has a vital role in drug development.

14.9   Appendix: fourier transform and some multi-pulse techniques

In the previous sections we have covered the basic parameters of the NMR spectrum. The subject of high-resolution NMR is a vast one, with a large number of sophisticated experiments available. Many of these are very useful in studying metabolism problems, and the aim of this appendix is to give an introduction to what is possible.

14.9.1   Why use pulse NMR?

We have already seen that the greatest drawback of NMR is its inherently poor sensitivity. This gives rise to lots of *noise* in the spectra. Using the traditional continuous-wave (CW) technique, varying the frequency and recording energy absorbtion, the sensitivity is simply not adequate with any practical size of magnet. In infrared or ultraviolet spectroscopy, the sensitivity is good enough for a single experiment, but for NMR, especially with low-abundance nuclei, such as ^{13}C or ^{15}N, we must use

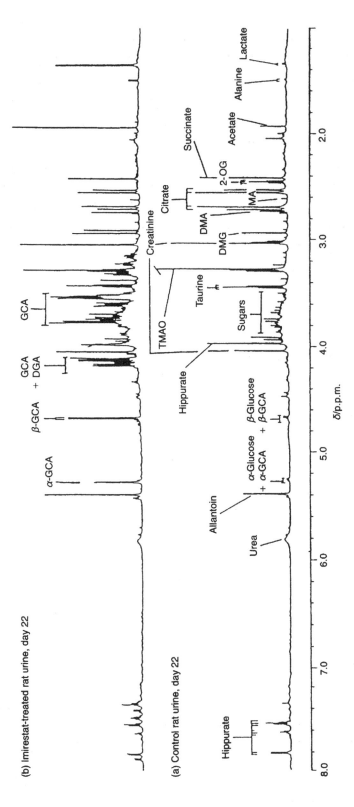

Figure 14.14 ¹H NMR spectrum (600MHz) of (a) control and (b) Imirestat-treated rat urine (50mg/kg p.o.). DGA, D-glucaric acid; GCA, D-glucuronic acid; IM, Imirestat.

some form of **signal averaging**. If we acquire repeated CW spectra, and add them together in a computer, the signal-to-noise (S/N) ratio improves by a factor of $N^{1/2}$ if N spectra are acquired. (If you want to check the maths of this, try any good statistics textbook.) Unfortunately, CW spectra take a long time to acquire – maybe hundreds of seconds – so to obtain the thousands of spectra necessary for ^{13}C NMR is prohibitive in time.

14.9.2   The pulse

In the slow CW experiment, the instrument spends most of its time sampling noise. If we could irradiate *all* the NMR signals at the same time and then unscramble the frequency information, we could speed up the experiment no end. Fortunately, it turns out that both these things are possible.

It is true to say that if a single frequency, F, is switched on and off rapidly, for a period of t seconds, then what results is a band of frequencies, centred at F. The bandwidth is given by the expression:

bandwidth $= F \pm 1/t$

Hence by applying a *pulse* with suitably chosen values of F and t, we can excite all the nuclei together. The value of t usually falls in the range 1–100μs.

14.9.3   Time and frequency

The measurement we make after a pulse is of signal amplitude as a function of *time* (this is called a **time domain** signal). The time domain and frequency domain forms of the data are interconvertible, by a mathematical function known as **fourier transform**. I do not wish to explain the mathematics of this, but here is the basic expression for those who might be interested:

$$f(v) = \int_{-\infty}^{\infty} f(t)e^{ivt}\,dt$$

where $f(v)$ is the frequency spectrum and $f(t)$ the time domain data. Luckily for us, very good algorithms are available to enable rapid transformation from time to frequency domain using computers. This is the reason why computers are an integral requirement in a high-field NMR spectrometer, and not a bolt-on extra as with some other instruments.

14.9.4   Multi-pulse experiments

I do not wish to go into the nuts-and-bolts of pulse NMR – many good textbooks are available to do this. However, it is useful to know that combinations of pulses and delays can transform the NMR spectrum to give us extra information in different ways. Without describing their physics, this section will briefly show some more advanced experiments.

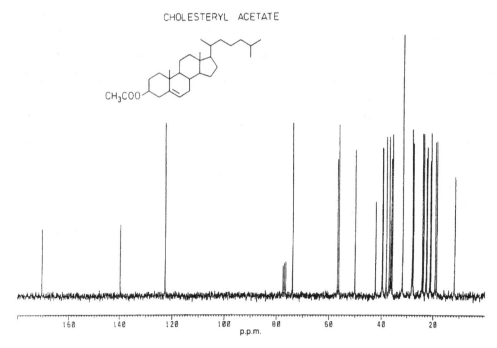

Figure 14.15 ^{13}C NMR spectrum of cholesteryl acetate.

Spectral editing

It is normal in ^{13}C NMR to 'decouple' proton signals by irradiating the whole ^{1}H spectral width. This gives 'stick' spectra, where each carbon gives a single signal. Couplings between ^{13}C and ^{13}C are not seen. Because the natural abundance is only 1%, the chance of two adjacent carbons being ^{13}C is only 0.01%. A typical ^{13}C spectrum, of cholesteryl acetate, is shown in Figure 14.15.

It would be very useful to know which signals belonged to CH_3, CH_2, CH or quaternary carbons. This is possible using an experiment known as DEPT (distortionless enhancement by polarisation transfer). We can choose a spectrum showing CH and CH_3 carbons only, or one where CH and CH_3 are in antiphase to CH_2. This is very useful for assignment. In addition, computer addition and subtraction of data can give us individual subspectra (Figure 14.16).

Homonuclear 2-dimensional correlation spectroscopy (COSY)

We can think of the fourier transformation in very simple terms as a process to convert a function of time into a function of frequency:

$$f(t) \xrightarrow{\text{FT}} f(v)$$

In a two-dimensional experiment, we introduce a second time variable, by putting a variable delay between pulses. We can then do a two-dimensional fourier transform:

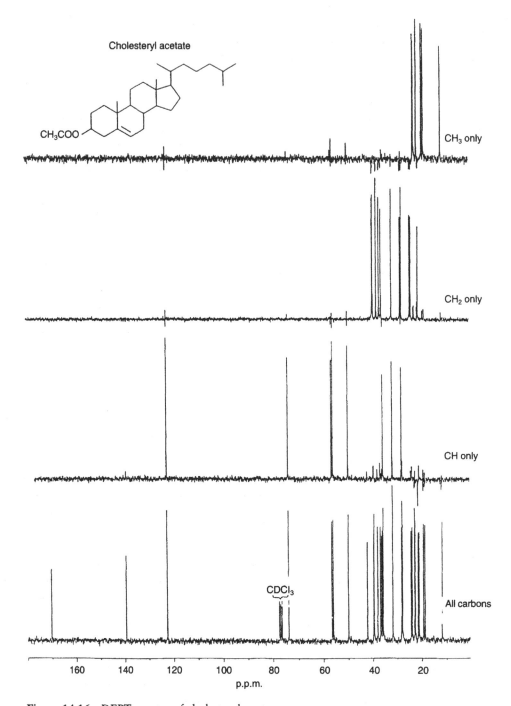

Figure 14.16 DEPT spectra of cholesteryl acetate.

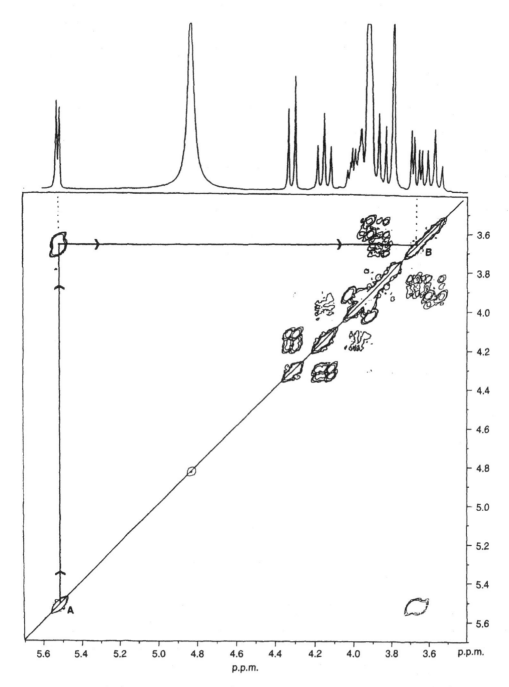

Figure 14.17 The ¹H COSY spectrum of sucrose. To interpret, imagine you are looking down from above on the spectrum, which is along the diagonal. From signal A at $\delta = 5.5$, follow a line to the cross-peak, then at right-angles back to the diagonal, arriving at A's coupling partner B at 3.65 p.p.m. You can then find B's couplings, and so on.

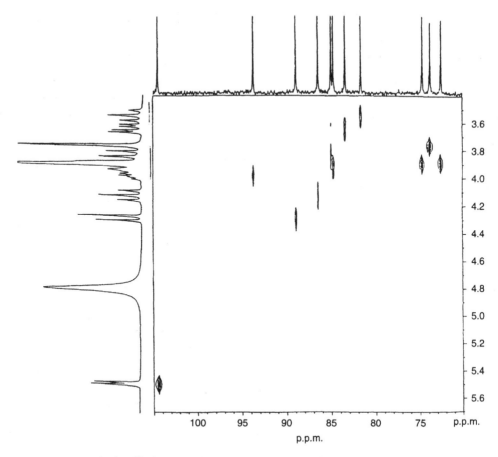

Figure 14.18 The ^{1}H–^{13}C heteronuclear COSY spectrum of sucrose. Each cross-peak represents a one-bond correlation between a proton and a ^{13}C. For example, the proton at 5.5p.p.m. is seen to be bonded to the carbon at 104p.p.m. on the ^{13}C spectrum.

$$F(t_1, t_2) \xrightarrow{\text{FT}} f(v_1, v_2)$$

The resulting spectrum has *two* frequency (chemical shift) axes. What do these represent?

In the simplest experiment, both axes represent proton chemical shifts. If proton A at shift v_A is coupled to proton B at shift v_B, then a **cross-peak** appears at coordinates (v_A, v_B) and at (v_B, v_A). We can imagine the normal spectrum as appearing along the diagonal. An example of this is given in Figure 14.17.

Here is a very powerful way of unravelling spectral overlap. This is particularly useful for the type of complex mixtures we often deal with when obtaining spectra of biofluids and drug metabolites. The main disadvantages are increased time to obtain the data and large data storage requirements. Both these drawbacks are easily overcome with modern high-field instruments.

Heteronuclear two-dimensional COSY

By a different choice of pulses and delays, we could make one of the frequency axes a chemical shift scale for a *different* nucleus, usually ^{13}C. In this case, each cross-peak represents a correlation between a proton chemical shift and a carbon chemical shift. This gives an even greater theoretical advantage in unravelling complicated spectra, as the ^{13}C shift range is over 200p.p.m. However, this experiment is inherently insensitive, and long acquisition times are needed. Again, though, recent advances in instrumentation (so-called 'inverse' probes) have dramatically improved the sensitivity of heteronuclear correlation. If you have access to one of these modern systems, it may just be possible to use ^{13}C–^1H COSY for drug metabolite identification. An example of ^{13}C–^1H COSY is given in Figure 14.18.

14.9.5 Conclusion

This brief appendix has given a very brief description of three of the most important types of multipulse experiment. I have included no technical or theoretical details; these are explained very clearly in a number of excellent textbooks.

The variety of available two- (and even three-) dimensional experiments is enormous and, given enough material, virtually any structure is obtainable – and I haven't even mentioned NOE difference, NOESY, INADEQUATE and many more acronyms. As with all subjects, the best way to learn is to try it. I suggest you take off your watches, remove your pacemakers and visit a friendly NMR spectroscopist right away!

14.10 Bibliography

Abraham, R.J. and Loftus, P. (1985) *Proton and Carbon-13 NMR Spectroscopy – An Integrated Approach*, New York: Wiley.

Derome, A.E. (1987) *Modern NMR Techniques for Chemistry Research*, Oxford: Pergamon.

Nicholson, J.K. and Wilson, I.D. (1987) High resolution nuclear magnetic resonance as an aid to drug development. *Progress in Drug Research*, 31, 427–479.

Preece, N.E. and Timbrell, J.A. (1990) Use of NMR spectroscopy in drug metabolism studies: recent advances. *Progress in Drug Metabolism*, 12, 147–203.

Sanders, J.K.M. and Hunter, B.K. (1987) *Modern NMR Spectroscopy – A Guide for Chemists*, Oxford: Oxford University Press.

Webb, G.A. (ed.) (1999) *Annual reports in NMR Spectroscopy*, Vol. 38, London: Academic Press.

Woolf, T.F. (ed.) *Handbook of Drug Metabolism*, New York: Marcel Dekker.

15

STRATEGY IN METABOLITE ISOLATION AND IDENTIFICATION

Hugh Wiltshire

Although the isolation of metabolites from biological material is usually difficult, there are a large number of techniques which can be applied. These make use of the different physico-chemical properties of the metabolites and their ionic, polar and non-polar interactions with solvents and solid supports. As there are so many endogenous compounds in biological media, each metabolite is bound to be contaminated with components that behave similarly in any particular system. Only by using a combination of extraction and chromatographic techniques will it be possible to design an efficient procedure for the isolation and identification of the metabolites produced by a complex catabolic process.

15.1   Stage 1: radiochemical synthesis

It is probably possible to carry out an *in vivo* metabolic investigation without the use of a radiolabelled form of the drug if the compound has a very characteristic UV spectrum which is unaltered by metabolism. *In vitro* studies with unlabelled material are more likely to be successful as there will less endogenous material to interfere with the isolation process. However, the chances of real success are remote and it is likely that the results obtained will be unreliable with major metabolites being overlooked and minor metabolites attracting undeserved attention.

15.1.1   Choice of label

Both tritium and ^{14}C-labelled drugs can be used but the carbon label is to be preferred because of its greater biological stability and ease of detection. Tritium does have advantages: cheapness, higher specific activity and sometimes ease of synthesis. The problems associated with the formation of tritiated water are less serious in a metabolite identification study than in, for example, the measurement of the plasma pharmacokinetics of total drug-related material, but the insensitivity of many detectors (linear analysers, solid scintillant HPLC detectors) to tritium is a major disadvantage.

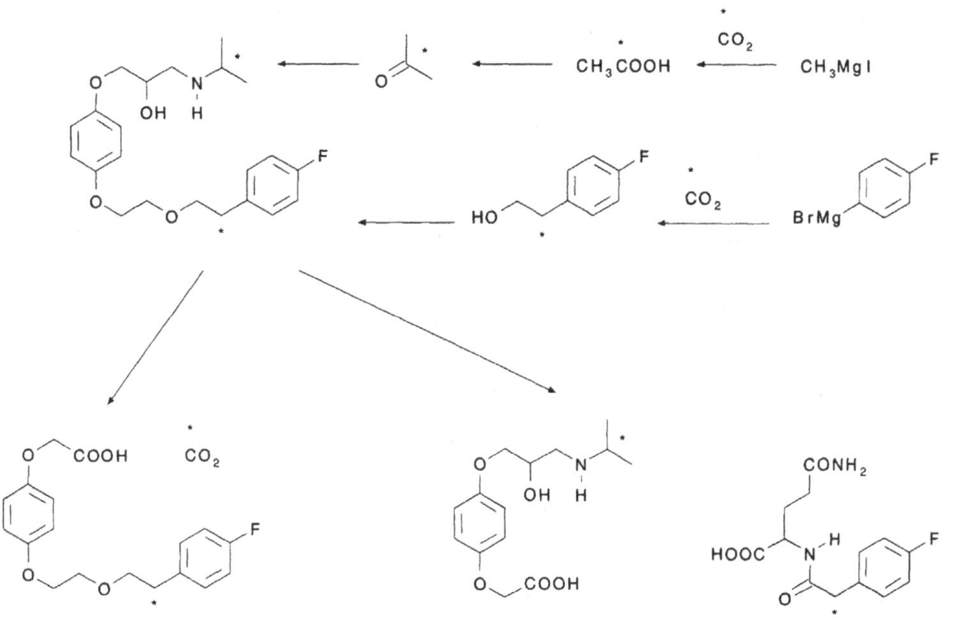

Figure 15.1 Radiosynthesis and major metabolic pathways of Ro 31-1411.

15.1.2 Position of ^{14}C label

There are three major considerations as to the positioning of a ^{14}C label in a molecule. First, it should be a metabolically stable carbon atom and not rapidly converted into carbon dioxide. Secondly, it should be near the groups of pharmacological interest, so that any potentially active metabolites still contain the label and can therefore be identified. Thirdly, it must be synthetically feasible. The ideal position is often next to a benzene ring; these carbons are rarely lost and are usually converted into conjugates of the corresponding benzoic acid. Carbon-14 labels adjacent to an aryl ring are often relatively easy to prepare from ^{14}C-labelled carbon dioxide (which is the cheapest available form of radiolabelled carbon) *via* a Grignard reagent (e.g. Figure 15.1).

Figure 15.1 illustrates some of the problems associated with ^{14}C labelling. One labelled form of the β-blocker, Ro 31-1411, was partially (~20% by the rat) converted to carbon dioxide, whereas the other was largely metabolised to the glutaminyl (primates) or glycyl (rat, dog) conjugates of *para*-fluorophenylacetic acid, metabolites of remarkably little interest. Ideally the label would have been incorporated into the hydroquinone ring, but the chemical synthesis would have been much more difficult – and expensive. In retrospect, the synthesis and metabolic investigation of the two radiolabelled forms of the drug was probably the best approach to the determination of its metabolic pathways.

15.2   Stage 2: animal experiments

Humans are usually to be avoided for initial metabolic studies as the doses of both drug substance and radioactivity must be kept low for ethical reasons. Comparative studies between humans and the species used for toxicological studies of the metabolic profiles in plasma, excreta and perhaps liver microsomes or hepatocytes will, however, be needed at a later stage.

15.2.1 Routes of excretion

Excretion balance studies after intravenous administration of radiolabelled drug will define the routes of excretion of drug-related material. It is much easier to collect urine than bile and urinary metabolites are generally easier to isolate than biliary ones. Even if substantial amounts of biliary excretion occur in the rat, it is possible that urinary excretion predominates in primates (as a result of the higher molecular weight 'cut-off' for biliary secretion), in which case the primate is likely to be more suitable for metabolic isolation studies. However, rats will often tolerate larger doses of drugs and could still be the preferred species.

15.2.2 Formulation and route of administration

It is usually desirable to give the maximum tolerated dose in order to maximise the concentration of drug-related material in excreta. It should, however, be remembered that the ratio of metabolites at high doses may not be identical to that at the pharmacological dose, as some metabolic steps may be saturated. In this case the recovery of the parent compound will often be disproportionately high. A comparison of the metabolic profiles at various dosage levels should, therefore, be made at a later stage.

If a drug is not readily soluble in water at the required dose, some other formulation will be needed. This problem will probably already have been addressed during toxicological studies. Suspensions in carboxymethylcellulose, solutions in organic-aqueous mixtures or wet-milled microsuspensions may well be suitable. Oral administration is usually to be preferred if absorption is high, but intraperitoneal dosing can be a good alternative.

15.2.3 Collection of urine and bile

Urine is readily collected using suitable metabolism cages. As it will normally be necessary to collect bile for 1–2 days in order to obtain a good recovery of metabolites, conscious animals will be needed together with suitable apparatus to eliminate any losses of bile.

15.3 Stage 3: metabolite isolation and characterisation

The isolation of metabolites can be divided up conveniently into five phases: enrichment, analysis, separation, purification and characterisation.

15.3.1 Enrichment

Having obtained a sample of excreta containing sufficient drug-related material, it is necessary to enrich the metabolites by removing as much as possible of the endogen-

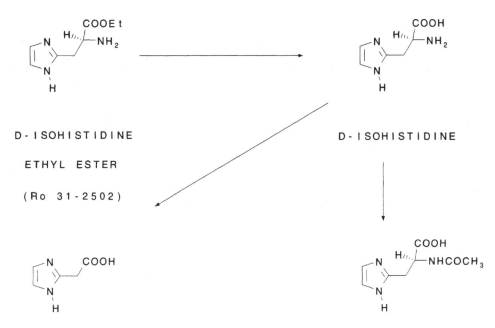

Figure 15.2 The metabolism of Ro 31-2502.

ous material, with minimal losses of the compounds of interest. There are three main ways of carrying out this **enrichment** step.

Solvent extraction is the simplest technique for concentrating metabolites and will give good recoveries and good enrichment of drug-related material. Thus if 80ml of urine containing 6g of solid is shaken with an equal volume of ethyl acetate at pH 9, only about 60mg (1%) will be extracted. At pH 3, approximately 300mg (5%) of endogenous material will be removed. The technique falls down if the metabolites are unstable at basic or acidic pH and if they are too polar for ready extraction.

Ion-exchange chromatography is especially suitable for the extraction of polar compounds, particularly zwitterions, which do not dissolve in organic solvents. Adsorption onto strongly acidic or basic polystyrene-based resins can cause problems resulting from the lack of stability of metabolites if basic or acidic conditions are needed for their elution. If the compounds under investigation are at all lipophilic, they will also be retained on the resin as a result of non-polar interactions and may be difficult to remove. Nevertheless, the sequence of adsorption to and elution from a strongly acidic resin, followed by a strongly basic one, can concentrate polar metabolites by a factor of 100. A good example of the successive use of a strongly acidic (DOWEX-50) and a strongly basic (DOWEX-1) resin to concentrate polar metabolites is illustrated in Figure 15.2, which shows the metabolites of Ro 31-2502 which were concentrated approximately 100-fold by this method.

The third technique available to the metabolist is that of solid-phase extraction. It is possible to adsorb fairly large amounts of drug-related material on Mega Bond-Eluts™ or else to use a column of XAD-2 polystyrene resin. A 30 × 2.5cm column of this will retain all but the most polar materials from at least a litre of urine. It is then possible to wash away polar endogenous material with water and to elute the compounds of

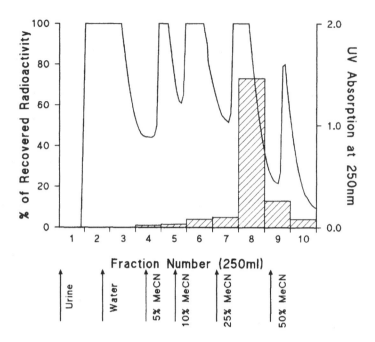

Figure 15.3 XAD-2 chromatography used for the initial isolation of the metabolites of cilazapril from the urine of the baboon.

interest with increasing concentrations of an organic solvent in water (Figure 15.3). Small increases of organic solvent will result in the elution of significant quantities of endogenous material. It is well worthwhile, therefore, to increase the concentration of organic modifier in steps, even if all of the metabolites are removed at a single concentration. In certain cases a clean separation of metabolites is possible using this method.

Better retention to reverse-phase material will be obtained if the ionisation of the metabolites is suppressed. Mild acidification will improve the isolation of acidic metabolites and will also stabilise acyl-glucuronides, common conjugates of carboxylic acids. However, acidification can cause precipitation of bile acids and may not always be worthwhile.

Although these three techniques are best applied to the enrichment of metabolites they also have limited uses for separation and purification, as discussed in Sections 15.3.3 and 15.3.4. In some cases a single fraction will be produced from the enrichment stage, but usually there are between three and six fractions, which will need to be analysed.

It must be stressed that the recoveries of every step must be determined by liquid scintillation counting of aliquots so that an idea of the relative abundance of the metabolites can be obtained when the study has been completed.

15.3.2 *Analysis*

Attempts to quantify the number and relative abundance of metabolites will be done at an early stage, but are unlikely to be very reliable. Now that the concentration of

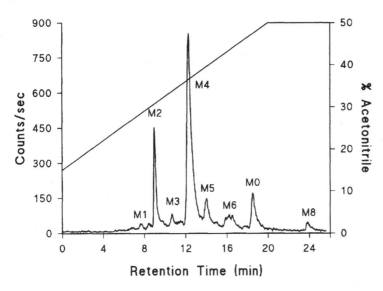

Figure 15.4 Separation of the urinary metabolites of Ro 31-6930 by gradient HPLC on a C18 column at pH 3 (0.1% AcOH).

the metabolites has been raised ten- to 100-fold it should be much simpler. Two techniques of **analysis** are available, both of which use radiochemical detection.

Analytical gradient HPLC is a powerful method for separating drugs and metabolites, and HPLC radio-detectors remove the drudgery of collecting fractions and measuring the radioactivity in aliquots by liquid scintillation counting. However, HPLC will probably be used for the separation and purification of the metabolites and the analytical data provided will become redundant.

Far more effective is the use of TLC for the analysis. This technique uses normal-phase silica and so quite different physico-chemical properties of the metabolites are employed in the separation (Figures 15.4–15.6). Metabolite M4 of Ro 31-6930 (Figure 15.6) was apparently a single compound (Figure 15.4), but TLC (Figure 15.5) easily resolved it into two distinct components.

A major advantage of TLC is that it is possible to run samples from several fractions side by side and this simplifies the comparison of the metabolic profiles. The shapes of the spots, which are often not perfectly circular, can be useful diagnostically and it is important to examine the chromatogram autoradiographically, not only because of the high resolution obtained but also to see the shape of the spots. The chromatogram should then be analysed with a linear analyser to obtain quantitative information on the relative amounts of each metabolite in the fraction.

Analysis should follow each enrichment/separation step until it is certain that a fraction contains a single metabolite.

15.3.3 Separation

The **separation** phase is designed to produce fractions containing individual metabolites.

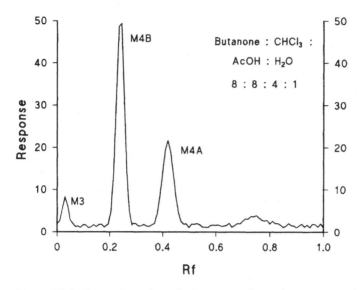

Figure 15.5 Separation of metabolites 4A and 4B of Ro 31-6930 by TLC.

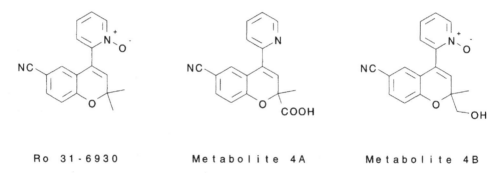

Ro 31-6930   Metabolite 4A   Metabolite 4B

Figure 15.6 The structures of Ro 31-6930 and metabolites 4A and 4B.

Ion-exchange chromatography on polystyrene is mainly applicable to highly polar compounds (Section 15.3.1) but cellulose-supported weakly acidic or basic ion-exchangers can, on the other hand, be useful for many ionisable metabolites. Polymers such as diethylaminoethyl (DEAE) and carboxymethyl (CM) Sephadex, suspended in aqueous media (pH 7.5 and 2.5, respectively), elicit a minimum of solid-phase adsorption. Elution with increasing concentrations of aqueous sodium chloride (0.1–0.6M) will separate metabolites, depending solely on their acidity or basicity.

The cardiotonic drug, dazonone (Figure 15.7) is metabolised to a number of phase 1 derivatives, which are excreted in urine either unchanged or as glucuronide or sulphate conjugates. Figure 15.8 shows how these three classes of compounds can be efficiently separated by ion-exchange chromatography on DEAE-Sephadex.

De-salting will usually best be effected by solid-phase extraction followed by elution with a polar organic solvent.

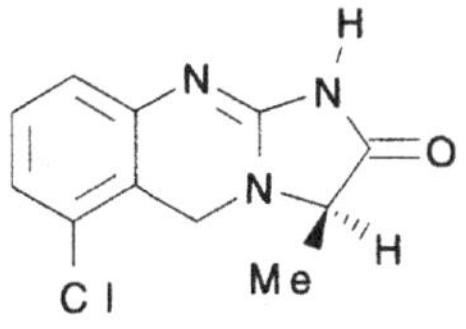

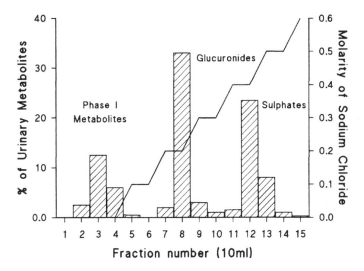

Figure 15.7 The structure of dazonone.

Figure 15.8 Separation of the metabolites of dazonone from human urine on DEAE-Sephadex A-25, using an aqueous sodium chloride gradient.

Solvent partition is a relatively crude procedure for the separation of metabolites and so is best used at an early stage. As shown in Figure 15.9, it can readily differentiate between acidic, neutral (both hydrophilic, i.e. amphoteric, and lipophilic) and basic metabolites. Care must be taken to check for the hydrolysis of labile metabolites (e.g. by TLC), particularly when using basic conditions.

Separation of glucuronide and sulphate conjugates is also sometimes possible by partition between water and organic solvents. The sulphate derivative (Figure 15.10) of Ro 15-1570 (Figure 15.22) is ionised at pH 3 and is soluble in water but the corresponding glucuronide is protonated at this value of pH and is readily extracted into ethyl acetate.

By far the most powerful method for the separation of metabolites is gradient reverse-phase HPLC. If there are not too many metabolites, it may be possible to use a single chromatographic step to prepare individual fractions containing single radioactive components. More often, however, it is preferable to use successive chromatographic systems employing different solid or mobile phases. The following variations are available:

- changes of mobile phase; for example, methanol, acetonitrile and tetrahydrofuran (Figures 15.11–15.13);

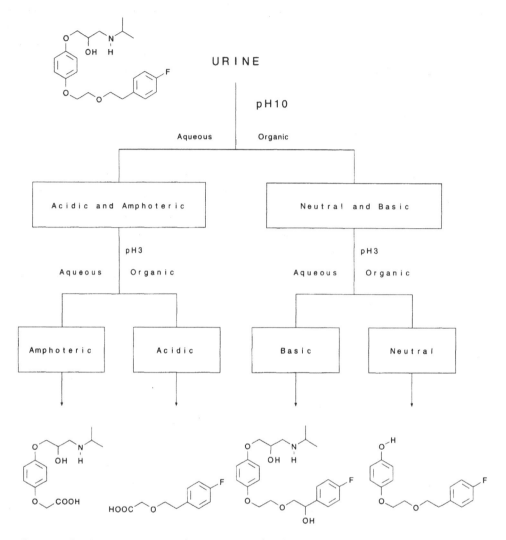

Figure 15.9 Aqueous/organic solvent partition for the separation of various classes of metabolite of Ro 31-1411 from urine.

- changes of stationary phase; for example, C18, phenyl, cyanopropyl, diol (Figures 15.14, 15.15); and
- changes of pH; usually between 3 and 7 (Figures 15.16, 15.17).

Changing the mobile phase

The use of acetonitrile rather than methanol improves the separation of the metabolites of cilazapril (Figure 15.11) because the retention times of the two ethyl esters (M0, M5) are greatly lengthened (Figure 15.12). (I have no explanation for the physico-chemical reasons behind this; retention times are usually shorter with acetonitrile!) The peaks of the esters are, however, much broader than when methanol is the organic modifier (Figure 15.13).

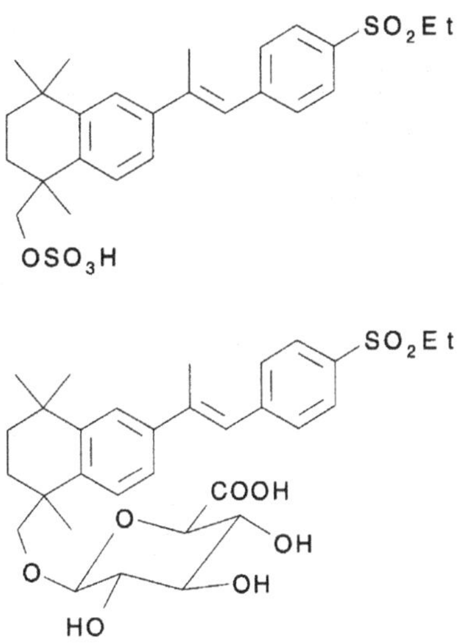

Figure 15.10 Sulphate and glucuronide conjugates of a metabolite of Ro 15-1570.

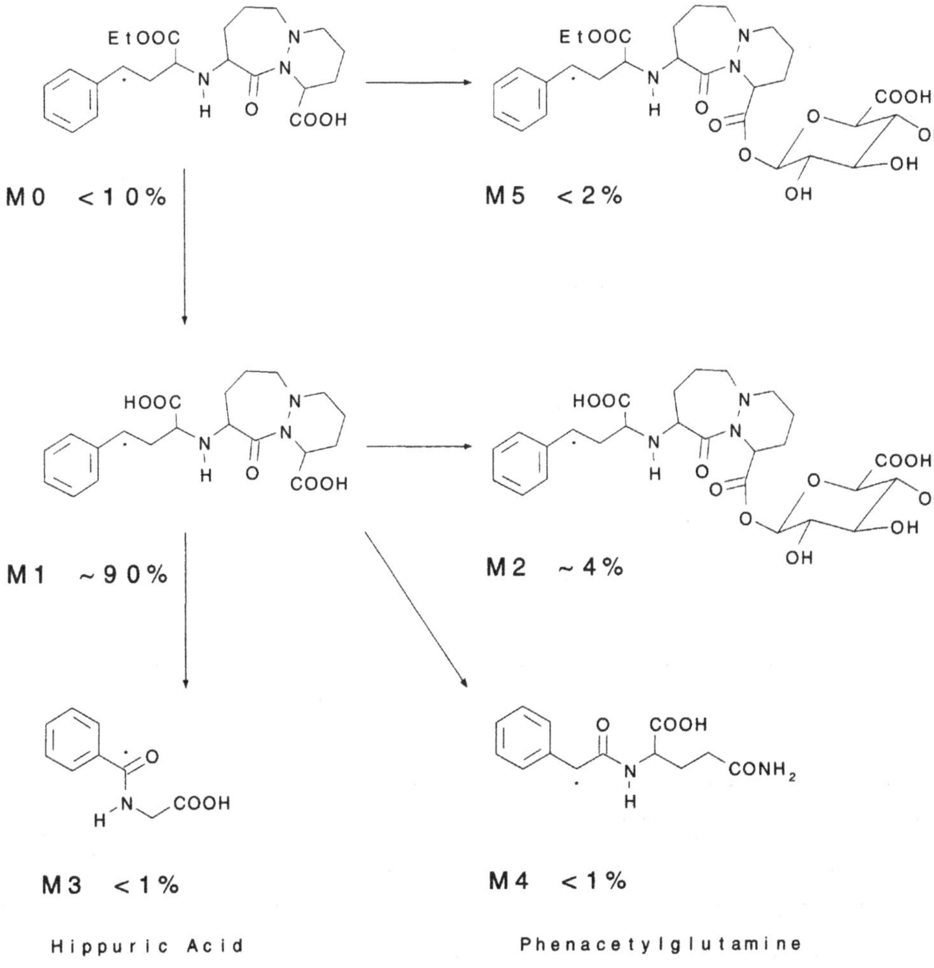

M0   < 1 0 %

M5   < 2 %

M1   ~ 9 0 %

M2   ~ 4 %

M3   < 1 %

Hippuric Acid

M4   < 1 %

Phenacetylglutamine

Figure 15.11 The metabolism of cilazapril.

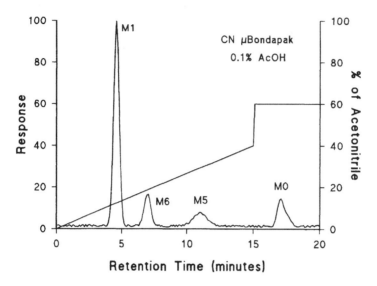

Figure 15.12 Separation of cilazapril and three of its metabolites by HPLC, using acetonitrile in the mobile phase.

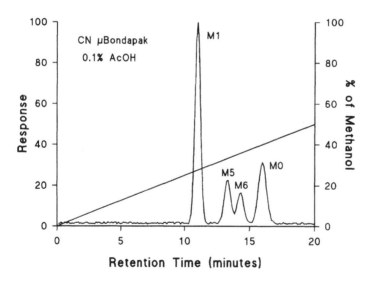

Figure 15.13 Separation of cilazapril and three of its metabolites by HPLC, using methanol in the mobile phase.

Changing the stationary phase

The first HPLC step in the isolation of the major metabolite of cilazapril from baboon urine (from Fraction 8, Figure 15.3) was carried out on a C18 μBondapak column at pH 3, using an acetonitrile gradient (Figure 15.14). The radioactivity attributed to this metabolite appeared to be associated with a single sharp UV peak. The subsequent use of a cyanopropyl μBondapak column separated the 'single' peak

312

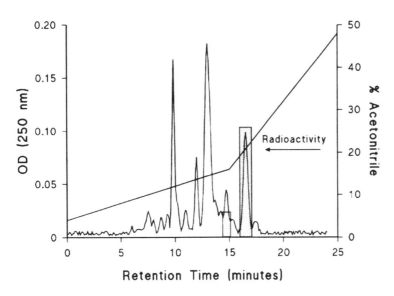

Figure 15.14 HPLC separation of metabolite 1 of cilazapril on a C18 μBondapak column, using an aqueous acetonitrile gradient at pH 3.

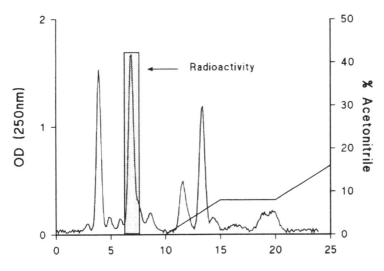

Figure 15.15 HPLC purification of metabolite 1 of cilazapril on a CN μBondapak column, using an aqueous acetonitrile gradient at pH 3.

into several components with very different retention times (Figure 15.15). (Note this is really a *purification* step, Section 15.3.4.)

Changing the pH

Metabolites 4A and 4B (Figure 15.6) co-elute after HPLC at pH 3 (Figure 15.4) but raising the pH to about 7 causes the acidic metabolite (4A) to ionise and hence to

313

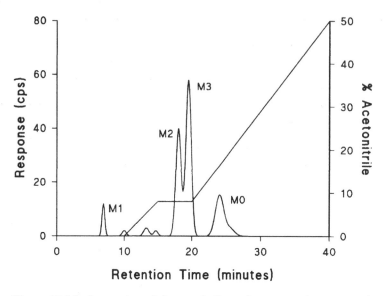

Figure 15.16 Separation of the metabolites of romazarit by reverse-phase HPLC on a CN μBondapak column at pH 3 (0.1% ACOH).

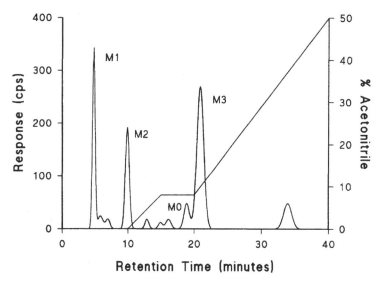

Figure 15.17 Separation of the metabolites of romazarit by reverse-phase HPLC on a CN μBondapak column at pH 6.3 (0.1% ACOH + NH₃).

increase in polarity and therefore its retention time is reduced relative to the neutral 4B. The effect of pH can be more subtle than this, as shown by the metabolites of romazarit in Figures 15.16 and 15.17. All of the compounds (Figure 15.18) are carboxylic acids and metabolites 2 and 3 are poorly separated at pH 3 (Figure 15.16). Raising the pH to 6.3 causes both to ionise, but the increase in polarity is

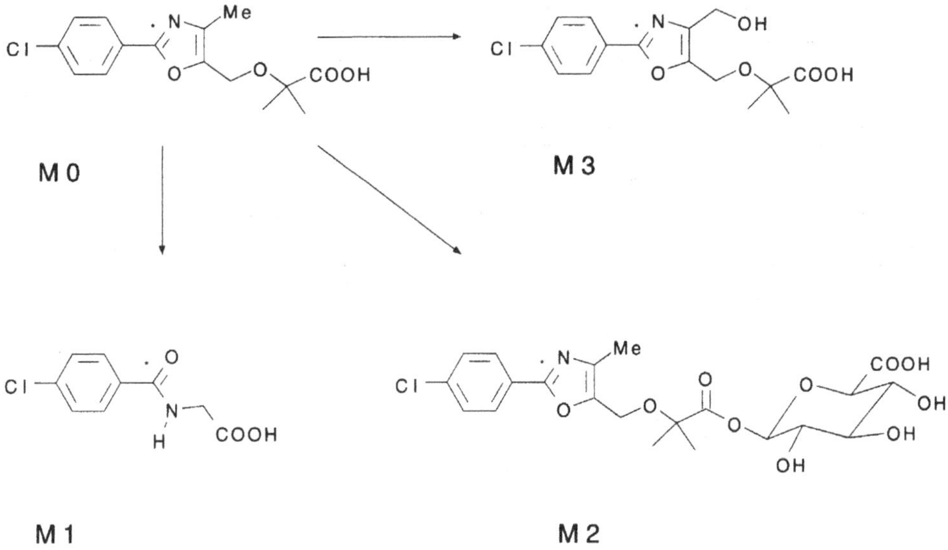

Figure 15.18 The metabolism of romazarit.

considerably more marked for the glucuronide (metabolite 2) than for metabolite 3 (Figure 15.17).

It is usually preferable to use volatile buffers such as acetic acid/ammonium acetate for ease of removal. However, solid-phase or solvent extraction can be used if necessary to remove the buffers from the separated metabolites.

Occasionally TLC will be useful for the isolation of less polar metabolites; the two amino-acid conjugates (M3 and M4) formed following the degradation of cilazapril (Figure 15.11) could not be separated cleanly by reverse-phase HPLC (Figure 15.19), but were readily resolved by TLC (Figure 15.20).

15.3.4 Purification

The **purification** stage is designed to remove any remaining endogenous material from separated metabolites, and there are obviously often overlaps between separation and purification. By this time the metabolites are likely to be contaminated only with compounds with closely similar chromatographic properties, and gradient chromatography will be unnecessary. It should be possible to purify the metabolites using an isocratic system, making use of differences in retention times between metabolites and contaminants resulting from changes in pH and stationary and mobile phases, as discussed above.

15.3.5 Characterisation

The **characterisation** of the metabolites is carried out in conjunction with the other processes already considered. Acids and bases may have been identified from their solubilities and chromatographic behaviour after changes in pH. Glucuronide and

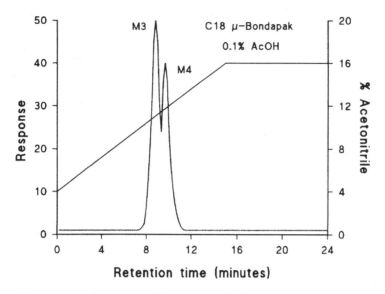

Figure 15.19 Separation of hippuric acid and phenylacetylglutamine by HPLC.

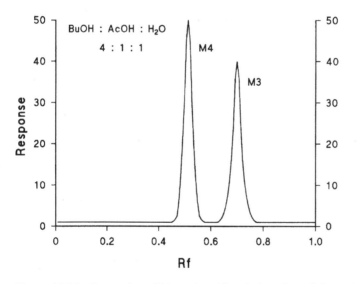

Figure 15.20 Separation of hippuric acid and phenylacetylglutamine by TLC.

sulphate conjugates will often have been recognised by their polarity, especially on TLC, and also by their susceptibility to enzymic hydrolysis (Section 15.4.3).

Two particularly useful methods of characterisation which do not require the actual isolation of pure material, and which can be carried out on the microgram quantities eluting from an analytical HPLC column, are diode-array UV spectroscopy and thermospray mass spectrometry. In both cases all that is needed is the separation

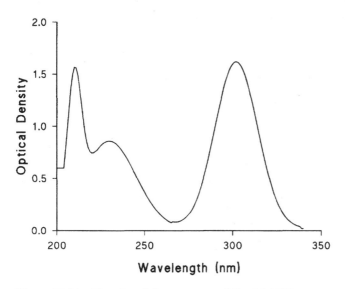

Figure 15.21 The ultraviolet spectrum of Ro 15-1570.

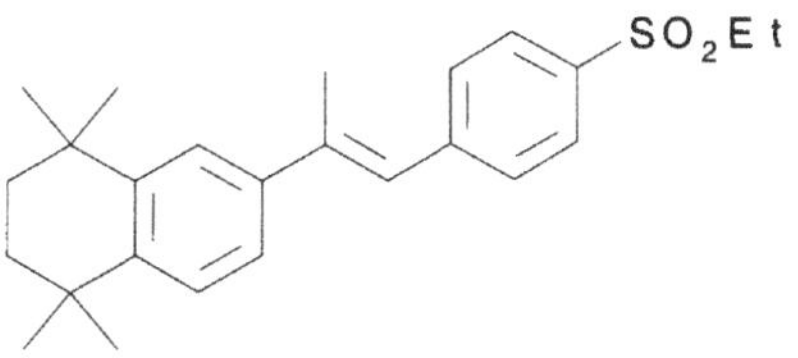

Figure 15.22 The structure of Ro 15-1570.

of the compound of interest from contaminants by HPLC, the eluent then being directed through the detector.

Figure 15.21 shows the characteristic UV spectrum of a retinoid (Ro 15-1570, Figure 15.22) which has dermatological applications. All of the 40 metabolites isolated from rat bile have the same highly characteristic UV spectrum and so no metabolism of the aromatic rings can have occurred.

The chromatogram shown in Figure 15.23, of one of the fractions isolated from rat bile, was obtained using a diode-array UV spectrometer. The intensity of UV absorption of both components at 300nm is approximately twice that at 253nm and 330nm. This demonstrates that the two compounds are metabolites of Ro 15-1570 and that no metabolism of the aromatic rings has occurred. The greater sensitivity of UV spectroscopy compared with radiochemical detection enabled considerably smaller amounts of metabolites to be recognised than if these compounds had not had such an unusual UV spectrum.

On the other hand, Figures 15.24 and 15.25 demonstrate how hydroxylation of the benzimidazole ring of mibefradil alters the UV spectrum and allows unambiguous identification of the position of oxidation of two metabolites. (The assignment of the UV spectra to the substitution pattern of simple hydroxybenzimidazoles was made by

317

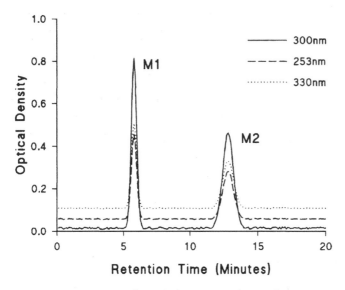

Figure 15.23 HPLC of metabolites M1 and M2 of Ro 15-1570.

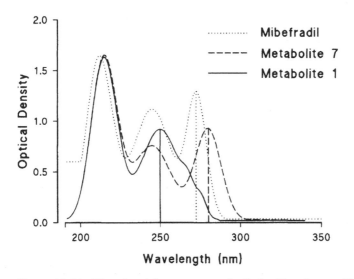

Figure 15.24 The ultraviolet spectrum of mibefradil and two of its metabolites.

chemical synthesis nearly 30 years ago; in this particular case, NMR spectra (Section 15.4.2) have confirmed the structures.)

Thermospray mass spectrometry provides quite different information. Knowledge of the molecular weight of a metabolite may be insufficient unequivocally to assign its structure but, when combined with other data, may be conclusive. The generalised fragmentation pattern of mibefradil and many of its metabolites is shown in Figure 15.26.

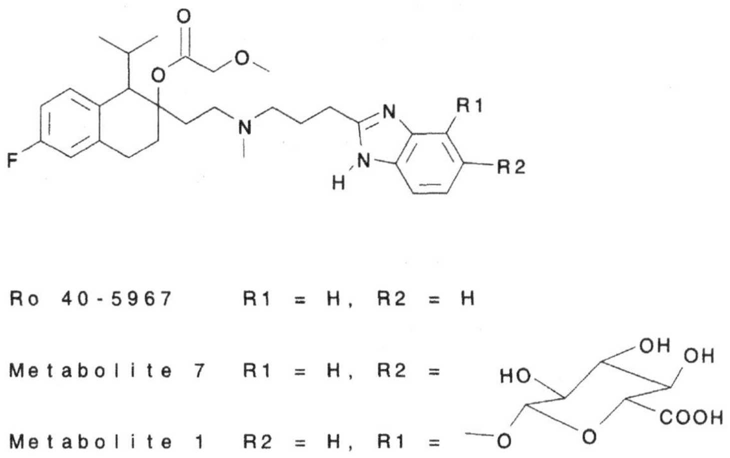

Figure 15.25 The structure of mibefradil and two of its metabolites.

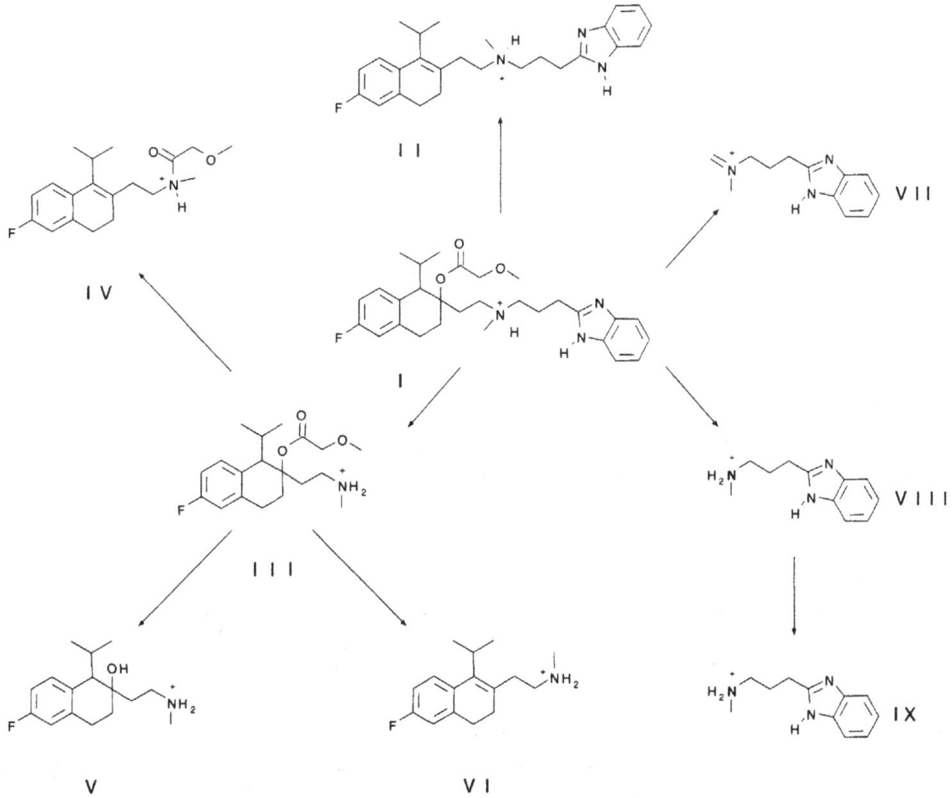

Figure 15.26 The mass spectral fragmentation pattern of mibefradil after thermospray isolation.

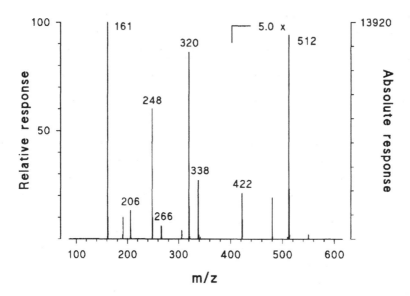

Figure 15.27   Thermospray mass spectrum of metabolite M22 of mibefradil.

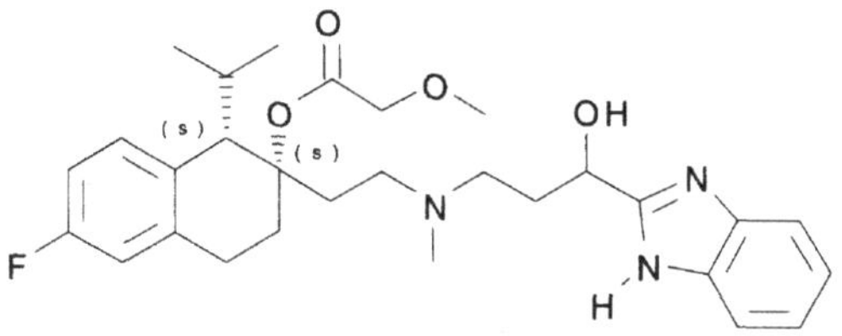

Figure 15.28   Probable structure of metabolite M22 of mibefradil.

The molecular weight of metabolite 22 (which was equivalent to approximately 0.1% of the biliary metabolites of mibefradil) indicates that it is a mono-hydroxy derivative (m/z for the protonated molecular ion = 512, m/z for mibefradil = 496) and the UV spectrum shows that no oxidation of the benzimidazole ring has occurred. The fragmentation pattern (Figures 15.26, 15.27) shows that the hydroxylation is to the right of the tertiary amine (ion VII: m/z = 206 for metabolite 22 but m/z = 190 for mibefradil; ions III, IV, V, VI: m/z = 338, 320, 266 and 248, respectively, for both drug and metabolite). The most likely structure is that of Figure 15.28, with a hydroxyl group α to the aromatic system (the alternative β-substituent is possible but would have involved oxidation at a less-reactive carbon atom).

The electrospray interface differs from thermospray in that much less fragmentation occurs. This has the disadvantage that less structural information can be obtained, but the advantage that the results are less likely to be ambiguous. Two applications are illustrated by ion chromatograms of the microsomal metabolites of mibefradil in Figure 15.29. First, ion chromatograms can clearly show the presence of isomers (e.g. five hydroxylated metabolites with m/z = 512). Secondly, overlapping peaks can be distinguished easily without resorting to background subtraction (metabolites E1

320

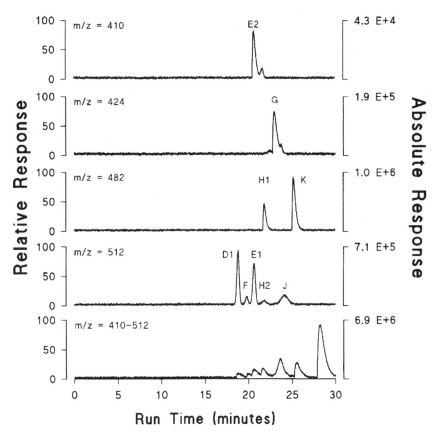

Figure 15.29 Ion chromatograms (electrospray interface) showing the major metabolites of mibefradil formed in human hepatic microsomes.

and E2 with protonated molecular ions of 512 and 410Da are distinct metabolites and not fragments from the same compound.

In some cases GC–MS can still be useful, particularly if the metabolites have poor UV chromophores. The methane chemical ionisation mass spectrum of the neutral metabolite of Ro 31-1411 (Figure 15.9) obtained from a direct insertion probe was complicated by dimeric ions and their accompanying cluster and fragment ions (Figure 15.30). After gas chromatography, a much simpler spectrum was obtained (Figure 15.31) and the confusing ions were readily explained (Figure 15.32; e.g. $553 = 2 \times 276 + 1$; $399 = 2 \times 276 - 153$).

15.4 Stage 4: identification of metabolites

As with the isolation and characterisation of metabolites, their identification also requires a combination of physico-chemical techniques. If less than 50μg are available, it is probable that only mass spectral and chromatographic techniques will provide useful information. If more material has been isolated, proton NMR spectra should help the identification process considerably.

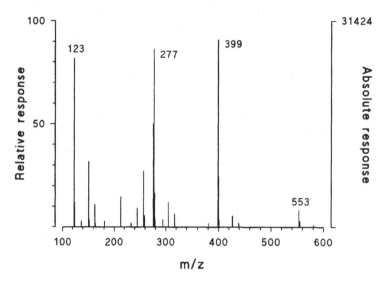

Figure 15.30 Methane chemical ionisation mass spectrum of metabolite 8 of Ro 31-1411 (direct insertion probe).

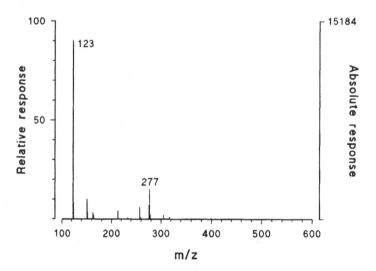

Figure 15.31 Methane chemical ionisation mass spectrum of metabolite 8 of Ro 31-1411 (after gas chromatography).

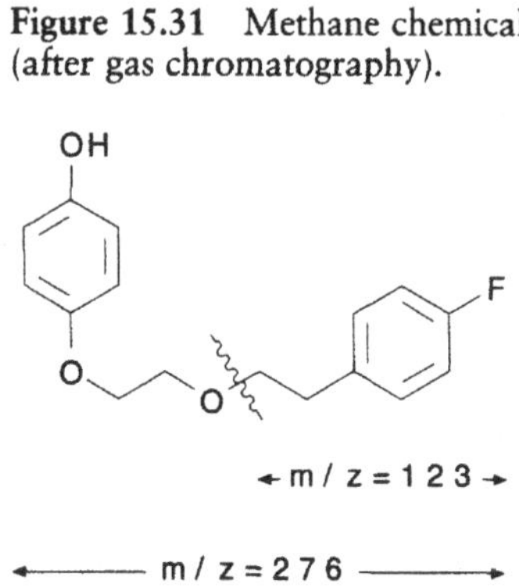

← m/z = 1 2 3 →

←——— m/z = 2 7 6 ———→

Figure 15.32 Structure and fragmentation of metabolite 8 of Ro 31-1411.

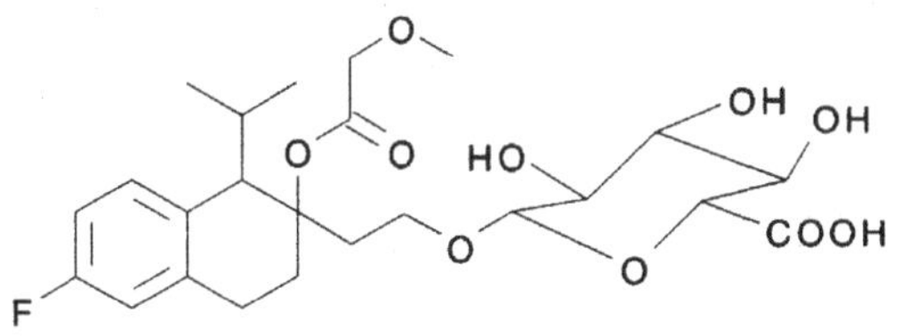

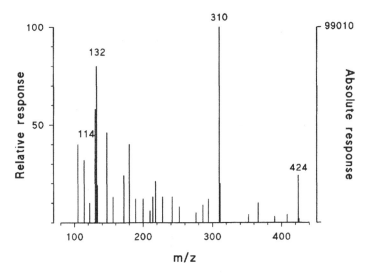

Figure 15.33 The structure of metabolite M9 of mibefradil.

Figure 15.34 Ammonia chemical ionisation mass spectrum of metabolite 1 of Ro 31-1411.

15.4.1 Mass spectrometry

Although the molecular weight will usually have been obtained by thermospray mass spectrometry, FAB mass spectrometry can sometimes be useful in resolving ambiguities. The thermospray mass spectrum of metabolite 9 of mibefradil consisted of a single peak with $m/z = 518$. The FAB spectrum, on the other hand, had the peak at $m/z = 518$ (10% of the base peak) together with one at $m/z = 523$ (25%). The two peaks were ammonium (M + 18) and sodium (M + 23) adducts, respectively, of the metabolite whose molecular weight was actually 500. Unlike most of the other metabolites of this compound, which retained at least one of the nitrogen atoms of the parent drug and which exhibited strong M + 1 ions, this metabolite (Figure 15.33) had no basic group and spectra were only obtained from adducts with anions. Negative-ion mass spectrometry might also have been useful in resolving the question of the actual molecular weight of the metabolite.

Chemical ionisation spectrometry with ammonia, isobutane and methane can also be useful as different fragmentation patterns, providing different information, can result (Figures 15.34–15.37). Thus the structure of the dihydroxylated metabolite of Ro 31-1411 has been unequivocally established by comparison of its mass spectra

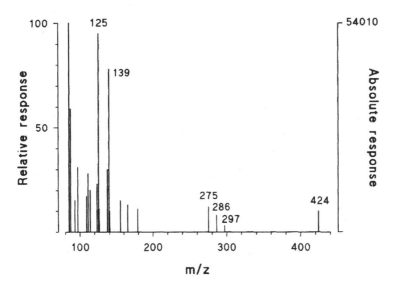

Figure 15.35 Methane chemical ionisation mass spectrum of metabolite 1 of Ro 31-1411.

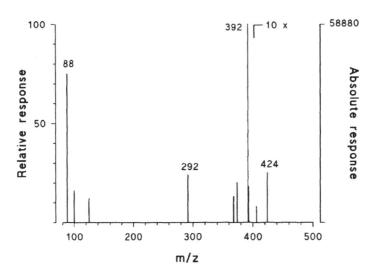

Figure 15.36 Electron impact mass spectrum of metabolite 1 of Ro 31-1411.

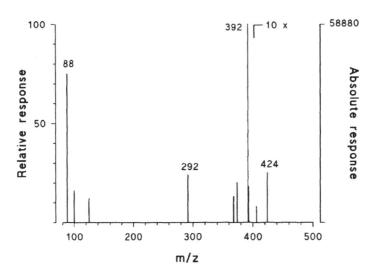

Figure 15.37 Mass spectral fragmentation of metabolite 1 of Ro 31-1411.

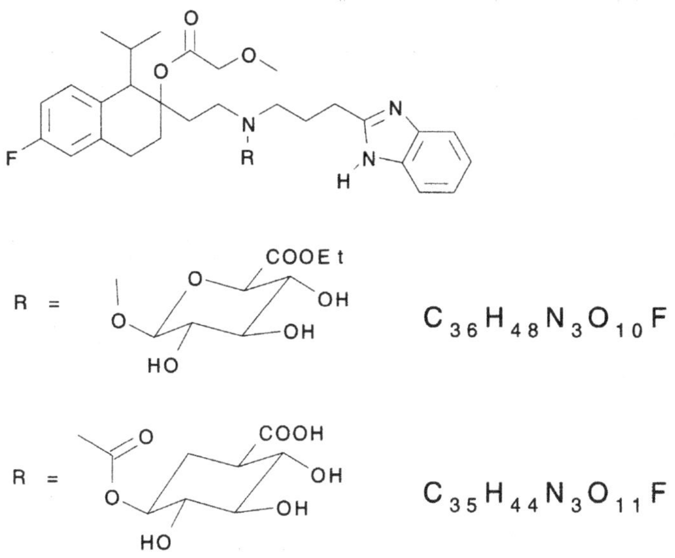

Figure 15.38 Possible structures of metabolite M3 of mibefradil.

with those of the parent drug and monohydroxylated metabolites. Ammonia chemical ionisation (Figure 15.34) gives a protonated molecular ion ($m/z = 424$) and loss of the hydroxylated β-blocker side-chain ($m/z = 310$, i.e. $292 + 18$ (NH_4^+), Figure 15.37).

Methane chemical ionisation also produces a protonated molecular ion (Figure 15.35), but the two fragments at $m/z = 139$ and $m/z = 125$ indicate that the second hydroxyl group was on the benzylic carbon atom (Figure 15.37). Finally, the electron impact spectrum (Figure 15.36), which had a very weak protonated molecular ion ($m/z = 424$) but no true molecular ion, demonstrated that hydroxylation of one of the geminal methyls of the isopropyl group had occurred by its intense ions at $m/z = 392$ and $m/z = 88$ (Figure 15.37). Loss of the β-blocking side-chain also occurred ($m/z = 292$).

High-resolution mass spectrometry can resolve ambiguities in complicated metabolites; see, for example, Figure 15.38. The molecular formula of the carbamyl glucuronide metabolite (M3) of mibefradil is $C_{35}H_{44}N_3O_{11}F$ (found 701.3044, theory 701.2960; $\Delta = -0.0084$) and not $C_{36}H_{48}N_3O_{10}F$ (theory 701.3323; $\Delta = +0.0279$), which could have been an artefact formed by transesterification of the corresponding N-glucuronide with ethyl acetate. Similarly, the empirical formula of the water-soluble metabolite (Figure 15.42) of Ro 31-4587 is $C_{22}H_{29}N_4O_8BrS$ (found 588.0915, theory 588.0899, $\Delta = -0.0016$) and not $C_{22}H_{29}N_4O_{10}Br$ (theory 588.1068, $\Delta = +0.0153$).

15.4.2 NMR

NMR spectroscopy is a powerful spectroscopic technique for the identification of metabolites and makes use of quite different properties of the molecules under investigation. It gives no idea of molecular weight but will generally enable the number and relationships of the protons to be obtained. Although it can be useful for the

325

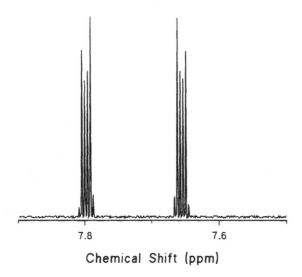

Figure 15.39 Part of the NMR spectrum of mibefradil, showing the protons of the benzimidazole group.

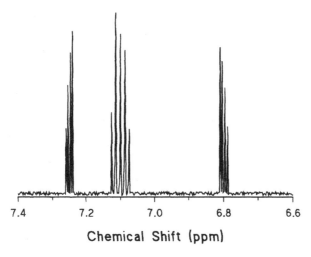

Figure 15.40 Part of the NMR spectrum of metabolite M1 of mibefradil, showing the protons of the benzimidazole group.

analysis of impure samples, it is often necessary to isolate 100µg of pure material to obtain unambiguous data. Confirmation of the structural assignments of the ring-hydroxylated metabolites of mibefradil (Figure 15.25) was obtained by NMR (Figures 15.39–15.41 show the resonances of the protons of the benzimidazole ring systems).

The assignment of the structure of the metabolite in Figure 15.42 as a cysteinyl conjugate was only possible with the help of the NMR spectrum (Figure 15.43). The phenacetyl group was obviously unchanged (four protons at $\delta = 7.2$–7.8p.p.m. and a

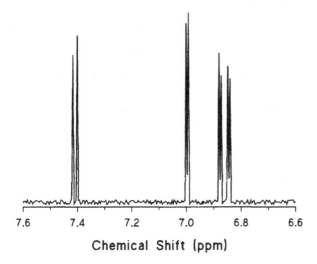

Figure 15.41 Part of the NMR spectrum of metabolite M7 of mibefradil, showing the protons of the benzimidazole group.

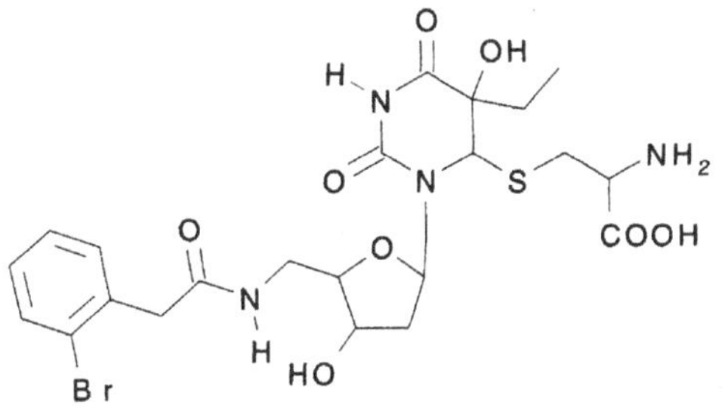

Chemical Shift (ppm)

Figure 15.42 The structure of the water-soluble metabolite of Ro 31-4587.

two-proton singlet at $\delta = 3.84$p.p.m.). The COSY spectrum showed that the α proton of the cysteinyl moiety ($\delta = 3.96$p.p.m.) was coupled to the two β protons at $\delta = 3.24$ and $\delta = 3.34$p.p.m. The five protons of the tetrahydrofuranyl ring were identified similarly and resonated at $\delta = 5.74$, 4.44, 4.02, 2.72 and 2.24p.p.m. This NMR spectrum is particularly complicated because the geminal protons of four of the five methylene groups are magnetically non-equivalent as a result of nearby asymmetric centres.

15.4.3 Degradation, derivatisation and comparison with authentic material

Degradation to known compounds was once the classical method for structural deter-mination. In metabolic studies it is usually confined to enzymic cleavage of conjugates. Thus, treatment of a metabolite of Ro 31-1411 with β-glucuronidase gave a mono-hydroxylated metabolite identical by HPLC to one of those that had already been identified by NMR and mass spectrometry (Figure 15.44).

327

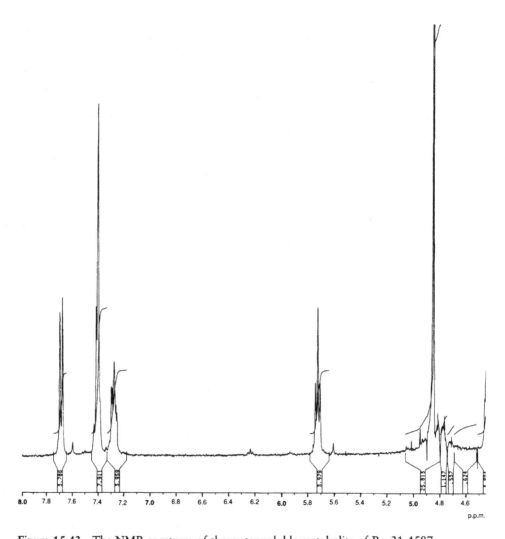

Figure 15.43 The NMR spectrum of the water-soluble metabolite of Ro 31-4587.

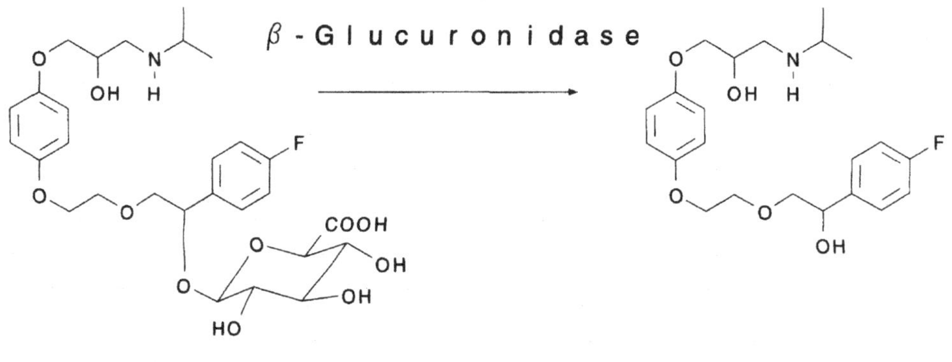

β - G l u c u r o n i d a s e

Figure 15.44 Hydrolysis of a glucuronide metabolite of Ro 31-1411.

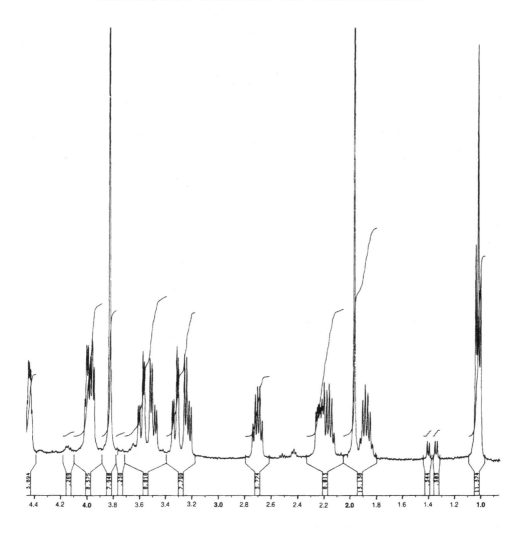

It should be remembered that not all conjugates are readily hydrolysed by the 'appropriate' enzyme. For example, acyl glucuronides are prone to rearrangement at neutral pH to glucuronic acid esters and only the 'natural' isomer is readily cleaved by β-glucuronidase. Thus the hepatic elimination of romazarit by the rat is largely as the acyl glucuronide which partially rearranges to three isomeric esters (Figures 15.45, 15.46). Treatment of the bile with β-glucuronidase results in the least polar isomer (the 'natural' one) being hydrolysed to the parent drug (Figure 15.47). Base-catalysed hydrolysis, however, results in all four compounds being degraded (Figure 15.48).

The mass spectra of impure acids are often difficult to interpret. An acidic metabolite of Ro 31-1411 was esterified with the boron trifluoride/methanol complex, purified by preparative TLC and its ammonia chemical ionisation mass spectrum obtained

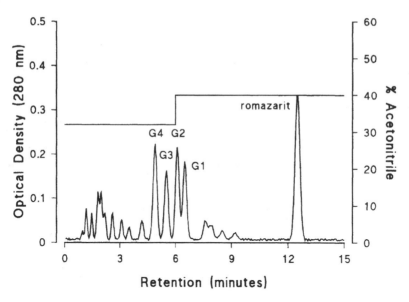

Figure 15.45 Reverse-phase HPLC of rat bile after oral administration of romazarit at 20mg/kg (C18 μBondapak, 0.1% CF₃COOH). (1) Before hydrolysis.

after gas chromatography (Figure 15.49). The M + 18 and M + 35 ions (Figure 15.50) were consistent with the methyl ester of a likely metabolite (Figure 15.51) and the assignment was confirmed by comparison with authentic material by GC–MS.

15.4.4 Ambiguities

Reliance on a single spectroscopic technique is unlikely to provide reliable structural data; it should always be the aim of the metabolic scientist to characterise metabolites by UV and thermospray mass spectrometry and then to isolate sufficient material of the major metabolites for NMR spectra. Confirmatory experiments involving degradation and derivatisation may follow if necessary. A thorough investigation of the metabolism of Ro 15-1570 using diode-array UV spectroscopy (Figures 15.21–15.23) and thermospray mass spectrometry characterised 36 metabolites. The two major primary metabolites had molecular weights corresponding to the gain of 16 and 30 atomic mass units and were therefore assumed to be an aliphatic alcohol and the corresponding carboxylic acid (Figure 15.52). There were also seven mono-hydroxy/mono-carboxy derivatives. Without NMR and authentic compounds it was quite impossible to deduce more about their structures.

15.5 Stage 5: quantitative aspects of metabolism

15.5.1 Quantification of excretion balance studies

The relative recoveries of drug-related material in the different excreta are determined by liquid scintillation counting.

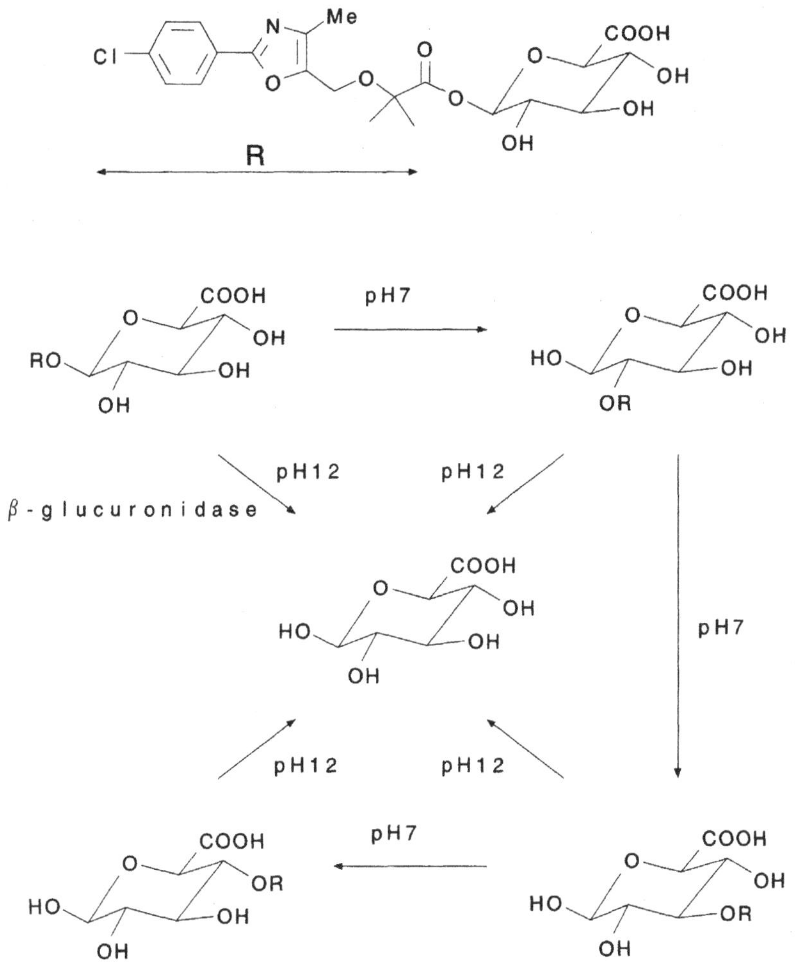

Figure 15.46 Rearrangement and hydrolysis of romazarit-glucuronide.

15.5.2 Quantitative aspects of metabolite isolation

The yields of each step are also determined by liquid scintillation counting as the sum of the recoveries of the individual fractions. The relative abundance of the metabolites should normally be estimated assuming that the losses at each stage are the same for each sub-fraction.

15.5.3 Quantitative measurement of metabolic profiles

Metabolic profiles in excreta are usually estimated from data obtained using a radiochemical detector after HPLC or TLC on the mixture of interest. If preliminary fractionation is needed to simplify the chromatograms, the assumption that losses are

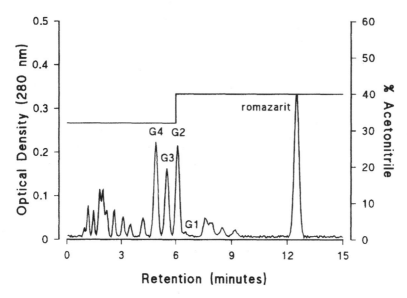

Figure 15.47 Reverse-phase HPLC of rat bile after oral administration of romazarit at 20mg/kg (C18 μBondapak, 0.1% CF₃COOH). (2) Hydrolysed by β-glucuronidase.

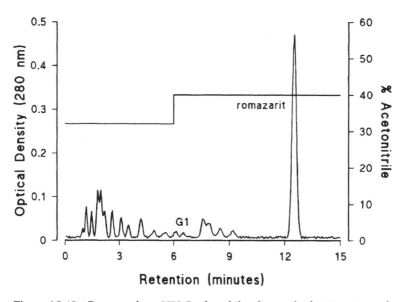

Figure 15.48 Reverse-phase HPLC of rat bile after oral administration of romazarit at 20mg/kg (C18 μBondapak, 0.1% CF₃COOH). (3) Hydrolysed by NaOH.

evenly distributed between the various metabolites will again have to be made. If it is not possible to analyse all of the samples, owing to lack of sensitivity, a second assumption – that the profiles in the minor samples are the same as those of the major ones – will be necessary.

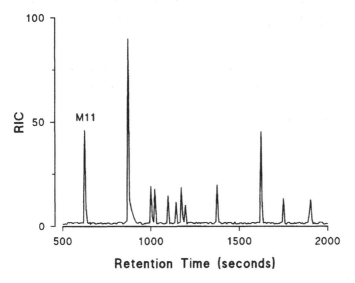

Figure 15.49 Gas chromatogram of metabolite 11 of Ro 31-1411 after methylation with BF$_3$/methanol.

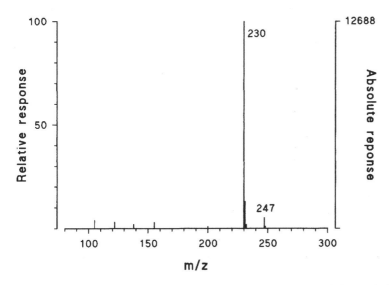

Figure 15.50 Ammonia chemical ionisation mass spectrum of the methyl ester of metabolite 11 of Ro 31-1411.

15.6 *In vitro* studies

In vitro studies range from simple measurements of the rate of hydrolysis of ester pro-drugs in plasma to the sophisticated use of isolated perfused organs. Compared with *in vivo* experiments, they offer the advantages of many fewer endogenous compounds and consequently easier analysis. Thus it is often possible to carry out

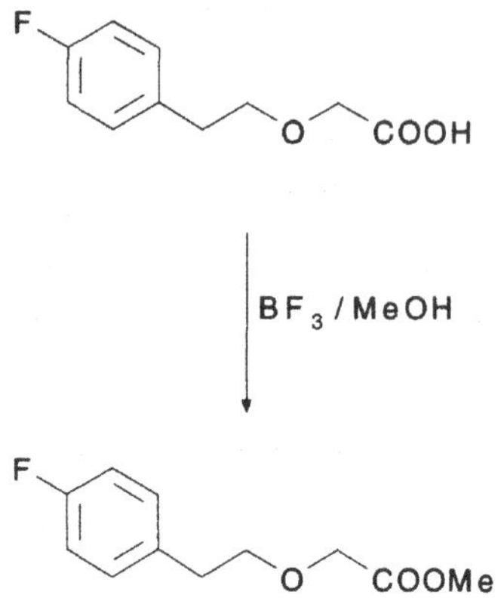

Figure 15.51 Methylation of metabolite 11 of Ro 31-1411.

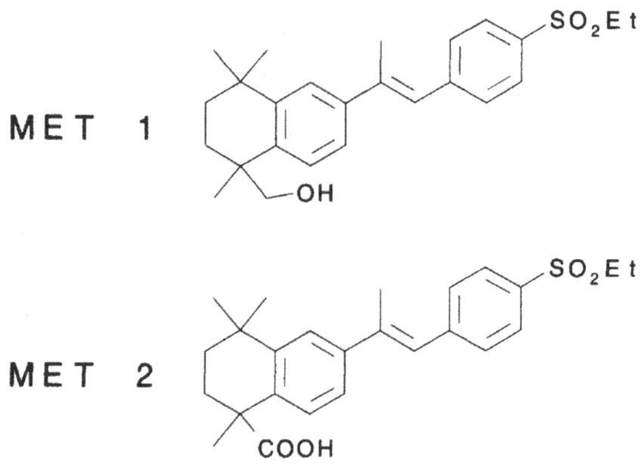

Figure 15.52 Possible structures for metabolites M1 and M2 of Ro 15-1570.

preliminary metabolic studies with liver microsomes or hepatocytes using drugs that have not been radiolabelled. (Nevertheless, the information obtained from such experiments may well be misleading as far as extrapolation to whole animals is concerned, and must be treated with extreme caution.) On the other hand, the viability of *in vitro* preparations is relatively poor and the metabolism of slowly cleared drugs may be negligible.

These types of experiments fall into three broad classes: metabolite isolation, cross-species comparisons and mechanistic investigations.

15.6.1 *Isolation of metabolites from* in vitro *incubations*

The techniques required for the isolation of metabolites from *in vitro* incubations are similar to those used from animal excreta, but the advantage of cleaner samples is often counterbalanced by the disadvantage of smaller amounts of material. The major

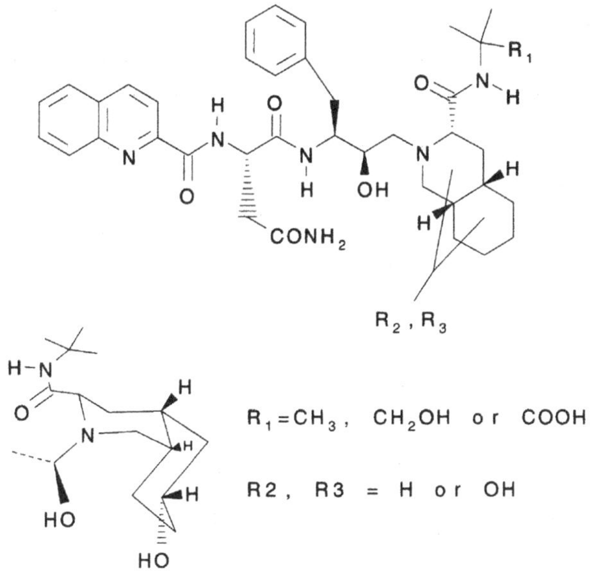

$R_1 = CH_3$, CH_2OH or $COOH$

$R2$, $R3$ = H or OH

Figure 15.53 The biliary metabolites of saquinavir produced by the rat: part structure of metabolite M-iv.

Table 15.1 The biliary metabolites of saquinavir produced by the rat

Metabolite number	R1	R2	R3	%
Saquinavir	CH₃	H	H	
M-i	COOH	H	H	14
M-ii	CH₃	OH	OH	12
M-iii	CH₃	OH	OH	9
M-iv	CH₃	OH	H	15
M-v	CH₃	OH	H	7
M-vi	CH₃	OH	H	5
M-vii	CH₃	OH	H	2

metabolites of saquinavir (Figure 15.53 and Table 15.1) have been isolated from both the bile (dose = 10mg/kg) and hepatic microsomes (drug concentration = 5µg/ml) of rats. A number of identical hydroxylated derivatives were characterised, but the carboxylic acid (M-i) was only found in bile; however, its precursor, the hydroxy-methyl derivative, was identified in the microsomal incubation.

Some of the dangers inherent in the isolation and identification of metabolites from *in vitro* preparations have already been mentioned. As an example, Ro 31-6930 (Figure 15.54) is almost completely metabolised by the rat to at least 20 metabolites. About half of the dose, of which less than 10% was the parent drug, was recovered in urine (Figure 15.4 shows a urinary metabolic profile of this compound). However, when incubated with the 'S9' supernatant from the livers of rats which had previously

335

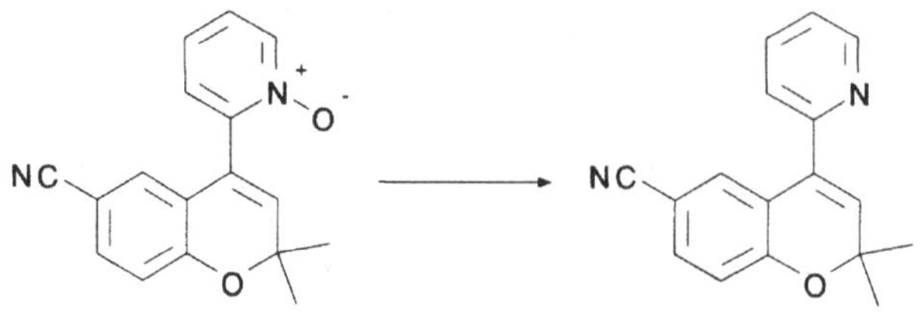

Ro 31-6930

Figure 15.54 The metabolism of Ro 31-6930 catalysed by rat 'S9' supernatant.

been dosed with inducers of hepatic cytochrome P_{450} (i.e. the method used in the 'Ames' mutagenicity test to simulate *in vivo* metabolism) only one metabolite was formed (Figure 15.54). This was shown by TLC/autoradiographic comparison with authentic material to have resulted from reduction of the N-oxide group to the simple pyridyl derivative – and no more than 1% of the drug was metabolised over a wide range of concentrations. This metabolite was not detected *in vivo*, although several metabolites resulting from its further metabolism (e.g. metabolite 4A, Figure 15.6) were identified.

The isolated perfused liver, however, will normally produce a similar biliary profile of metabolites to that of the intact animal; no differences were found in the case of romazarit (Figure 15.18). If a drug is selectively toxic (e.g. cardiotoxic), it may well be possible to employ larger quantities with the isolated organ than the whole animal, and hence to increase the ease of the isolation and identification of the metabolites. It would then be possible to compare *in vivo* metabolic profiles, obtained using small doses of radiolabelled material of high specific activity, with the *in vitro* results.

15.6.2 Cross-species comparisons of metabolic profiles

Cross-species comparisons of metabolism are necessary for the assessment of safety margins. If the exposure of the drug and its metabolites to the species used for toxicological investigations is not significantly higher than that experienced by humans, it may be difficult to obtain regulatory approval. These comparisons are of two types: investigations of metabolic profiles in excreta, usually urine (Sections 15.3.2 and 15.5.3) and measurements of plasma AUCs for relevant metabolites.

If a drug is highly metabolised and if it is not eliminated in urine, cross-species comparisons are often best carried out with hepatic microsomes. Although these will generally only convert the parent drug into its primary metabolites, this is probably the most useful information for determining the suitability of the chosen toxicological species. Human microsomes are commercially available and the metabolic profiles of drugs can usually be determined quite readily. An example of the HPLC/fluorescence comparison of the metabolism of mibefradil by humans, rat, cynomolgus monkey and marmoset is included in Figure 15.55.

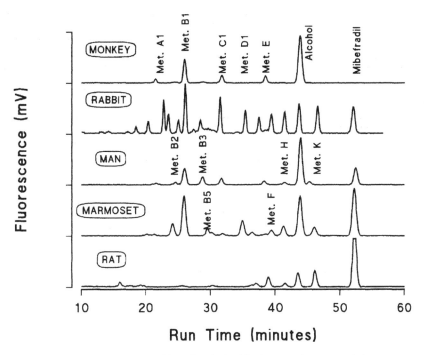

Figure 15.55 Metabolic profiles of mibefradil after incubation with hepatic microsomes from humans, cynomolgus monkey, rabbit, marmoset and rat.

In this case the primates (particularly the cynomolgus monkey) form substantial amounts of the alcohol by hydrolysis of the ester group (Figure 15.25) and metabolise it further by means of cytochrome P450-catalysed reactions. The major reactions in the rat, on the other hand, are demethylation (metabolites H and K) and hydroxylation of the benzimidazole ring (metabolite E).

15.6.3 *Mechanistic studies*

Mechanistic experiments to investigate the kinetics of the disappearance of the parent drug and the appearance of metabolites in, for example, microsomal incubations can be valuable. It is possible to determine kinetic parameters such as the Michaelis constant (K_M) and the maximum rate of the reaction (V_{max}) for a drug, to determine which of the cytochrome P450 isoenzymes are responsible for metabolism and to investigate the drug's metabolic interaction with other compounds for extrapolation to the clinical situation.

15.7 Identification of plasma metabolites

The identification of metabolites in plasma can present serious problems. Unless a drug has a particularly small volume of distribution, the concentration of total drug-related material is unlikely to be greater than a few micrograms per millilitre. Characterisation by radiochemical detection (although not in humans), diode-array UV and thermospray mass spectrometry after reverse-phase HPLC is usually feasible, but the results are often inadequate and it is often impossible to be certain which peak corresponds to which metabolite.

If the metabolites in excreta are known, or if microsomal metabolic studies have been completed, the primary metabolic pathways can be deduced. It may then be possible to identify the plasma metabolites by chromatographic comparison with authentic derivatives or by selected ion monitoring via a suitable HPLC–MS interface. These derivatives could be synthetic, metabolites isolated from excreta or *in vitro* systems, or primary metabolites prepared by deconjugation of isolated glucuronides, sulphates or amino-acid amides. An example of the identification of a number of metabolites of mibefradil circulating in human plasma after oral administration of the drug using gradient HPLC with electrospray mass spectral detection is shown in Figure 15.56.

There are a number of pitfalls to this approach and it is necessary first to determine the importance of the circulating metabolites by radioactive means. For example, the majority of romazarit is excreted as metabolites (Figure 15.18), but the plasma levels of the parent drug are always greater than 90% of those of total drug-related material (i.e. radioactivity). The metabolites are excreted so rapidly that high plasma levels are never reached.

The similarity between the biliary and microsomal metabolites of saquinavir (Figure 15.53) suggested that all the problems of metabolism had been solved. Plasma kinetic studies with radioactive material showed that the parent drug accounted for up to half of the drug-related material in rats and humans. However, specific ion-monitoring failed to show the presence of any of the known metabolites in human plasma. In addition, the metabolites detected in rat plasma by radio-chromatographic means were considerably more polar than those identified; their structures are, as yet, unknown.

It may not always be possible to identify an important plasma metabolite by chromatographic comparison with microsomal incubations or excreta. In this case you will have to try to isolate and identify the metabolite from plasma. Good luck!

Ideally, one would wish to compare the plasma 'areas under the curve' of humans and animals for each metabolite. In practice this is usually difficult, as radiochemical methods are seldom suitable in human studies because of the restricted dose of radioactivity that can be employed, and other analytical techniques which would allow the quantitative determination of the concentrations of more than one or two metabolites would be very difficult to develop. A reasonable approach is to compare peak areas using some suitable method such as HPLC/UV or fluorescence for each significant metabolite and hence to obtain semi-quantitative profiles and AUCs. If radioactive metabolites have been isolated, a correction factor can then be applied by measuring the extinction coefficient relative to the specific activity. A cross-species comparison of the exposure of humans, cynomolgus monkey, rat and marmoset to the major circulating metabolites of mibefradil after oral administration of the drug is shown in Figure 15.57.

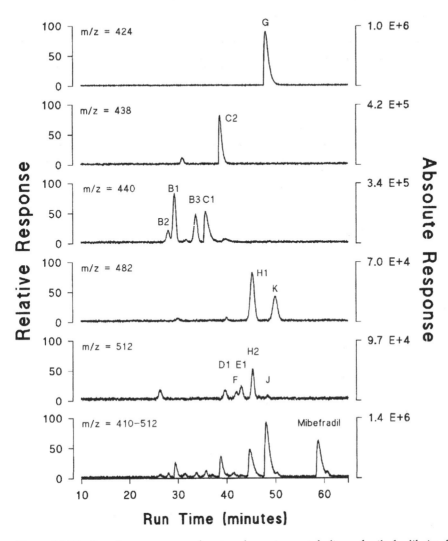

Figure 15.56 Ion chromatograms showing the major metabolites of mibefradil circulating in human plasma after daily administration of a 100mg tablet.

15.8 Good laboratory practice

It is not mandatory to carry out metabolite isolation/identification studies to GLP standards. Nevertheless, the complexity of metabolic experiments means that such GLP principles as the complete documentation of objectives, observations, results and conclusions are essential. Before each stage of the isolation process it will be necessary to list the rationale, objectives and methods to be used.

For example: 'TLC (page . . .) shows that Fraction I.2. (equivalent to 5% of drug-related material recovered in urine, 4% of dose, page . . .) contains three radioactive

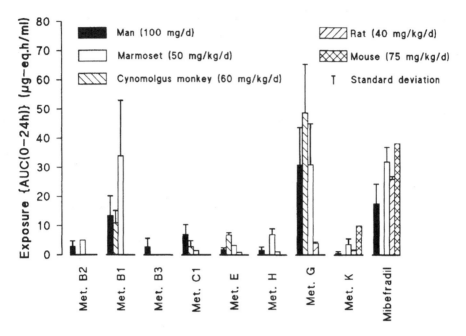

Figure 15.57 Cross-species comparison of the exposure of humans, marmoset, rat and cynomolgus monkey to mibefradil and its major metabolites after oral administration of the drug.

compounds (ratio 1 : 1 : 2). These will be separated by gradient HPLC on a C18 μBondapak column (7.8 × 300mm, ref . . .) using an aqueous methanolic mobile phase containing 0.1% acetic acid (pH ~3). HPLC equipment = . . .'

For each HPLC run, details of the actual conditions of the method and observations must be recorded with the traces or computer print-outs. For example: 'Run 1, analytical, injection 10μl out of 3ml (50 000dpm, 25μg drug-related material, ~500μg total weight). Linear gradient 10%–50% methanol over 30 minutes, UV detection 220nm at 0.1OD, radiochemical detector 50cps.'

Observations:

'Three major peaks: 6.9min (% MeOH = 19.2) ~25% of total 14-carbon;

9.1min (% MeOH = 22.1) ~20% of total 14-carbon;

19.6min (% MeOH = 36.1) ~50% of total 14-carbon;

minor peak? 14.2min (% MeOH = 28.9) ~ 5% of total 14-carbon.

Minor compound not seen on TLC, possibly contamination or an artefact arising from degradation of a major metabolite or maybe TLC resolution inadequate ???'

Conclusions:

'System unnecessarily long. Try 10–45% methanol over 25 minutes and stop after last endogenous peak (expected at ~23.5 minutes) comes off. Increase injection to 50μl (250 000dpm) and check that first two peaks still resolved. Retain all fractions'

Signature/date

15.9   Conclusions

Despite the inherent difficulties of isolating small amounts of metabolites from biological material and of identifying complicated structures, the many techniques available for both processes provide the metabolist with the tools for his or her work. Provided that he or she remembers the importance of using a variety of methods for the isolation and characterisation of the metabolites, no problem should be insoluble.

16

STRATEGY FOR THE DEVELOPMENT OF QUANTITATIVE ANALYTICAL PROCEDURES

David Bakes

16.1 Introduction

Analytical method development is the process of creating a procedure to enable a new chemical entity of potential therapeutic value to be quantified in a biological matrix. Such methods are required to demonstrate exposure of the test system to the compound for toxicological assessment, and as an aid to understanding its pharmacokinetic and pharmacodynamic behaviour. This chapter outlines the major steps that may be applied in order to establish a new quantitative analytical method and is intended to be read together with other chapters that deal with each of the aspects in more detail.

> Any description of this process is bound to be subjective by its very nature since it describes a sequence of observations and subsequent decisions. Although this chapter describes a general approach, the principles will be universally applicable; where there appears to be a significant difference between these and your laboratory, then obviously your own laboratory policy should prevail.

The development process falls into three main categories: detection, separation and quantification, each of which will be discussed in general terms. It is prudent, at the start, however, to list six of the major characteristics of a developed method that must be borne in mind during its development.

1 *Accuracy*: the ability of an assay technique to measure the analyte concentration correctly at the time of sampling. This is perhaps the most important aspect of all assay development and the one factor that alone can lead to an assay being accepted or rejected. The accuracy of a method may be influenced by any of the following criteria.
2 *Precision*: the precision of an assay is a measure of its ability to give the same result for the repeated analysis of samples at the same concentration. It is *not* the same as accuracy.

3 *Recovery*: the recovery of a method is a measure, usually expressed as a percentage, of that proportion of analyte present in the original sample which is retained throughout the extraction process and is available for direct quantification. The minimum generally acceptable recovery of an extraction process is 60%, although a target value of at least 80% is usually applied as this will often improve accuracy, precision and, of course, sensitivity.

4 *Sensitivity*: the sensitivity of a method is the lowest concentration of analyte that can be detected or quantified, referred to as the limit of detection (LOD) and the limit of quantification (LOQ) respectively. The LOD is that concentration which gives a signal equivalent to three times the background signal at the retention time of the analyte. The LOQ is the lowest concentration to be measured with an accuracy and precision better than 20%. It therefore follows that an analyte may be detected but not quantified.

5 *Specificity/selectivity*: these two terms are often used interchangeably although they have different meanings. A specific assay is one in which only the analyte produces a response at the detector. This is often impossible, with many co-extracting compounds also giving a response in optical or electrochemical detection. In this situation the assay must demonstrate selectivity in as much as it is able to discriminate one detector response from several present, and to be able to quantify that signal alone without interference from the others. A specific assay, where the analyte alone produces a response, is the ideal but where more than one peak is produced selectivity is essential.

An alternative understanding of the term specificity relates to the positive identification of a chromatographic peak, e.g. by its mass/charge (m/z) ratio, as is achieved by mass spectrometry.

6 *Robustness*: this is a measure of the ruggedness of the system. Ideally a method will to be able to tolerate small changes in the reagents, different batches of columns and different operators, etc. without compromising the analytical results. Where there are critical conditions or parameters, these need to be identified and suitable limits imposed; a system suitability established.

Having given a basic outline of some of the requirements of a developed method, the flow chart in Figure 16.1 illustrates the major steps to be taken in reaching these objectives.

16.2 Preliminary requirements

There is much information that would be useful to the development analyst when faced with the task of assay development, such as a published method for an analogous compound for example. Unfortunately, this is not often the case with new chemical entities for which a quick assay for toxicological purposes is required within a few days. Later in the drug's development, when assays with the appropriate sensitivity for pharmacokinetic evaluation are required, more information is available. However, even at the earliest stages some information is very useful, if not essential. The chemical structure gives an indication of reactive groups and possible detection modes, such as ionic species for isolation by ion-exchange, or carboxyl, amino, hydroxyl and sulphonyl groups whose dipolarity suggests isolation by electrostatic

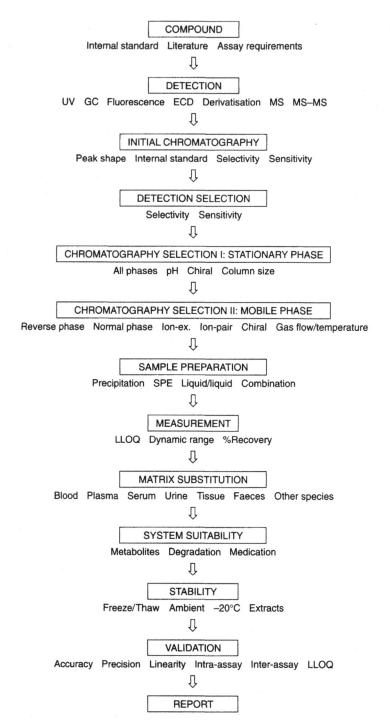

Figure 16.1 Flow path of the development of an analytical method.

interaction, or *cis*-diols where covalent bonding may be the method of choice. The presence of unsaturated bonds is an indication of UV absorbance and conjugated ring systems suggest that fluorescence detection may be suitable. Whatever the structure, it will be necessary to screen the drug for a response by mass spectrometry through flow injection analysis to determine the applicability of LC–MS–MS as the mode of analysis. If volatility is indicated, then GC evaluation is also appropriate.

So the structure of the analyte provides almost indispensable information at the start of a development procedure. The structure may also indicate potential solubility through an estimation of its log *P*, and potential stability difficulties through the presence of ester or peptide linkages, for example.

Most chromatographic analyses in a pharmacokinetic environment involve quantification by peak response ratios with internal standards, where a structural analogue of the analyte is added to the samples and standards prior to the sample extraction process. This necessitates the availability of such an analogue and, again, the structure and related information may suggest possible candidates for the internal standard. If no suitable internal standards are available, then quantification will have to be by what is often referred to as external standards, where the response of the analyte alone is plotted versus concentration, to generate a calibration line.

Other essential information concerns the requirements of the assay: Is it to be for toxicokinetic or pharmacokinetic work? What is the anticipated concentration range? What matrix and species are to be investigated, with their concomitant restrictions on sample volume? Does the assay require automation or is it to be contracted out? Are there any known metabolites or breakdown products that also require quantification? Is the analyte chiral? What is the time deadline?

Once armed with this information, or as much of it as possible, the practical side of method development may begin.

16.3 Detection

The detection of the analyte, and perhaps internal standard, is the first aspect of chromatographic method development to be considered. The rationale behind this being that unless the analyte can be detected by the instrumentation it is impossible to assess any further development work.

As a first step solutions of the compound at, say, ~10µg/ml may be prepared in a methanol/water mixture under acidic, neutral and basic conditions, assuming solubility. Solutions of the proposed internal standard should also be prepared. These solutions are then assessed against their corresponding blanks for UV absorbance on a UV spectrometer and for fluorescence on a fluorimeter. When using the fluorimeter in this manner, it is prudent to allow the sample to be irradiated for a short time and then re-scanned to ensure fluorescent stability, as a number of molecules become fluorescent, or alter their λ_{max} and/or quantum yield through UV-mediated reactions. If a GC assay is indicated, then the structure of the analyte will indicate whether the detection is to be by electron capture (halides), the thermionic detector (nitrogen/phosphorus) or by flame ionisation as a universal detector without specificity, or the GC–MS interface as a universal detector with specificity. The 10µg/ml standard may also be used to tune the ionisation parameters on the mass spectrometer in either the ESI or APCI mode to assess fragmentation patterns and potential suitability for analysis by LC–MS–MS. For example, do the internal standard and analyte possess the same parent and product ions? If so, time resolution will be required. Ideally, a compound should produce a single parent and product ion, as, if there is extensive fragmentation to many small ions, the sensitivity is compromised as well as selectivity, through the need to discriminate between low m/z ratios in the presence of sample impurities and mobile-phase additives.

The results of the spectrometric and fluorimetric scans may indicate a likely detection mechanism and wavelength(s). If both UV absorbance and fluorescence are observed, then the latter would be the detection mechanism of choice, due to its inherently greater sensitivity and selectivity. If no, or very little, signal is observed from these scans, and analysis by LC–MS–MS is not possible, it may be appropriate to consider derivatisation with dansyl chloride for primary and secondary amines or 4-bromomethyl-7-methoxy-coumarin for carboxylic acids, for example, to attach a fluorophore to the analyte. Coulometry may be assessed in a similar manner to the LC–MS–MS, through flow injection analysis to produce a voltamogram, with a solution in the low µg/ml range prepared in a 50% methanol in 50mM phosphoric acid solution.

After these selective investigations have been carried out, the most appropriate detection method for the analysis may be selected. This will depend upon both the physico-chemical properties of the analytes and the requirements of the assay. For example, UV absorbance is the easiest and most widely available detection mechanism, is of moderate sensitivity and selectivity, while being applicable to a very wide range of compounds. Fluorescence detection has comparable ease of use but usually an order of magnitude greater sensitivity and selectivity, consequently greater care is needed with the selection of an internal standard. The electrochemical detector may have comparable sensitivity to the fluorimeter, providing that the specific functional groups are present, and therefore also demonstrates good selectivity. It does present serious problems with stability, particularly with high-sensitivity work, where a change in chromatographic conditions may result in unstable detection for up to 5 days. It also suffers from pump pulsations, requires a higher system pressure and may be prone to electrode desensitisation. The mass spectrometer provides the highest sensitivity, with low femtomole levels achievable with the appropriate compounds, and since it measures mass, a fundamental property of the analyte, it offers a specificity higher than others that record an effectual property; the ability of the analyte to absorb light, for example. Full details of all detectors to be used in HPLC are presented in Chapter 5, which should be consulted in addition to this chapter.

In considering UV detection, double bonds in the analytes would tend to suggest UV absorbance, with, generally speaking, greater absorbance at lower wavelengths. At, say, 210–220nm almost all the eluted compounds will register a response at the detector, as will methanol and THF, in addition to many other modifiers in the mobile phase, such as triethylamine, acetic acid and ammonium acetate. Although these wavelengths will usually produce larger chromatographic peaks for the analytes, the signal-to-noise ratio will often be less when compared to higher wavelengths. Usually a wavelength is selected at, or close to the λ_{max} of the analytes in the region of 225nm upwards. A word of caution here. If the detector wavelengths are selected from a stock solution of a salt or other impure solution, then the indicated λ_{max} may reflect a property of the salt more than the analyte. This is particularly true if a low wavelength, say 200–220nm, is indicated and the salt present is a formate. It is recommended that once a tentative wavelength has been chosen and used to set up an initial chromatographic system, the peak of interest should be scanned *in situ* to determine its true λ_{max}. This is done either on a diode-array detector or by stop-flow scanning or by repeated runs at various wavelengths.

If a simple binary isocratic system of acetonitrile/water is to be used in conjunction with an efficient sample clean-up step, then there is the greatest choice of wavelengths. If the method necessitates the use of, say, methanol with triethylamine, or acetate/formate buffers, then wavelengths below about 240nm are not really feasible for high-sensitivity work due to high background absorbance produced by these modifiers. If the analytes possess a chromophore at 300–400nm then, because so little endogenous plasma material absorbs at these wavelengths, the assay will usually demonstrate very good selectivity. The analyte peak may not be pure but, because any isoretentive material does not produce a response at the detector, that is immaterial for the purposes of quantification. This situation does not hold when the object of the development is to isolate pure analyte for structural elucidation, as any co-eluting material may interfere with this process. A diode-array UV detector may be more useful in this case since it will detect all UV-absorbing eluates regardless of their detection wavelengths, although this will probably compromise on sensitivity. Even so, there may be UV-transparent impurities that go undetected.

This principle also applies to fluorescence, where the high degree of selectivity inherent in this detection mode may obscure non-fluorescent isoretentive material. The fluorescence signal intensity may be pH dependent, particularly if the conjugated ring system conferring fluorescence contains, or is adjacent to, a charged moiety such as carboxylic or amino groups. In this case, in order to maximise the sensitivity, the pH will require adjustment to minimise the ionisation on such groups. This may be done with the mobile phase, provided that the range of pH 2.5–6.5 is not exceeded when most silica-based columns are used. There are some that will tolerate a pH as high as 10 but not many. Outside of this range it will be necessary to use a polymeric column which is stable in the pH range 1–13, or a post-column mixer with a second pump for the addition of either the acidic or basic reagent as appropriate. With both UV and fluorescence detection a post-column photoreactor may be beneficial in the enhancement of the analyte signal.

With the electrochemical detector there are additional considerations. First, the detection mechanism involves a chemical reaction, of either reduction or oxidation, in the analyte, so any post-chromatographic structural elucidation is not possible. Another point concerns the use of additives in the mobile phase. Although the optical detectors are generally tolerant of additives such as triethylamine to restore an ideal peak shape to the chromatography of basic analytes, electrochemical detectors can not be used with these amine modifiers as they dramatically increase the background current. Control of peak shape with electrochemical detectors is best achieved by adjustment of the mobile-phase pH, within its limits, to attempt to neutralise the charged groups on the analytes, or by the utilisation of an analytical column which is designed for such basic analytes. The presence of dissolved oxygen in the mobile phase will cause significant problems with high background current and drifting baselines with electrochemical detection. The problem is exacerbated at higher potentials and sensitivities, and, when used in the reductive mode in particular, it is essential that *all* the dissolved oxygen is removed from the mobile phase by continuous helium sparging or on-line vacuum degassing. Indeed *with all HPLC, regardless of detection mode, the best results will be obtained with thoroughly degassed mobile phases.*

So a number of detection modes are available for analysis by LC and GC. The method of choice will be a combination of assay requirements, analyte property, operator expertise and instrument availability, although the tendency for initial toxicokinetic work with high analyte levels to be done with optical detection, and pharmacokinetic analysis to be done through mass spectrographic analysis with either a liquid or gaseous coupling, will no doubt shift as quantitative mass spectrometry continues to become more established.

16.4   Separation

This is the most difficult and time-consuming part of the method development process and falls into two sections: sample preparation and chromatographic separation. The objective of the former is to selectively enrich the biological sample to such an extent that quantification down to the appropriate level is possible using the selected chromatographic method. The objective of the latter is to resolve and quantify a property of the analyte, any potential metabolites/degradation products deemed appropriate, and the internal standard in the presence of endogenous material in as short a time as possible, using appropriate reagents. These two aspects of method development are, of necessity, intrinsically linked, and neither can be optimised in the absence of the other. However, in an attempt to clarify the development process they will be dealt with here as independently as possible.

Chromatographic optimisation is the process of finding a set of conditions that adequately separates and enables the quantification of the analytes of interest in the presence of endogenous material with acceptable accuracy, precision, sensitivity, specificity, cost, ease and speed. The processes of optimising methods by HPLC or GC are covered in the appropriate chapter, with a few general points applicable to both techniques given here. Briefly, in order to optimise this process some measure of 'separation' is needed. This is given by the term 'resolution'. The higher the resolution of the system the more selective, sensitive and rugged the analytical method.

Resolution (R) may be calculated thus:

$$R = \tfrac{1}{4}\,[(\alpha - 1)/\alpha]\sqrt{N}[k_2/(k_1 + 1)].$$

For baseline separation, an R value of at least 1.5 between the peak of interest and its closest neighbour is usually necessary. This is generally the minimum resolution required for a good analytical chromatographic method. These terms are defined in Chapter 4.

Chromatographic optimisation aims to achieve this resolution between the analyte(s) and *all endogenous material generating a response at the detector*, with all peaks of interest eluting from the column such that $3 < k < 20$. This applies to all separation modes, whether they use UV, NPD or MS detectors. This is true of MS even though it is a specific detector responding to an inherent property of the analyte. Co-eluting material can alter or quench the ionisation of the analytes at the MS ion source and produce an effect that appears as low recovery, yet when analysed by UV or fluorescence, for example, the recovery is high. Although this material generates no response on the mass spectrometer, it can have a direct effect on the detector response to the analytes and may well vary between standards and samples, or within sample groups from different subjects. In such a case the chromatography will require adjustment to resolve the 'invisible' interference from the analytes.

In regard to the chromatography of pharmaceutical compounds from a biological matrix, there exists a plethora of extraction/enrichment procedures, a variety of detection modes and advanced integration capabilities. None of these is any substitute for good chromatography. Getting the chemistry and the separation right, *and appropriate for the intended application*, are indispensable characteristics of modern analytical techniques whether they are HPLC, GC, CE, MS or NMR.

16.5   Sample preparation

Sample preparation is necessary for at least two reasons: to remove as many of the endogenous interferences from the analyte as possible and to enrich the sample with respect to the analyte, thus maximising the sensitivity of the system. It also serves to ensure that the injection matrix is compatible with the selected column and mobile phase, or of the appropriate volatility for GC.

This is perhaps the single most intricate and time-consuming aspect of the whole development. Although there are computer packages available to assist with chromatographic aspects of the development there is, at present, only a variety of manufacturers' literature and an understanding of chemistry to aid the development analyst. If detection is by mass spectrometry, then the extraction optimisation may be somewhat simpler. There are those who maintain that the only extraction necessary for LC–MS–MS is protein precipitation, the so-called 'dilute and shoot' approach. This may be appropriate for single samples of certain analytes, but where the instrument is to be used on a routine quantitative basis, especially at high sensitivities, good sample preparation is necessary to maintain ion-source efficiency and to remove potential sources of quenching from the injection matrix. It is good practice to develop a thorough sample clean-up for routine mass spectrometry with the emphasis on good recovery (80% +), low non-volatiles and minimal quenching.

16.6   Solid-phase extraction

This is dealt with thoroughly in Chapter 2 and a summary of the various types of solid-phase separation mechanism that may be used are shown in Table 16.1. The degree of enrichment achievable for a particular sample is dependent upon:

1   the selectivity of the bonded phase for the analyte; and
2   the relative strength of that interaction.

The ideal is to create a 'digital' chromatography system on the SPE cartridge such that the analyte is at first fully bound, then the interferences are completely eluted, and then the analyte is entirely eluted; it is either all on or all off the cartridge. The separation mechanisms given in the table function due to the intermolecular interactions between analyte and the functional groups of the sorbent, discussed in detail in Chapter 2. These forces have been described in other chapters and can be categorised as: hydrogen bonding, dipole–dipole, dipole–induced dipole, dispersion forces, ionic and covalent.

16.7   Extraction sequence

Chromatographic analyses that require sample pre-treatment provide the highest selectivity, and hence specificity, when differing retention mechanisms are used in the sample extraction and analytical phases.

This is a basic rule of bioanalytical method development and should be observed wherever there is sufficient variety of functional groups on the analyte to permit it. For example, most chromatographic analyses are by reverse-phase HPLC, i.e. based on dispersion forces, consequently the greatest potential for a selective system will be achieved using an ion-exchange or covalent sample preparation mechanism. However, it may, of course, be appropriate to consider screening the full range of cartridges as described in Chapter 3 as the most applicable may not be apparent immediately.

As an example, a possible sequence for an ionisable compound will be considered. The first step is to examine the analyte and any available data regarding its pK_a(s). If such data are not available, then an experienced analyst may be able to estimate an approximate pK_a value from a consideration of the structure. Then select the appropriate class of either cation or anion exchangers and attempt to retain and selectively elute the analytes from buffer solutions by the adjustment of pH, analysing each step of the procedure for analyte elution. The target should be more than 90% of the analyte retained from the loading step and more than 90% of that retained being eluted in the final elution. If this is not achieved, then buffer strengths, pH and solvents will be optimised so that these conditions are met. The next step is to substitute the matrix under investigation for the buffer. This will usually be plasma but may also be serum, whole blood or urine. It will be necessary to adjust the pH of the matrix accordingly, and, if a low pH is required, the sample will need to be centrifuged to remove precipitated proteins which would otherwise block the cartridge. It may well be that no analyte is seen in the final eluate, which would suggest that there is a matrix interference with the bonded phases or that the analyte was highly protein bound. If matrix interference is suspected, it will not usually be observed to the same degree for all the bonded phases of a particular class and, to some extent, may be overcome by adjustment of pH or by further dilution of the sample with buffer or water.

If protein binding is suspected, then protein precipitation prior to sample extraction may be considered. Reagents to evaluate include perchloric, trichloracetic and tungstic acids, and organic solvents such as acetonitrile or methanol. With all of these it is necessary to bear in mind the lability of the analytes and the matrix requirements of the extraction procedure. If protein binding is believed to be through a covalent linkage, then there is very little chance of breaking it since this is the strongest of the intermolecular forces.

With each set of extraction conditions a positive and a blank matrix sample should always be run, with the positive being towards the top end of the anticipated calibration range as a check on recovery and analyte peak shape. It is not unknown for the extraction process to cause some analyte degradation, which may not be apparent at the lower concentrations. The development of the extraction, while usually occurring after the development of the analytical chromatography, may require that the latter be slightly modified to improve the resolution between minor endogenous peaks and the analytes. Therefore, once a tentative extraction method has been established, standards at the lower end of the concentration range should be evaluated. It is anticipated that, at this stage, there may well be more than one extraction process under evaluation, consequently the method that affords the greatest selectivity for the analytes will be chosen for further optimisation.

As already mentioned, another parameter that needs evaluation is recovery. This may be done by a comparison of the peak heights or areas of extracted standards to chemical standards, or by the use of a radiolabel. Many compounds can be made with either a ^{3}H or ^{14}C label. In this case the recovery in the final eluate may easily be calculated after scintillation counting. This technique has the added advantage of enabling the assessment of analyte loss at each stage of the extraction process. This is particularly useful if overall recoveries are low, enabling the steps with significant losses of analyte to be identified.

If there is no usable retention of the analytes on the ion-exchange columns, then it will probably be necessary to use the non-polar or reverse-phase mode. The same criteria apply: start with water or buffer and select those phases that exhibit retention and subsequent elution and then incorporate the biological matrix. It will be observed that there are often differences in the retention characteristics of the analyte between the water/buffer and the plasma matrices. This is because of additional factors such as protein binding, competition for the bonded phase, higher salt concentration and/or different pH, etc.

All of these non-polar bonded phases retain some residual silanol groups that are free to react with the analyte and matrix, giving them a secondary cation-exchange mechanism for retention, the propensity for which is generally inversely proportional to the chain length of the bonded phase. Although this dual functionality may be useful in particular instances, it should not be relied upon as the basis of an extraction since the degree of residual silanols is not usually subject to the manufacturer's quality controls and hence is inclined to show substantial batch-to-batch variation. To overcome this, the columns may be saturated with 10ml of 0.1M ammonium acetate prior to sample loading, or, alternatively, a small amount of aqueous ammonia may be included in the elution solvent, the exact quantity of which will need to be determined for each application.

There are further types of cartridges, designated 'mixed mode'. These utilise the two mechanisms of ionic (either anionic or cationic) and dispersion forces to achieve their selectivity and they can provide excellent extractions in certain applications.

One type of column with a unique functionality is the phenylboronic acid column (PBA). These are specific for aromatic co-planar hydroxyls and exhibit retention through strong covalent bonds. A major application of this type of column is in the

Table 16.1 Summary of separation mechanisms for solid-phase extractions

Separation mechanism	Analyte type	Loading solvents	Eluting solvents
Normal phase (silica)	Slightly to moderately polar	Hexane	Methanol
Normal phase (polar-bonded phase)	Moderately to strongly polar	Hexane	Methanol
Reverse phase	Non-polar	Water, buffer	For non-polar analytes: hexane For polar analytes: methanol
Anion exchange – SAX, NH$_2$, etc.	Ionic acid	Buffer: pH = pK_a + 2	Buffer (pH = pK_a − 2) pH where sorbent or analyte is neutral Solvent with high ionic strength
Cation exchange – SCX, CBA, etc.	Ionic base	Buffer: pH = pK_a − 2	Buffer (pH = pK_a + 2) pH where sorbent or analyte is neutral Solvent with high ionic strength
Covalent – PBA	Co-planar hydroxyls	Basic, pH > 8	Acid, pH < 2

isolation of glucuronide conjugates, which are retained under basic conditions and eluted with acids.

Table 16.1 provides a summary of separation mechanisms for SPE with fuller information on their use to be found in Chapter 2.

16.8 Liquid/liquid extraction

Samples received for analysis are usually in an aqueous biological matrix and, as has been shown, the analytes may be removed selectively from such matrices using solid-phase extraction methodologies based on a number of chemical interactions.

A traditional, and cheaper, alternative is liquid/liquid extraction, where the analytes partition between two imiscible solvents. This has been discussed in detail in Chapter 1, so there are just a few points here.

The properties of an ideal solvent for liquid/liquid extraction are:

- environmentally acceptable or easily recoverable;
- non-toxic and not highly flammable;
- imiscible with water;
- of convenient specific gravity;
- of suitable volatility;
- of high chemical stability and inertness;
- not prone to form an emulsion; and
- dissolves the neutral but not the ionised form of the analyte.

Although these properties are desirable, a compromise is usually required with some of them. Generally the choice of a solvent is governed by two main considerations:

> *At equilibrium the solvents form two imiscible layers. The analytes preferentially partition into one phase and the endogenous material into the other.*

The stoichiometric ratio of the analyte in the organic phase compared to that in the aqueous phase is known as the **distribution ratio**, D. Ideally this ratio should approach 100% in order to minimise losses through the effects of small changes in sample composition, temperature and pH. Reproducibility also increases with increasing extraction efficiency, although a consistent low recovery may be acceptable if an internal standard is used to compensate for changes in efficiency. One point of note is that liquid/liquid systems, unlike solid-phase systems, are more likely to give consistent results year after year, as there is usually less batch-to-batch variation with solvents; diethyl ether will always be diethyl ether, whereas the bonded phase coverage of a solid-phase extraction cartridge may show significant differences.

It is, however, important to realise that the solvents are not necessarily stable. An opened bottle of solvent has a finite life span. In particular, ethers are prone to oxidation through reaction with atmospheric oxygen to produce highly reactive peroxides, easily capable of oxidising certain amino groups, and, at high levels, are potentially explosive. To prevent this, fresh ether is used and it is purged with nitrogen before sealing. Chlorinated solvents usually have stabilisers (such as 1% ethanol to inhibit free-radical formation, or an olefin as a free-radical scavenger) added by the manufacturers but, even so, they are still prone to degradation through free-radical formation. Ketones, too, are susceptible to degradation. For acetone, the reaction is with the enol tautomer, which is always present at trace levels. When exposed to heat, the reaction occurs, and the by-products (dimers that may further react to form large polymers) may exhibit a differing selectivity of extraction to the monomer. So using old solvents for extractions or mobile phases is walking into a trap that could be so easily avoided simply by using fresh solvents and observing the manufacturer's expiry dates.

It may be appropriate at times to combine the two techniques of solid-phase and liquid/liquid extraction when a particular extraction selectivity is required. In one example, whole blood was mixed with a slurry of C8 solid-phase material as it was too viscous to be applied to the cartridges. After centrifugation the supernatant was discarded and the drug extracted from the solid-phase material by mixing it with diethyl ether, which was subsequently aspirated, evaporated and the reconstituted material injected to the chromatographic system. In this case an SPE isolation mechanism is applied but uses a liquid/liquid extraction procedure with the sorbent in the place of the imiscible solvent. This method has the advantage of allowing a greater contact time than is possible in conventional solid-phase extraction and, when compared to the routine analytical method, was seen to be more selective.

In the second example, the same analyte was extracted by liquid/liquid extraction from whole blood into diethyl ether. The ether was then loaded onto a normal-phase extraction cartridge of the aminopropyl type, which retained the polar constituents. The analytes were very selectively eluted in 5% acetonitrile in ether, which was evaporated, reconstituted and injected to the analytical system. This method has been shown to be the most selective yet for the analytes concerned and illustrates the benefits that may be gained through the use of a mixed extraction process.

16.9 Quantification

16.9.1 Rule of one and two

Once the extraction and chromatography have been established, the next step to consider is quantification. Ideally the method will yield an extraction efficiency of at least 80%, with good resolution between the analytes and endogenous material, symmetrical peaks and $3 < k < 20$ for all analyte peaks. By no means will these criteria always be satisfied and a compromise will be called for in one or more respect. It is important to remember that just because the chromatography and detection have been established it does not mean that they can not be altered should the results from the biological matrix indicate such. There are two basic rules that apply in this situation where one is changing conditions to enhance system performance: the rule of one and the rule of two.

The **rule of one** states that in altering any chromatographic parameter, and this includes the extraction step since this may justifiably be considered as 'digital' chromatography, only *one* parameter is changed at any *one* time. If this rule is not adhered to, then it can make the interpretation of results difficult.

The **rule of two** states that a chromatographic observation is not valid until it has been repeated. If a parameter is changed which results in an improvement in, say, recovery or peak shape, then it should be repeated to confirm the finding.

16.9.2 Standardisation

In chromatography there are two types of standardisation, classically referred to as external and internal standardisation. Views on how and when they should be used are varied, so here are a few observations.

In external standardisation the analyte is added at known concentrations to control samples which are then subjected to the same sample preparation procedures as the unknowns, and the detector response to the analyte is plotted versus concentration to generate a calibration curve. Although this is the classical interpretation of the term 'external standard', it may be understood to mean that a functional analogue of the analyte is added to the standards and samples after the sample preparation, to be used as a retention time marker only.

Quantification by internal standards requires a functional analogue of the analyte, which is added to standards and samples prior to sample pre-treatment. The detector response is then recorded as a ratio of the signal of the analyte to that of the internal standard. The assumption for the use of an internal standard is that the partition characteristics of the analyte and internal standard are very similar. This can be a false assumption; the only appropriate uses of non-isotopic analogue internal standards possibly being to serve as qualitative markers, to monitor detector stability and to correct for errors in dilution and pipetting. However, internal standards are usually beneficial for classical instrumentation and manual sample preparation. Thus the internal standard technique should, as far as possible, be adopted for manual sample preparation, which covers most analytical work. With LC–MS–MS the most appropriate internal standard is the deuterated analyte, with a *m/z* ratio of 3 or more than the analyte. This will have identical chemical properties to the analyte and so extract and chromatograph exactly as the analyte, producing a single peak. The internal standard is mass-resolved due to the difference in the *m/z* ratio of the parent ions and, if the labels are on the product ions too, at that stage as well. Time resolution is not necessary. It is important that the internal standard is pure and that there is no conversion through exchange. It is worth noting that quantitative HPLC may be achieved without internal standards, but that quantitative GC, GC–MS and LC–MS (or LC–MS–MS) are best achieved by their inclusion.

Modern equipment and automation, however, can provide extremely reproducible results, the quality of which can be adversely affected by the use of an internal standard due to the added variability of the internal standard measurement. The internal standard technique will not inevitably improve, nor will it always adversely affect, the precision and accuracy of an analytical method. Both internal and external standards may be evaluated when this issue is in question, with the understanding that for a fully automated method, external standardisation may, perhaps, be the most appropriate.

16.9.3   Peak height and area

Quantification in chromatography is by the measurement of a detector response to a physico-chemical property of the analyte over time, and is usually represented as a peak on a chart recorder. This peak has two main properties: height and area, either of which may be used for quantification. If the chromatographic peaks are well resolved from each other, symmetrical and indicative of good column efficiency, then height or area may be used. If, as is often the case in bioanalysis, there is incomplete separation of the peaks, particularly at very low levels where the chromatographic system itself may generate small interferences, or the analyte shows some degree of tailing, then the parameter of peak height should be selected as this has less potential for error in such situations, as can be seen in Figure 16.2. This is because small changes in the position of the start/stop integration signals affect peak area more than peak height. If enantiomeric ratios are to be determined, however, then the parameter of peak area has to be used.

16.9.4   Calibration check

Once the extraction and chromatographic techniques have been developed, the first part of the validation process is the calibration check. To do this a series of chemical standards are prepared over the required concentration range and injected to the chromatograph. From the results the linearity limits of the detector will be found. This is particularly useful when assessing the limit of detection (LOD), which is defined as that concentration producing a peak of a signal intensity of at least three times the background at the retention time of the analyte, and may be expressed as xg/ml or xg on column.

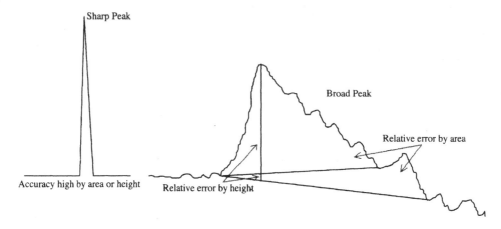

Figure 16.2 Relative errors by peak height and area.

16.10   Validation

There exists a plethora of information regarding the details of assay validation, with many companies having their own validation, protocols; however, some basic points are suggested here.

The accuracy and precision check should be carried out in the appropriate biological matrix, starting at the LLOQ (lower limit of quantification) and increasing three orders of magnitude higher in logarithmic steps comprising at least five data points in duplicate plus a blank. If this range does not cover the anticipated concentration range of the test samples, then a higher calibration range may also be produced, or dilutions of the samples may be considered with appropriate accuracy checks.

The calibration and validation standards may be produced independently and from different stock solutions. It may be appropriate to use disposable glassware, or plastic if adhesion of the analyte to glass is suspected.

The analyst may then extract and chromatograph, in duplicate, the calibration standards and six replicates of the validation standards. The linearity of the extraction is indicated by the correlation coefficient of the calibration standards and the precision from each group of six replicates of the validation standards, with the accuracy calculated from the mean of the six replicates compared to their target values as determined by interpolation from the calibration line. The linearity of the assay should be better than $r = 0.9900$, and accuracy and precision should show a deviation from target of less than 15%, although a precision and accuracy deviation of less than 20% may be accepted at the lower limit of quantification.

16.11   Support work

16.11.1   Matrix substitution

The majority of the samples received for analysis are plasma. However other matrices, such as serum, whole blood and urine from human, rodent, primate and canine sources, are used. It is therefore necessary to compare the effects of the various matrices on the assays to be used in the toxicokinetic and metabolic profile studies. The new chemical entities under development are intended for a particular therapeutic area, so in phase II, samples from patients in the relevant disease state will be presented for analysis. It therefore becomes necessary to evaluate subjects' plasma from the anticipated patient pool as the disease state may be reflected in altered metabolism, sample composition and concomitantly administered medication. Any of these may affect the assay performance.

Particular attention needs to be paid to the fact that the relative recoveries of the analyte and a non-homologous internal standard may differ between matrices, thus producing different regression lines of peak height ratio against concentration. In such a case it will be necessary to revalidate the assay in the new matrix, rather than attempt to combine the regression lines. If the substituted matrix shows unacceptable interference with the analyte or internal standard peaks or variable recoveries, then either the analytical chromatography or the extraction procedure, or both, will need re-development and validation.

16.11.2 Stability

Freeze–thaw cycles

One factor that affects stability of the samples is the process of freezing and thawing, and it is necessary to show that the measured analyte concentration does not change after the sample has been re-frozen and subsequently re-analysed.

Stability on storage

This is assessed by preparing bulk standards in the biological matrix at low, medium and high points on the calibration line. They are assayed in duplicate. An aliquot of each is then frozen for a specified time, then thawed and re-analysed against fresh standards. Another aliquot is stored at +4°C for a certain length of time and then re-analysed against fresh standards.

Stability of extracts

Analyses are often carried out on extracted samples in bulk. It follows that the samples at the end of the run will have been waiting in the autosampler perhaps overnight or, in the event of instrument failure, over the week-end. It is therefore necessary to determine the stability of these sample extracts. To this end a range of calibration standards are prepared with internal standard, extracted in duplicate and analysed. At the end of the trial all the extracts are re-analysed and the individual peak heights, as well as peak height ratios, are compared to the results of the original analysis. Based upon these results recommendations about the acceptable storage time of extracts may be given.

16.11.3 *Metabolites*

Many of the compounds undergoing development have impurities and breakdown or metabolism products. It is essential that these can be deselected from the quantification of the analyte. The first step, if these products are available in pure form, is to assess their detectability. If they do not cause a detector response under the analytical conditions of the assay, then they will not cause any interference. However, if the analytes are to be collected for structural elucidation, then it is important that analyte peaks are as pure as possible. If analyte-related material is detected on the chromatogram, it is important to check that there is no interference with the analyte peak ($R > 1.5$) and, should this condition not be satisfied, further chromatographic optimisation will be necessary at either the extraction or analytical stage or both.

Occasionally the situation arises where there are no analyte-related products available during the development phase of an assay but they appear as metabolites during the analysis of authentic samples. In this case, the chromatographic conditions may be modified to increase the selectivity of the system, and if this proves unsuccessful, then the extraction may be altered. It is worth restating that the majority of metabolites, and especially glucuronides and sulphates, are more polar than the parent material, so changes in the method to deselect for the more polar compounds may prove beneficial.

If it becomes necessary to measure the metabolites as well as the parent material, then the same process of detector selection, chromatographic separation and extraction/enrichment will apply. However, due to the differing properties of these related compounds, it may not be possible to either extract or chromatograph them all on the same system. It may be necessary, for example, to combine a liquid/liquid extraction with a solid-phase method, or to use ternary/quaternary elution, or to change a reversed-phase system into an ion-pair method, or to use different detectors in series. If LC–MS–MS is available, it is the technique of choice. In its absence column switching may be an appropriate way of solving this difficulty in certain circumstances.

As a last resort it may be necessary to split the sample and analyse each portion under different conditions in order to quantify the products of extensive metabolism.

16.12 Conclusions

This method development strategy gives only a brief outline of the major steps that may be taken in the development of an analytical chromatographic assay. It is not intended to be a definitive document, as changes and adaptations will always be appropriate, depending upon the nature of the compound, the assay and the assay requirements, as well as individual company procedures. This chapter has dealt primarily with analytical HPLC as this is the main analytical procedure in use at the time of writing, although the principles are widely applicable.

Although this chapter is of value in its own right, its use will be maximised if it is considered to be a part of the entire book. Bearing this in mind, this chapter should help to avoid many potential pitfalls and dead ends and facilitate an efficient and structured approach to assay development.

INDEX

Accuracy 342, 356
Acetic acid, as TLC solvent 158
Acetonitrile 79, 81, 89, 347
  aqueous molecular structure 4
  dipole moment 6
  in immunoassay 182
  in mixed solvents 21
  as mobile phase 310, 312
Acylation, GC 148
Administration of drug in excretion balance
  studies 304
Adsorption chromatography 64–5
Affinity 172
Alcohols
  miscibility with water 8
  in mixed solvents 20
Alumina 88
  stationary phase, TLC 149
Ammonia, as TLC solvent 158
Analysis, of metabolites 306
Animal, choice for excretion balance studies
  304
Anion exchange 86
Antibody 171
Antigen 171
Antiserum 171
APCI see Atmospheric pressure chemical
  ionisation
Applied field strength 163
Artefact peaks 61
Atmospheric pressure chemical ionisation
  (APCI) 267
Atomic nucleus 278
Autosampler 140, 197

Band broadening 47, 60–1, 62
Band focusing 99
Band spread see Band broadening
Bases 12
Beer's law 111, 112
Benzene, as solvent 20

Bile 170
  collection in excretion balance studies 304
Binders, in TLC 150
Bioanalysis 138, 143, 148
Biochemical changes, NMR observation of
  294
Boiling point 133
Bonded phases 64–7
Bonds
  dipole–dipole 8, 32
  dispersive 8
  energy 31
  hydrogen 5, 8, 32
  intermolecular in solvents 3
  ionic 8
  strength 8, 31
Buffers 16
Butanol, as TLC solvent 158
Butyl chloride, as solvent 20

Calibration 355
  mass-spectrometry 235
Capacity factor 54–7, 62, 66
Carbon disulphide, dipole moment 6
Carbon-13 NMR 278
Carbon-14, as label in metabolite studies
  302
Carboxymethylcellulose, for formulating
  drugs 304
Carrier gas (GC) flow rate 133
Carrier protein 179
Cassette dosing 241
Cation exchange 86
Cellulose, stationary phase, TLC 149
Characterisation of metabolites 315
Chemical ionisation 267, 268
Chemical shift 294
Chemiluminescence 120
Chloroform 20
  hydrogen bonds 6
  as TLC solvent 158

CI *see* Chemical ionisation
CID *see* Collision-induced dissociation
Collision cell 256
Collision voltage 245
Collision-induced dissociation (CID) 256
Column ovens, HPLC 50
Column switching 97
Cone voltage 245, 246
Continuous flow FAB 267
Counter ion 85
Coupling in NMR, *see* Spin–spin coupling
Cyclodextrin 164
Cytochrome P450 isoenzymes 337

Daughter ion 256
Degassing 348
DELFIA 180, 183
Derivatisation
    of drugs 24
    for electrochemical detection 122
    for GC
        acylation 148
        esterification 147
        silylation 147
    in identification of metabolites 329
    post-column 127
    pre-column 126
Desalting, in separating metabolites 308
Detection 345
    electrochemical 78, 346–7
    fluorescence 345–7
    mass-spectrometric 103, 345–6, 348
    UV 89, 345
Detector, CE
    conductivity 168
    electrochemical 169
    fluorescence 169
    mass spectrometric 169
Detector, GC
    electron capture 146
    flame ionisation 140, 145
    mass selective 147
    thermionic 145
Detector, HPLC
    amperometric 121
    bulk property 108
    coulometric 121
    diode-array 113
        effect of fluorescence 113, 114
    electrochemical
        dynamic 121
        equilibrium 122
    flow-cell 107, 114
    fluorescence 115, 200
    infrared 125
    light-scattering 125

linearity 109
nitrogen 125
noise 128
optically active 124
radiochemical
    liquid scintillant 124
    solid scintillant 124
resolution 109
selectivity 108, 127
sensitivity 114, 115, 117, 119, 120
solute property 107
time-constant 110
universal 108
UV 200
    fixed-wavelength 111, 112
    variable-wavelength 113
Deuterated solvents 287
Diastereoisomers 94
    Dalgleish rule 95
Diazomethane 147
Dichloromethane
    solvent 20
    TLC solvent 158
Dielectric constant 3, 20
Diethyl ether, as solvent 20
Di-isopropyl ether, as solvent 20
Diode-array UV spectroscopy, in metabolite
    characterisation 317
Dipole moment 6, 20
Dipoles, non-permanent 7
Direct on-column injection 140, 143
Direct probe 268
Disequilibrium 184
Displacement 88
Distribution coefficient 17
Distribution ratio 353
Drugs, miscibility in water 8
Drugs, pH of salts 12
Dynamic FAB 271

Eddy diffusion 60, 62–3
Effective plate number 53
EI 267, 268
Electrical double layer 162
Electrochemically active groups 123
Electrochromatography 165
Electrokinetic sampling 167, 168
Electron capture detector 146
Electron ionisation 268
Electro-osmotic flow 161, 165
Electropherogram 161
Electrophoretic mobility 161
Electrospray 267, 273
Emulsions in drug extraction 21
Enantiomer 90, 164
Endcapping 67, 77

Enhancers, use in TLC 153
Enrichment
    of drugs 1
    of metabolites 1, 304
Enthalpy 3
Entropy 3
Enzyme labels 176
Enzymic cleavage, in identification of
    metabolites 327
Epitope 171
Equilibrium 172, 173
Equilibrium distribution coefficient 51, 54
ESP 267, 273
Esterification, GC 147
Ethyl acetate
    hydrogen bonds 6
    solvent 18
    TLC solvent 158
Excretion balance studies 304
Exposure, cross-species comparison 338,
    340
External standard 76, 345, 354

FAB see Fast atom bombardment
Fast atom bombardment 259, 260, 267,
    270
Ferrule 139
Field amplification 167
Film thickness 133, 136
Filter 117
First law of metabolism 26
Flame ionisation detector 140, 145
Flow cell 113, 114
Flow rate 137
    effect on HPLC 62
    effect on SPE 30
Fluorescence 113
    temperature dependence 120
Fluorescent groups 120
Fluorescent label 174, 176-7
Fluorine NMR 291, 293
Formulation, of drug in excretion balance
    studies 304
Fourier transform 279, 286, 296, 297
Fragmentation patterns 238
Free solution capillary electrophoresis 160,
    162
Frit FAB 271
FSCE see Free solution capillary
    electrophoresis
Functional groups 343
Fused silica capillary 166

Gas filter 135
Gas-phase ionisation 223
Gas chromatography and MS 269

Gaussian peak 52, 58
GC column
    diameter 136
    dimensions 133, 136
    film thickness 133
    length 136
    packed 131
    PLOT see Porous-layer open tubular
        column
    SCOT see Support-coated open tubular
        column
    WCOT see Wall-coated open tubular
        column
GC/MS 267, 269
    in characterisation of metabolites 321
Gibb's free energy 2
Glucuronide 352
Good laboratory practice 339
Gradient 112
Gradient reverse phase 68-70

Hapten 171
Heated nebuliser LC/MS APCI 270
Helium 135
Henderson-Hasselbach equation 13
Hepatocytes, in metabolic studies 334
Hexane 89
High Performance TLC 149
HPLC
    analytical gradient, in analysis of
        metabolites 307
    apparatus 45-50
    applications 44-5
    basic principles 50-64
    chiral 90
    column 45-7
        amino 88
        C18 77, 80
        C8 80
        care 73-4
        cavity 92-3
        chiral 91
        conditions 100
        cyano 80, 88
        diol 88
        macrocyclic antibiotic 94
        phenyl 80
        protein 92
        three-point interaction 92
    gradient 100
    gradient reverse phase, in separating
        metabolites 312
    injectors 49-50
    ion-exchange 85
    ion-pair 82
    normal phases 87

optimisation 75
origins 44
plumbing 47
pumps 48
reverse phase 76, 81–2
HPLC–NMR 291
Hydrocarbons, miscibility with water 8
Hydrogen 135
Hydrogen bonds *see* Bonds, hydrogen
Hydrophilic functional groups 9
Hydrophobic functional groups 9

Identification of metabolites 321
Immunisation 179
Immunoassay separation
  charcoal 177
  immobilisation 178
  second antibody 178
Immunogen 171
Ion traps 274
In vitro studies 333
Injection, GC
  direct on-column 140, 143
  split 140
  splitless 140
Injection volume 102
Inlets, mass-spectrometry
  direct liquid 214
  moving belt 214
  particle beam 214
Integrals (NMR) 286
Integration 110, 111
Internal standard 76, 98, 345, 353–4
  in mass spectrometry 355
Iodine, as spray reagent in TLC 158
Ion evaporation 221
Ion exchange 72, 82–3
Ion exchange chromatography
  in enriching metabolites 305
  for separating metabolites 308
Ion trajectory 230
Ionisation constants 11
Ion-pairing 69–72
Ion-suppression 69–72
Isocratic separation 48
Isotherm 58
Isotope labelling 265

Jablonski diagram 116
Joule heating 163, 164

K_a 11

Laser 117, 125, 226
  desorption, matrix-assisted 226
Linear flow velocity 45, 62–3

Liquid scintillation counting, to determine recoveries 330
Liquid/liquid extraction 352
Liver, perfused isolated, in metabolic studies 336
Localisation 88
Log P 87, 345
Longitudinal diffusion *see* Molecular diffusion

Magnetic field strength 279
Mass chromatography 239
Mass selective detector 147
Mass spectrometry 78, 81, 93, 103
  chemical ionisation, in identification of metabolites 323
  electrospray, in characterisation of metabolites 320
  FAB, in identification of metabolites 323
  high resolution, in identification of metabolites 325
  in identification of metabolites 321
  ion source 104
  negative ion, in identification of metabolites 323
  thermospray
    in characterisation of metabolites 318
    in identification of metabolites 323
Mass transfer 60–3
Matrix substitution 356
MECC *see* Micellar electrokinetic capillary electrophoresis
Megabore GC 136
Metabolic profiles 331, 336
Metabolic profiling, use of TLC 151
Metabolites 358
  plasma 337
  purification 315
Methanol 79, 346–7
  creation of artefacts using acidic methanol 23
  hydrogen bonds 4
  mobile phase 150, 158, 310, 312
Method development 342
Methylethylketone, as TLC solvent 158
Micellar electrokinetic capillary electrophoresis (MECC) 160, 164, 170
Micelle 161
Michaelis constant 337
Microsomes, in metabolic studies 334, 336
Migration time 161, 168
Mobile phase 135
Molecular diffusion 60
Molecular forces 51
Molecular imprinting 42, 193
MS–MS 256, 257

Neutral loss 258, 264
Neutral loss mass spectrum 262
Ninhydrin, as spray reagent in TLC 158
Nitrogen detector *see* Thermionic detector
NMR spectroscopy, in identification of
    metabolites 325
Normal phase 64–5

Organic modifier 164, 165

Parent ion 256
Particle beam LC/MS 267, 269
Partition 132
Partition chromatography 65–7
Partition coefficient 9, 17
Peak
    area 96, 355
    broadening *see* Band broadening
    fronting 58–9
    height 96, 355
    shape 58–61
    tracking 103
    volume 98
    symmetry 58–9
    tailing 58–9
PEG 266
Peroxides 89, 353
pH
    in CE 164, 166
    changes to improve separation of
        metabolites 313
    effect on EOF in CE 161
    effect on fluorescence 347
    effect on HPLC 77, 347
    HPLC optimisation 80–3, 86
    problems with extreme values in solvent
        extraction 21
    in solvent extraction 21, 350, 353
    in SPE 350
    in titration 12
Phenylboronic acid 33
pK_a 12, 77, 83, 350
Plasma 170
    metabolites 337
Plate height 53
Plate number 136, 76, 102
Pneumatics 139
Polar compounds, enrichment in metabolic
    studies 305
Polyethylene glycol 134, 266
Polyimide coating 139
Polysiloxanes 134
Porous layer open tubular column 134, 136
Precision 342, 356
Precursor ion mass spectrum 261
Precursor ions 256, 257

Pressure 101
Product ion mass spectrum 258
Product ions 256, 257
Programmed temperature vaporisation 140
Programmed temperature vaporisation
    injection 141
Protein binding 350
Protein precipitation 21, 196, 197
Protonated molecule 258
Pulse NMR 294, 296
Purification, of metabolites 315

Quantification 76, 354
    see also External standard; Internal
        standard
Quantitation, calibrationless 114
Quantum efficiency 116

Radial compression columns 47
Radiochemical detection in TLC 150, 151
Radiochemical methods in determining
    recoveries 26
Radiochemical synthesis in metabolite
    identification 302
Radiolabel 175–6
Recoveries in extraction 26
Recovery 343, 351, 354
Reduced plate height 53
Resolution 54–8, 76, 161, 348
Retention factor 76, 85
    see also Capacity factor
Retention gap 143
Retention volume 54
Reverse phase
    chromatography 65–7
    HPLC for isolation of metabolites 1
    TLC 149
Rf 149
Robustness 343
Run time 101

Safety margins 336
Salting out 3
Sample introduction
    CE
        differential pressure 167
        electrokinetic 167, 168
        gravimetric (syphonic) 166
    GC 138, 140
Sample preparation 75, 349
    for NMR 287, 288, 291
Sample volume 64
Scattering
    Raman 119
    Rayleigh 119
    second-order 119

Selected ion monitoring 235
Selectivity 343
  see also Separation factor
Selectivity factor 54–8
Sensitivity 343
Separation factor 76, 85, 89, 91, 348
Separation of metabolites 307
Sephadex, DEAE and CM, ion-exchange
  chromatography 308
Silanols 34, 77, 85, 351
Silica 88
  stationary phase, TLC 149
Siloxanes 134
Silylation 147
Size exclusion chromatography 72
Sodium chloride, for salting out 5
Solid-phase extraction 288, 349
  in enriching metabolites 305
  of faeces 37
  in pipette tip 206
  sorbent capacity 30
  of tissues 37
  toxic solvents 29
  urine 36
  96-well format 29, 39, 40, 41, 198
  problems 206
Solubility 132
Solvent extraction 18
  in enriching metabolites 305
Solvent partition, in separating metabolites
  309
Solvents
  mixed 20
  table of properties 20
SPE *see* Solid-phase extraction
Specificity 343
Spectrum reference (NMR) 280
Spin 278
Spin–spin coupling 281–5
Split injection 140
Splitless injection 141
Spray reagents, TLC 158
Stability 357
Stable isotopes 293
Starch/potassium iodide, as spray reagent in
  TLC 159
Stationary phase 133
  TLC 149
Support-coated open tubular column 134
Surfactant 164

Tailing peak 143
Tandem mass spectrometry 256
Temperature
  effect of in HPLC 50, 61–2
  GC 137

gradient in GC 137, 141, 141, 143
HPLC detectors 112, 120
HPLC optimisation 80, 85–6
Tetrahydrofuran 79
Theoretical plates 52–3
Thermal ramp *see* Temperature, gradient in
  GC
Thermionic detector 145
Thermospray 271
Time-resolved fluorescence 176–7
Titration curves 12, 13, 14, 15
Titre 171, 175, 179, 180
TLC
  for analysis of metabolites 307
  comparison with HPLC 155
  preparative 150
  to separate metabolites 315, 316
  two-stage analytical technique 153
Toluene, as solvent 20
Total ion current 235, 236
Triethylamine 78, 85, 347
Trifluoroacetic acid 79
Triple quadrupoles 256
Tritium
  detection in TLC 153
  label in metabolite identification 302
TSP 267, 271
Two-dimensional NMR 297, 301

Urine 36, 170
  collection in excretion balance studies
    304
UV light, detection in TLC 150
UV-absorbant groups 115
UV-irradiation 126

Validation 356
Van Deemter equation 62–3, 137
Van der Waal's forces 6
Vapour pressure 132
Viscosity 63–4
V_{max} 337
Void volume 54

Wall-coated open tubular column 134,
  136
Water
  dipole moment 6
  hydrogen bonds 4
  ionisation constant 11
  as solvent 5
96-well format 29, 198

Zwitterions 13
  enrichment 305
  solvent extraction 26

T - #0207 - 071024 - C0 - 254/178/21 - PB - 9780748408436 - Gloss Lamination